LAND OF MOUNTAIN AND FLOOD

THE GEOLOGY AND LANDFORMS OF SCOTLAND

Alan McKirdy, John Gordon and Roger Crofts

Previous pages. The Eildon Hills from Scott's View

This edition published in 2009 by
Birlinn Limited
(in association with Scottish Natural Heritage)
West Newington House
10 Newington Road
Edinburgh
EH9 1QS

2

www.birlinn.co.uk
www.snh.org.uk

ISBN: 978 1 84158 626 7

British Library Cataloguing-in-Publication Data
A catalogue record for this book is available from
the British Library

Designed and typeset by Mark Blackadder

Printed and bound in China

CONTENTS

FOREWORD

I first fell in love with Scotland more than twenty years ago. I came up from London to Edinburgh by train and then drove north through the Highlands, across to the West Coast and back down to Glasgow. With every passing mile, I knew that I had found my home. As a geographer growing up in the south of England, mountains, glens and lochs were something I'd only ever read about in textbooks – but suddenly I was right in the middle of the most incredible landscape I had ever encountered. I was mesmerised by the Linn of Dee, left speechless by Glencoe, and by the time I had reached Oban I had vowed to move here forever.

I was lucky: I'd had years of training in 'reading' the landscape to help me understand the awesome nature of what was now in front of my eyes – an understanding which added real depth and wonder to what I was seeing. But *sharing* that with others was harder to achieve, and it's for this reason that I'm thrilled to have this fabulous account of a landscape I love so much.

For me, Scotland is the most beautiful country in the world. But the beauty which characterises our landscape goes more than skin deep. The mountains, glens, islands and lochs which give Scotland its character come from the rocky 'bone structure' that lies beneath the surface. Scrape away the purple heather, the machair grasslands or shimmering water down to the underlying landforms and bedrock and the hidden wonder of Scotland is revealed like a fabulous journey through time.

This stunningly produced book tells the story of that journey – of how Scotland came to be over the millions of years of its history. At times, it's a detective story, at others an action thriller; parts will read like a travelogue, while some sections are pure romance. It's a story that will take you all over the world, through fire and ice, powerful meetings and devastating separations.

But perhaps the best thing about this book is that everyone can join the journey of discovery. The clearly written text and wonderful illustrations bring the magnificence of the landscape home to both novice landscape detectives and experts alike. Whether you are lucky enough to get out into the hills or if you're just sitting at home as an armchair traveller, poring over the breathtaking photographs, this book is guaranteed to reveal the hidden magic of Scotland's landscape.

Vanessa Collingridge
January 2007

Buachaille Etive Mór from Rannoch Moor.

PREFACE

The Stacks of Duncansby, Caithness.

Our aim in writing this book is to reveal the story of Scotland's geological past and its rich Earth heritage to an audience beyond the academic world. In comparison with many other subjects, geology is relatively little known among the wider public, yet Scotland has some of the most remarkable rocks and landforms on the planet. Moreover, geology and landforms are directly relevant to the way we live our lives. They have influenced not only our history, culture and economy, but also our understanding of many natural phenomena that impact on society today, such as climate change, rising sea levels, floods and landslides. They also give us an insight into how our environment may change in the future and help inform how we might manage or adapt to such changes. There is now growing recognition, too, of the considerable value of geology and landforms as an asset for tourism and sustainable rural development. This is highlighted, for example, in the recent award of European Geopark status to the North-west Highlands – an award endorsed by UNESCO.

In this book, we explain the formation of Scotland's landscapes and landforms and how they are closely linked to the underlying rocks and the natural processes that have shaped the surface of the Earth. This is woven around the story of shifting continents as Scotland journeyed across the surface of the globe from near the Equator, towards the South Pole and then north to its present location. We also recognise and celebrate the many important contributions that Scots, or those working in Scotland, have made to Earth science through fundamental discoveries about how the Earth works. Above all, we hope that readers enjoy the book and discover more about Scotland's Earth heritage through exploring its varied landscapes.

In bringing together the story of Scotland's rocks and landforms, and how they have formed our scenery, we owe a great debt to Archibald Geikie's *The Scenery of Scotland Viewed in Connection with its Physical Geology* (3rd edition, 1901) and Brian Sissons' *The Evolution of Scotland's Scenery* (1967). Both of these books, with their detailed and perceptive observations, have influenced our own reading of the landscape. In addition, *The Geology of Scotland* (2002), edited by Nigel Trewin, has been an invaluable and authoritative source. Our understanding of the geology, landforms and landscapes of Scotland, and further afield, has also been strongly influenced by the inspired teaching, both in the class and in the field, of Chalmers Clapperton, the late Sandy Mather, Bill Ritchie, the late John Smith, David Sugden, the late Francis Synge, Nigel Trewin and the late Ken Walton. We have also drawn inspiration from the enthusiasm and insight of Aubrey Manning and his appreciation of the landscape.

In addition, Roger Crofts would like to thank Terry Driscoll, John Jones and Derek Woodward.

This book would not have been possible without the assistance of Scottish Natural Heritage. The authors gratefully acknowledge its support in the production of this volume.

Many people assisted in the preparation of the book. We particularly thank Alastair Dawson, Jim Hansom, John McManus, Clive Mitchell, Stuart Monro and Nigel Trewin, who reviewed whole sections of the manuscript. We also thank Colin Ballantyne, Adrian Hall, David Jarman, Lesley Macinnes and Richard Tipping for insightful discussion on particular topics. From SNH, Stewart Angus, Patricia Bruneau, Andrew Coupar, Noranne Ellis, Colin Galbraith, George Lees, Colin McLeod and Alistair Rennie provided helpful comments or discussion. Elaine Dunlop, formerly of SNH, helped to shape the book from its inception. Her advice and comments assisted us to find the appropriate language to describe the complex and unfamiliar concepts for a lay audience. We are particularly grateful to Lorne Gill, SNH's photographer, who provided many of the photographs that accompany the text. Special thanks are extended to Betty Common from SNH's photographic library and to Pam Malcolm, who helped source much of the illustrative material. We also acknowledge the assistance Andrew Bachell during the latter stages of the project and we thank Jim Lewis and Robert Nelmes for drawing many of the diagrams.

Numerous individuals and organisations provided photographs and other illustrative material for the book. Their assistance is recognised in the copyright acknowledgements, but we also record our gratitude here.

Special thanks are also extended to Andrew Simmons, Birlinn's Managing Editor, who gave us patient and helpful advice throughout the project. Hugh Andrew, Birlinn's Managing Director, saw the potential of the book from the outset and we are grateful for his support. Mark Blackadder developed the design and layout for the book and we thank him for his patience in accommodating our changes to the various drafts.

As well as this assistance, we also owe a debt to the work of the many academics whose results we have drawn on and interpreted or synthesised. However, we remain responsible for any omissions, inaccuracies or errors.

Alan McKirdy, John Gordon, Roger Crofts
January 2007

'The result, therefore, of our present enquiry is, that we find no vestige of a beginning, – no prospect of an end.'

James Hutton, *Theory of the Earth*, 1788

'The story of the origin of our scenery . . . leads us back into the past farther than imagination can well follow . . .'

Archibald Geikie, *The Scenery of Scotland Viewed in Connection with its Physical Geology*, 1901

INTRODUCTION

SCOTLAND'S ROCKS AND LANDFORMS: A JOURNEY THROUGH TIME

The mind seemed to grow giddy by looking so far into the abyss of time.
John Playfair, *Transactions of the Royal Society of Edinburgh*, 1805

Scotland is justly famed for its wonderful scenery of mountains, lochs, islands, wild rocky places and sandy beaches. Whether you fly over Scotland, drive around it, walk its mountains, or just enjoy reading about it, you cannot have failed to notice its great diversity of landscapes and landforms. Think about the mountains of the North-west Highlands, the rocky pass of Glencoe, the wide, open spaces of Rannoch Moor or the beaches of East Lothian and the Hebridean islands. Even our towns and cities have distinctive landscapes – Edinburgh with its seven hills and Glasgow with many more. Look at the photograph of Scotland taken from space. The landscape is remarkably varied, from the spectacularly mountainous north and west with its deeply indented coastline and fringe of islands and peninsulas facing the Atlantic from the Clyde to Shetland, to the high, dissected plateaux of the eastern Grampians rising above the coastal lowlands of the Moray Firth and the North-east. Or from the Central Lowlands with their straight-edged boundaries to the north and south seemingly cut into the landscape with a knife and studded with distinctive hill masses, to the rolling Border Hills and the relatively smooth lines of the east coast interrupted by the Forth and Tay estuaries. How did such a small country acquire such a diversity?

The answer to this question lies in Scotland's rocks and landforms. They tell a remarkable story: a story of 3,000 million years of Earth history and a journey through time, which truly stretches the imagination. The land that makes up Scotland today has travelled the world from near the Equator to the South Pole, and then north to its present location. The record of this journey, and the events and forces that have shaped the landscape of Scotland, are preserved in our rocks and landforms. They reveal shifting plates, continents splitting and colliding, ancient volcanoes, mountain building and ice ages. By 'reading the landscape' and exploring the clues in the rocks and landforms, we can retrace Scotland's journey through the great enormity of geological time which was first appreciated by James Hutton, the father of modern geology. We can look back into the very "abyss of time", so aptly described by John Playfair when he visited Siccar Point in Berwickshire in 1788 with Hutton. However, we can only look so far back, and there is no clear evidence of the very beginnings of Scotland. And, although we may speculate about the future, we cannot be sure how long Scotland will remain in its present form, or where its continuing journey will lead. Even now, we are only at a staging post in Scotland's journey through time, and that journey provides the central theme of this book.

Our aim is to tell the story of Scotland's Earth heritage in a widely accessible way and to help others to 'read the landscape'. This is not a definitive geological textbook, and to the specialists who may feel we have oversimplified, we make no apology – there are plenty of other technical publications available. We have deliberately provided many photographs, diagrams and maps to help explain Scotland's Earth history in a new light. We hope you will enjoy it, be enthralled by the story and be stimulated to go out and look at Scotland in an entirely different way and better appreciate its fascinating history and landscapes. The story told in this book has many different strands. In the following paragraphs we introduce some of the main highlights.

There are many reasons why Scotland's rocks, landforms and landscapes are so diverse. Geologically speaking, Scotland is not a single country. It comprises a number of separate pieces

The major geological features and landforms of Scotland are clearly revealed in this satellite image.

of the Earth's crust that have been joined together in the distant past through the movements of the Earth's plates. You can see some of the joins if you look carefully at the satellite image of Scotland. Each of the pieces has a very different geological history and we shall explain in more detail the story of how they came to be assembled like a giant jigsaw.

Scotland has a long geological history and has been continually on the move as the Earth's plates have shifted. Scotland started life in the Southern Hemisphere near the Equator and then moved polewards to a position considerably south of the present antipodean landmasses. At that time, Scotland was part of a much bigger continent and was initially linked to what are now parts of North America and South America. On its subsequent journey northwards, this continent gradually broke up into smaller pieces. Remarkably, Scotland and England were, for a

Foinaven, Arkle and Ben Stack rise above an ice-scoured landscape of Lewisian gneiss in North-west Scotland. The Lewisian rocks are among the oldest in the world.

Geology and landforms feature prominently in many of Scotland's urban landscapes. The skyline of Edinburgh is dominated by Arthur's Seat and Salisbury Crags, where James Hutton first demonstrated the existence of molten rocks and past volcanoes.

time, on separate continents divided by a vast ocean, Scotland forming a small part of North America. Then the movements of the plates joined them together. Later, Scotland parted company with North America as the North Atlantic Ocean began to open up and volcanoes erupted along the west coast.

Plate movements have also given rise at different times to great upheavals in the Earth's crust, folding and deforming the rocks and uplifting them into vast mountain ranges. These upheavals were accompanied by upwelling of molten rock through weaknesses in the crust, leading to great bursts of surface volcanic activity. Today, the long-eroded roots of these mountains form the core of the Highlands, and remnants of the volcanoes occur at Ben Nevis and Glencoe along with many other locations. The ripples from more distant plate movements also buckled the later rocks in Scotland. Thus the movements of the Earth's plates have been fundamental in shaping the structure of Scotland and its geological diversity.

During the course of its remarkable geological journey, Scotland has experienced all of the Earth's climates, from polar to tropical. Consequently, many different environments are represented in the rock record. The rocks are historical archives revealing the former presence of tropical rain forests, deep oceans, shallow seas, sandy coasts, huge deserts, vast rivers, arctic frosts and great ice caps. The remains of plants and animals from these different environments are also preserved as fossils in the rocks and tell the story of the evolution of life on Earth.

Many different natural processes have fashioned the rocks and landforms of Scotland. Volcanoes have erupted frequently and we can

Tantallon Castle stands on a headland of Carboniferous volcanic rocks on the coast of East Lothian. Defended by cliffs on three sides, it was built in the fourteenth century from the local red sandstone seen in the foreground.

still see the remains of their interior plumbing, now stripped bare by the forces of erosion. Huge streams of lava have poured out over the land surface in parts of the Hebrides, Lochaber and the Central Lowlands. Glaciers and ice caps have repeatedly covered the country and moulded the landscape. The waves of the sea and large inland lakes have also left their mark in many places. Together, the processes of weathering and erosion have cut down entire mountains and revealed the underlying rocks at the ground surface.

Reflecting its long and varied geological history, Scotland is made of many different types of rocks, with varying resistance to the forces of weathering and erosion. Some are relatively hard, like the granites of the Cairngorms, and some are relatively weak, like the sandstones found in many parts of southern Scotland. Some display a strong resistance to the elements and remain upstanding, such as the ancient sandstones of the Torridon Hills, whereas others have been more readily worn down by rain, rivers and glaciers to form the lower parts of the Moray Firth coast and the Central Lowlands. Moreover, the rocks have reacted differently to the huge changes in climate, and hence weathering processes, that have taken place. In a geologically short period of time, Scotland experienced a tropical climate like that of northern Nigeria, an arctic climate like that of northern Canada and some very rapid fluctuations between warmer and colder conditions. Some of the harder rocks disintegrated through chemical weathering in the warm, humid climates, while frost action under the arctic conditions readily split others apart. Whole slopes have even collapsed downhill and rivers have carried vast amounts of material downstream and out to sea.

In many parts of Scotland, the Ice Age glaciers and their meltwater rivers significantly eroded and reshaped the landscape. In the north and west, particularly, they scraped the landscape bare, deepened the glens, formed the basins that now hold our lochs, and created the corries. In the glens and lowland areas, they left a thick blanket of stony debris across the ground surface, often sculpted into a variety of mounds, ridges and terraces. They also provided the materials from which our soils have formed. The growth

During the Ice Age, glaciers played a significant part in shaping the present landscape of Scotland. They carved the spectacular ridges, glens and cliffs in the granite mountains of north Arran, now enjoyed by hillwalkers and climbers.

and decay of the glaciers was also accompanied by changes in the relative level of the land and the sea. These changes have shaped our coast, with its diversity of cliffs, beaches and sand dunes.

Together, all these aspects have enriched the detailed story of Scotland's journey and help to explain Scotland's diverse Earth heritage. Scotland is really quite exceptional in that rocks of most geological periods are represented somewhere in the country. However, the story is not without its gaps. The rock record is massively incomplete, with some time periods poorly represented. In some places, the gap in the record is really quite difficult to conceive; for example, in parts of North-west Scotland, glacial deposits formed only 20,000 years ago lie directly on top of rocks over 2,000 million years old.

Although geology can open a window on the past, it is far from a dead subject – in fact it is crucially relevant today. The record of the rocks and landforms clearly demonstrates continuing environmental change. Looking around Scotland,

The dynamic nature of the landscape today is evident along Scotland's sandy coasts. At St Cyrus, north of Montrose, there has been a progressive northwards movement of sand during the last few decades and the edge of the dunes has built seawards by up to 40 metres.

many might think that the landscape is quiet. This is far from the case – change continues to take place, although much less dramatically than at certain times in the past. Scotland experiences mild earthquakes – very gentle compared with those in the past and with those in other parts of the world. Slopes are moving under the effects of gravity, the continued freezing and thawing of the ground, and heavy rainstorms. Rivers continue to carry sediments downstream, dumping them on their banks and beds or out at sea. Coasts are continually changing; some are building outwards, and others are retreating. Some of these changes are easy to see, as the result of a particular storm on land or at sea; elsewhere the change is gradual and less easy to detect, but it is happening nevertheless. Along with these natural forces, during the last 6,000 years or so, human activity has increasingly become a force for change through deforestation, agriculture, urban expansion, coastal and river management activities, and, most recently, through the effects of greenhouse gas emissions on the climate. Earth scientists have a key part to play in helping to understand such changes, to place them in context and to manage or mitigate their impacts.

And what will happen next? During the remainder of this century, as a result of greenhouse gas emissions from human activities, we can expect the climate generally to get warmer, wetter and possibly stormier, the sea to rise and engulf some lower-lying parts of the coast, and rivers to flood more frequently. If such emissions continue uncontrolled, they carry as yet

Geodiversity: foundation of Scotland's Earth heritage

Geodiversity is the variety of rocks, minerals, fossils, landforms, sediments and soils, together with the natural processes which form and alter them. It provides the foundations for life on Earth and for the diversity of natural habitats and landscapes, and it has had a profound influence on economic activities and on the history, culture and settlement patterns in many parts of the country.

For a relatively small country, Scotland has a remarkable geodiversity, reflecting a long and varied geological history that spans some 3,000 million years of the Earth's existence. It is the richness of this geodiversity which gives Scotland the wonderful Earth heritage described in this book. For example, Scotland's geodiversity provides a compelling testimony, written in the rocks and landforms, of how colliding continents, ancient volcanoes, powerful glaciers and changing climates have shaped the present landscape, how different life-forms have evolved and how rivers, floods and sea-level changes are continuing to alter the land surface. Scotland's geodiversity has therefore played an important part in our understanding of the history and workings of the Earth.

Scotland's Earth heritage is also part of our wider scientific and cultural heritage. The modern science of geology was born in Scotland a little over 200 years ago when James Hutton investigated the rocks around Edinburgh and further afield. Interpretations of the rock record have changed over time, and Scotland and Scottish geologists have made major contributions to the study of geoscience. Many of the fundamental ideas and principles have been developed in Scotland and applied worldwide. Today, many of Scotland's rocks and landforms are an asset of national and international importance. They have provided crucial evidence for interpreting geological processes of global significance, such as volcanism and ice ages. In addition, some of Scotland's rocks contain a rich variety of fossils, which have greatly elucidated the evolution of the plant and animal kingdoms. Many sites and areas therefore have great historic value through their crucial role in the development of geoscience and their association with key historical figures or ideas. Where possible, we have highlighted this historical dimension of Scotland's Earth heritage and tried to recognise the key places and the contributions of many figures who have largely been forgotten outside scientific circles, but deserve to be more widely known.

Society has benefited in so many ways from Scotland's geodiversity: building stones and aggregates, raw materials for industry, the productive soil, the awe-inspiring landscapes, the myths and legends and the historical truths that attach to certain key places. Without our natural resources we would not have been able to feed ourselves, or provide fibre and fuel and many of the other items that we take for granted. But there have been activities in the past, and many today, that are unsustainable and risk reducing the value and longer-term benefits that we gain from our rocks, landforms and soils. Scotland has a priceless Earth heritage and we have an obligation to conserve it for our own and future generations' benefit and enjoyment.

unquantified risks of more dramatic global changes over the next few centuries, such as melting of the Greenland and West Antarctic ice sheets and weakening of the Gulf Stream, which would have significant effects on sea levels and climate in Scotland. In some tens of thousands of years' time, the climate will probably get colder again with the possibility of the Ice Age returning and glaciers and ice caps reappearing in the uplands. Even further ahead, tens of millions years on, we can expect the Atlantic Ocean to continue to widen and then, eventually, begin to close. Who knows – maybe Scot-land in the long-distant future will become wedded with other countries as the Earth's plates continue to rearrange the pieces of the global jigsaw?

We have already touched on the great array of Scotland's rocks and landforms, perhaps the most diverse on Earth for a country of Scotland's size. In most parts of Scotland, this rich geodiversity is readily apparent. Look in the right place and you will be able to see evidence of continental collisions and splits, discover the huge gaps in time between older and younger rocks, and find rocks from deep in the Earth's interior and the floors of the oceans, as well as rocks and landforms from the hottest, coldest and wettest environments. Throughout the text we have identified places and sites where particular features can be seen. In addition, we have selected twenty places to visit (see chapter 6), deliberately spread around Scotland, which we hope will encourage you to go out and 'read the landscape' and to appreciate and enjoy Scotland's outstanding Earth heritage.

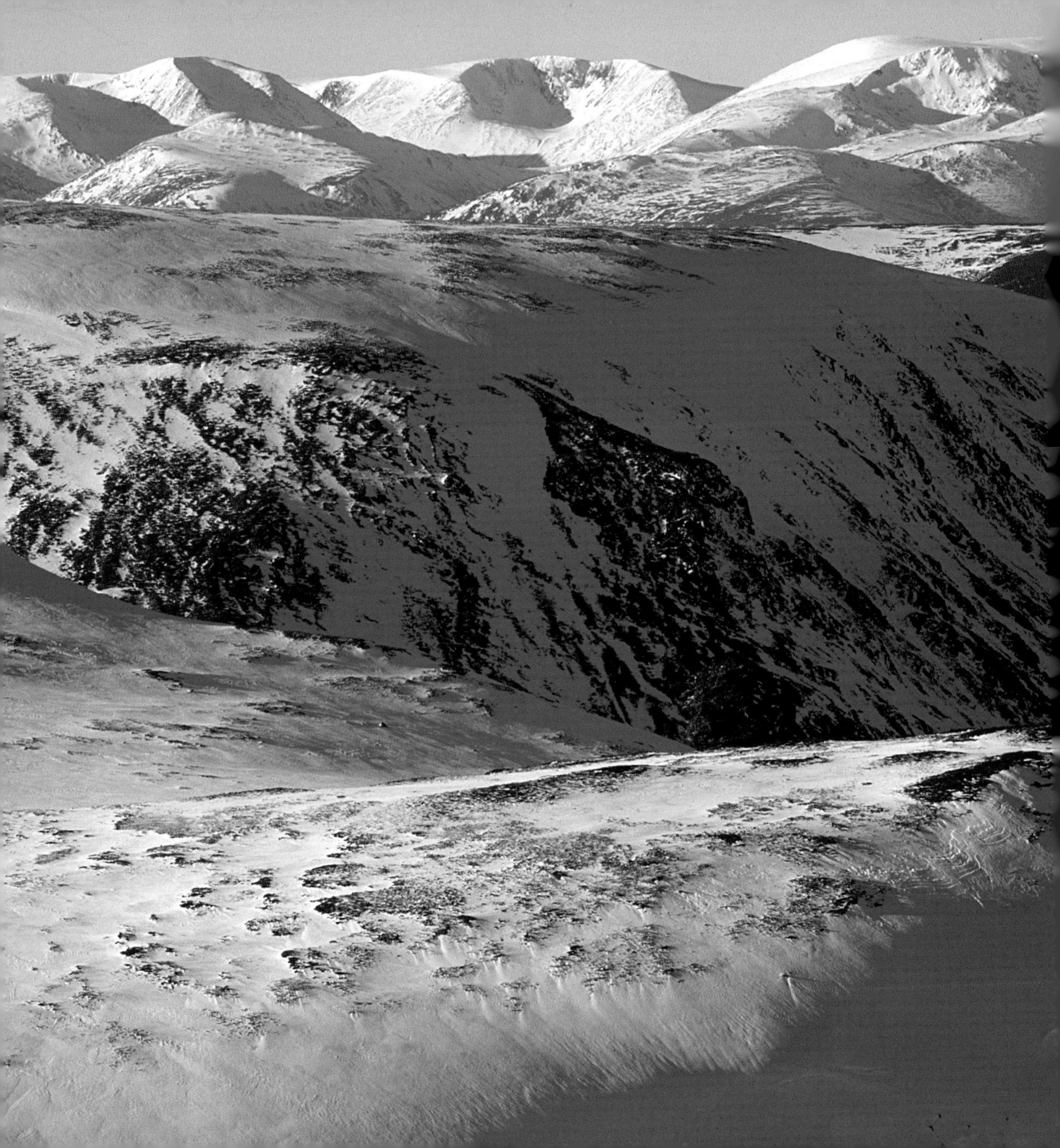

CHAPTER 1

GEOLOGY ENRICHES ALL OF OUR LIVES

The Standing Stones pointed long shadow-shapes into the east, maybe just as they'd done of an evening two thousand years before when the wild men climbed the brae and sang their songs in the lithe of those shadows while the gloaming waited there above the same quiet hills.

Lewis Grassic Gibbon, *Sunset Song*, 1932

Geology enriches all of our lives in a variety of ways. The rocks, landforms and soils of our country affect where we are able to build our homes and settlements, what crops we can grow and even the jobs we do. They also provide the raw materials for us to live our lives and the places where we take our leisure. The underlying rocks, landforms and soils also partly determine the nature and diversity of the ecosystems that the landscape is able to support. Geology is therefore one of the key determinants of our quality of life and the nature of the environment in which we live. It is important that this resource is managed effectively and that the power of the natural processes that continue to shape our landscapes is understood and respected. Latterly, we have also recognised the need to conserve the best and most representative elements of the rocks and landforms for teaching, research and demonstration purposes, as well as for their wider Earth heritage value. This chapter introduces some of the obvious, and also some of the more unexpected, ways in which geology, landforms and soils influence and enrich our lives.

Opposite. The high granite plateaux, steep-sided glens and deep corries make the Cairngorms one of the most renowned and distinctive mountain landscapes in Britain. They form the largest area of high ground in the country and have a strong 'arctic' character. Along with the climate, the rocks landforms and soils of the Cairngorms have exerted a powerful influence on the landscape, wildlife and land use of the area. Thousands of people visit the Cairngorms each year to walk, climb and ski, and to appreciate the magnificent scenery and wildlife. Few, perhaps, are aware of the Earth heritage which underlies this unique environment and its exceptional qualities.

Right. A core through a peat bog reveals a great deal about an area's recent geological history. It contains a pollen archive that allows the vegetation history to be accurately reconstructed.

GEOLOGY AND HUMAN HISTORY

Peat bogs reveal much about prehistory

Peat bogs, most rather unprepossessing in appearance, are nonetheless invaluable storehouses of information about past changes in vegetation and climate and, indirectly, about the activities of early human settlers in Scotland. A rich postglacial history, extending back over the last 10,000 years, can be deduced from these peat bogs. Pollen grains from plants that lived nearby became trapped in the layers of peat as they built up over the millennia. As a rough guide, up to a metre of peat accumulated over a thousand-year period. Sampling the pollen archive from the various layers in the peat allows a picture of the changing vegetation cover to be established from the earliest deposits at the base of the bog through to the younger layers at the top. This pollen record tells the story of the vegetation that formerly clothed the country. Some of the plants recorded in this archive are now rare or extinct in Scotland.

After the Ice Age glaciers melted, pioneer herbaceous plants gained a toehold in the desolate moonscape of glacial debris, softening the harshness of the landscape. Juniper and birch scrub with abundant tall herbs followed. In the early Holocene, the period of time after the ice melted, trees spread back into Scotland and extensive areas of woodland thrived on the lower ground. Those in the south of Scotland were dominated by oak, hazel and elm. In the Grampian Highlands and North-west Highlands, pine and birch were the main tree species. Further north and in the Northern and Western Isles, the woodlands largely comprised birch and hazel. Human influence on the natural ecosystem is also picked up in the pollen records. There are marked declines in elm pollen about 5,800 years ago, and around 4,500 years ago further major reductions occurred in woodland pollen. These changes reflect the forest clearances during the Neolithic and Bronze Age, although other factors may have contributed. There is also a commensurate rise in the occurrence of cereal-type and heather pollen, clearly indicating the influence that agricultural activity was already having on the landscape. It was probably at this time that Scotland ceased to be a place of true wilderness and increasingly over the years, the countryside, even including the most remote areas, could no longer be described as having an entirely natural landscape.

Geology and the early settlers in Scotland

The natural landscape exerted a powerful influence over the early inhabitants of the British Isles. Scotland was initially populated by hardy communities of Mesolithic hunter-gatherers living mainly around the coast. They made use of the abundant natural resources there, including stone for tools and caves for shelter. The occupation sites of these early inhabitants are rarely preserved. However, Skara Brae on Orkney is one of the best examples of a later Neolithic habitation in Scotland. And it was because of its coastal position that this monument of national importance survived in good condition. After its abandonment, the buildings were completely covered by sand for many millennia, only to be rediscovered when the site was partially exposed in 1850 by a great storm.

The woodlands that covered much of Scotland were gradually cleared to make way for a more agrarian existence, as faithfully recorded in the pollen record. So, over time, people modified

Traprain Law in East Lothian was occupied by an early tribe that dominated much of South-east Scotland. This site provided all their basic needs.

Human activity in early Scotland

500,000–10,000 BC	Old Stone Age (Palaeolithic)
10,000–4,000 BC	Middle Stone Age (Mesolithic)
4,000–2,500 BC	New Stone Age (Neolithic)
2,500–700 BC	Bronze Age
700 BC–AD 79	Iron Age
AD 79–367	Roman occupation

their surroundings to make their immediate environs conducive to a more comfortable and productive way of life.

Archaeological evidence suggests these early settlers used flints as tools and arrowheads. An important source of flint for local use was from a gravel deposit at the Den of Boddam, near Peterhead. Other local sources of stone that had flint-like qualities, such as the pitchstone from Arran and the Rum bloodstone, were also fully exploited.

In these early years, the requirements of the first settlers were simple. A location that could be easily defended against natural predators and rivals was essential. So too was access to land for foraging and, later, for growing crops. A ready source of clean water was also needed. Traprain Law, in East Lothian, fulfilled all of these criteria and it is no surprise that this blister of chilled magma that dominates the East Lothian landscape was occupied for many generations. It had its heyday during the Iron Age, when it was probably the headquarters of a tribe that ruled over much of South-east Scotland. From earliest prehistory to the post-Roman period, its location clearly made it an important strategic site.

There is also clear evidence of human occupation of Arthur's Seat in Edinburgh. A flint arrowhead and knife have been found there, as has the site of a presumed Bronze Age village. Evidence is clearly visible for cultivation terraces near Dunsapie Loch, where crops were grown on the steep slopes. Fortifications dating from at least the Iron Age have also been found, suggesting that despite the strategic advantages that this defunct volcano conferred upon the early inhabitants, additional security measures were required to ensure their safety and well-being.

A pulse of magma, long-chilled, also partially builds the Lomond Hills just to the east of Kinross, creating another impressive strategic site fashioned by geology. The cap of hard igneous rock that covers much of the Lomond Hills resisted erosion by the Ice Age glaciers and, as a result, the hills now tower over Loch Leven. An Iron Age hill-fort has been found on the summit of East Lomond, along with an enormous prehistoric cairn that has not been fully excavated in modern times. Again, geology provided an important strategic location that allowed the early inhabitants of Scotland to live their lives in relative safety.

Standing stones

The people of late Neolithic and Bronze Age Scotland remain with us in the form of the stone monoliths they left behind. Callanish, on the Isle of Lewis, is arguably their finest work and is known as Scotland's Stonehenge. Fourteen

The stone circle at Callanish, Lewis, is built of massive blocks of Lewisian gneiss.

slabs of the local stone, Lewisian gneiss, point heavenwards. Another twenty-nine stones form an avenue leading to the circle. Although its actual purpose is enigmatic, it is likely to have been some kind of ritual or religious centre where symbolic ceremonies were performed. The stones could even have been some kind of early astronomical observatory. Legend has it that the stones represent giants who refused to convert to Christianity and were turned to rock as a punishment. Although the stones are all of local derivation, the effort to quarry, transport and erect them would still have been considerable. A sophisticated social organisation and unity of purpose would have been required to manage and execute this early civil engineering project.

Kilmartin, in Argyll, is another important location where standing stones are found. Some of the stones show 'cup' marks, which are frequently associated with 'ring' markings. This type of art is quite common across northern Britain, although experts are still unsure as to the precise significance of these markings. They are also found carved into natural rock outcrops, often in prominent locations. Machrie Moor, near Blackwaterfoot on Arran, is another impressive standing stone location. One of the stones has a number of parallel grooves cut into its surface. A nearby stone burial chamber or cist contained the skull of a young man, along with two flint arrowheads.

There are many other carved stones across

Craw Stane, near Rhynie, is a block of granite with sculpted symbols that would have been important in the everyday lives of the early inhabitants of Scotland.

Scotland, some bearing the carvings of elaborate symbols and devices. The indigenous Pictish people erected many of these monuments. They harried the Romans during their occupation of the area of Scotland south of the Antonine Wall, a defensive structure which ran from the Forth to the Clyde. The warring clan was named Picti by the Romans, which means 'the painted people'. By studying their stone carvings, historians have discovered a great deal about the lives of the Pictish people. That they were a warlike and disputatious people is beyond doubt! Their carvings demonstrate them to be armed to the teeth with sword and shield. But they were also farmers and fishermen and they buried their dead with respect.

Early cave-dwellers

Sea caves and rock shelters have been used by humans from earliest times. Again, nature provided a secure place for them to live and practise their crafts. At East Wemyss in Fife, waves cut caves in the solid sandstone cliffs when relative sea levels were considerably higher than today. As the land later emerged, the caves became habitable. The walls are decorated by symbols dating from Pictish times and later. Animals and everyday objects such as mirrors, swords and cauldrons are all represented in the crudely chiselled wall art that survives to this day. It is possible that the Vikings also left their mark on the walls of these caves.

King's Cave, in the south-west corner of

St Kilda – life on a rock at the edge

St Kilda is the most spectacular of all Scotland's offshore island groups. But even this most distant and inhospitable outpost was visited by people around 3,000 years ago. The discovery of a variety of tools is testament to this early presence and to later Iron Age occupants. Vikings were the next to settle these rocks in the middle of the ocean. Accounts of the more recent human history of the islands date back to the sixteenth century.

But geologists can trace an even more ancient history. St Kilda first appeared as an active volcano some 65 million years ago at the beginning of the Palaeogene and, as with the other eruptions of this age, it was subsequently planed down to its roots. Only the eroded stump of the volcano remains, comprising granites and gabbros in almost similar measure. It sticks out of the sea like a sentinel, west of North Uist in the Western Isles. The islands are now a World Heritage Site and owned by the National Trust for Scotland. They are currently the site of a radar tracking station for a missile range on Benbecula.

What marks out this archipelago on the edge from other island groups is the height of the cliffs. Conachair, on the northern coast of the largest island, Hirta, is the highest sea cliff in Britain at 430 metres. The sheer rock face plunges into the foam of the Atlantic breakers that batter the cliffs incessantly. The coastline comprises a wide variety of coastal landforms of outstanding quality, including sea caves, geos, arches, tunnels, stacks and blowholes.

Despite the remoteness and associated hardships, the islands were permanently inhabited until 29 August 1930 when the remaining St Kildans finally succumbed to the ravages of endemic illness and the relentless battering of the elements. There can be no place in Scotland where the quality and nature of the inhabitants' lives were so completely determined by the vagaries of the weather and their harsh physical surroundings. The small population of islanders lived around Village Bay – a haven in relative terms, in part protected from the relentless westerly winds. The thin soils and short growing season restricted the crops that could be grown. Fresh meat was also in short supply, so the islanders harvested a resource that was plentiful – the bird life. Gannets, fulmars and puffins were to the St Kildans what reindeer are to the Sámi. Boys were taught to climb from an early age and negotiated the precipitous cliffs of Dùn and Oiseval with ease. This acquired skill allowed them to catch seabirds to fill their larders with enough protein to see them through the harsh winter months. The St Kildan way of life was not so much a triumph of humanity over nature, as a case of people eking out a living in one of the most hostile physical and climatic environments that Scotland has to offer.

The cliffs on Hirta are amongst the highest and most spectacular in Scotland. The highest cliff towers 430 metres above the Atlantic Ocean.

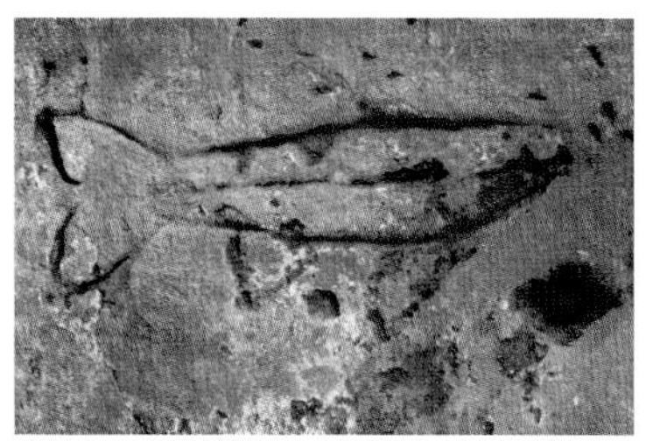

Wemyss Caves wall art – carving of a fish

The Iron Age broch on Mousa in Shetland is one of the best preserved in Scotland. It offered protection from attack and also a few home comforts. These structures were largely constructed from local stone.

Arran, is cut into the soft sandstone of Permian age. Both Christian and Pictish symbols have been carved into the walls of the cave – a horse and a cross amongst others. It is also thought to be the place where Robert the Bruce, that great Scottish patriot of much later times, hid whilst he planned his campaign to gain the Scottish crown.

Stone brochs and houses

During the Iron Age, great defensive fortresses were constructed in many parts of Scotland. These broch towers, as they are known, served as a living space for families that was relatively secure from attack. The stone materials used for these constructions were, for the most part, locally derived. Although brochs were double-walled, windowless structures, excavation of one on the island of Rousay on Orkney revealed that there were some home comforts to be had. The extended family that occupied this fortified farmhouse had a water cistern, a fireplace and partitioned rooms.

Geology provides strategic locations for castles

Edinburgh and Stirling Castles, two of Scotland's most iconic buildings, are located on prominent geological features. The ancient volcanoes of Carboniferous times that created both strategic locations were active around 350 million years ago. In more recent times, the action of Ice Age glaciers stripped away the surrounding softer rock to form precipitous cliffs that could not be easily scaled by an enemy. The

The site of Edinburgh Castle has been continuously occupied for over a thousand years. Precipitous cliffs on three sides have made it virtually impregnable from attack.

site for Stirling Castle also lay near the highest navigable point of the River Forth and overlooked the crossroads between the principal east–west and north–south routes. These natural attributes created the ideal location for a castle that dominated Scottish history from 1124, in the time of King Alexander I. An associated volcanic ridge also provided the space for the town of Stirling to develop.

The site of Edinburgh Castle has similar attributes. This place has a record of almost continuous occupation for over a thousand years. It was initially chosen because the location could be easily protected from warring neighbours. It was then progressively fortified until, in 1603, it was firmly established as the most significant strategic and military building in Scotland. Edinburgh Castle was described as "the first and principal strength of the realme . . . where His Majesty's jewels, moveables, munitionis and registaris are kepit."

There are many other castles, such as Tantallon in East Lothian, Dunnottar in Kincardineshire and Roxburgh near Kelso, that owe their strategic strength, and therefore their historical importance, to their underlying geology. Tantallon and Dunnottar are both perched high on impregnable coastal cliffs and Roxburgh Castle sits on top of a prominent glacial landform.

The Stone of Destiny

The Stone of Destiny has a special place in the history of Scotland. The kings of Scotland have been crowned on this lump of sandstone since time immemorial. Its origins are shrouded in myth and legend. This famous stone was said to come originally from the Holy Land, transported to Scotland via Egypt, Sicily and Spain. There are also strong Irish connections. Saint Patrick was said to have blessed the stone and legend has it that it was used in the coronation of the kings of the Kingdom of Dalriada. The stone was then moved to Scotland. Its temporary home was on Iona until it was finally moved to Scone, near Perth. It was then used for the installation of the kings of Scotland. The Stone of Destiny was last used in 1292 when John Balliol was proclaimed as monarch of Scotland. During one of the most turbulent periods in Scottish history, the Stone was then removed from its ancestral place by Edward I, who carried it off as a trophy to a new resting place in Westminster Abbey. This act of theft robbed the Scots of one of their most potent symbols of nationhood, a wrong that was not put right for 700 years. The Stone of Destiny was repatriated on St Andrew's Day, 30 November 1996, installed in Edinburgh Castle and placed alongside the other honours of Scotland.

But why has this lump of rather unprepossessing, mica-rich, dullish red sandstone exerted such a hold on the nation over many centuries? The authoritative view from Historic Scotland is that the Stone of Destiny is probably a Pictish crowning stone.

All that can now be determined with a degree of scientific precision is its geological provenance. A number of eminent scientists have attempted to pinpoint its geological origins over the years. Sir Archibald Geikie from the Geological Survey examined the Stone and concluded that "it had been quarried out of one of the sandstone districts between the coast of Argyll and mouths of the Tay and Forth, but there is no clue in the stone itself to fix precisely the source." In 1998, scientists from the British Geological Survey attempted to determine where it came from by matching the mineral content of the Stone of Destiny with various quarries and bedrock strata from around Scotland. Scone seemed a good place to start and the match with local strata was found to be strong. Survey geologists were assisted by the discovery of minute samples from the Stone of Destiny in their archive. A small cardboard pillbox labelled 'Coronation Stone S17850' contained three chips of pink sandstone from the Stone itself. There were also six previously prepared microscope slides that allowed a detailed examination of the mineral content of the Stone to be made. This more detailed study allowed the conclusion to be reached that "the Stone of Destiny is a block of Old Red Sandstone from near Scone itself . . . and moreover [to] strongly suggest that accounts of the Stone coming from the Middle East through Spain and Ireland to Scotland are no more than legend." This harsh scientific reality debunks some of the more extravagant myths that surround the Stone, but its historical significance as a symbol of royal coronation and celebration can never be denied.

GEOLOGY PROVIDES THE RAW MATERIALS FOR LIFE

The bedrock of the country has provided Scots over the centuries with the raw materials they needed to develop an industrial infrastructure and thus improve their lives. The distribution of the natural resources of coal, oil shale, building stones and minerals helped to determine the settlement patterns across the country and also 'selected' the areas that would be further developed during the Industrial Revolution. It is no accident that the Forth and Clyde valleys were reinforced as the main population centres during this period of rapid industrialisation: reserves of coal, iron ore and limestone were close at hand to bankroll and power the emerging smokestack industries. The Clyde estuary was an ideal location for the shipyards that sprang up along its shores. The steel plate required to build the ships was produced locally in the flat hinterlands. The physical characteristics of the land combined perfectly with the mix of natural materials that were available locally to produce an industrial powerhouse. Glasgow soon became the 'second city of the Empire' largely on the back of nature's bounty and the sweat of the local labour force.

Coal mining bankrolls the Industrial Revolution in Scotland

Scotland's coal reserves are a legacy of the country's epic journey across the globe. During Carboniferous times, the land that was to become Scotland sat astride the Equator and tropical rain forests flourished. Thick organic layers were formed as the forests waned and became buried under layers of sand and mud. Over time, as this process of burial continued, the organic layers gradually turned to coal. These rich reserves sat undisturbed until they were systematically exploited during the Industrial Revolution.

At this time, coalmines were sunk across the Central Lowlands. The coal reserves were seemingly inexhaustible and coal output increased to a peak of 40 million tonnes in the early twentieth century. Scottish mining engineers took their skills to Yorkshire and other mining areas in England to help develop the coal resource south of the border. This exploitation of a natural resource affected the way in which many people across the country lived their lives. The rural idyll, if indeed it ever existed, was gone and replacing it were tightly knit mining communities, united in their grinding poverty and social deprivation.

Environmental degradation of these industrial areas also followed. Mountainous coal bings, the waste from the coalmining process, reared up across the country. River courses were polluted as the by-products of the various heavy industries were discarded without a second thought. Today, many complain about environmental bureaucracy and red tape stifling development, but in those unregulated days of the Industrial Revolution, parts of the Central Lowlands were heavily polluted and became dangerous places to live and work.

Deep coalmining in Scotland has now ceased. But coal continues to be won from opencast sites in the Central Lowlands, although in nothing like the prodigious quantities of previous centuries.

Peat – a timeless source of energy

Peat has been used as a source of energy for millennia in Scotland. It is readily available in many parts of the country and, best of all, it is a sustainable resource, if managed properly for local use. Traditionally, crofters throughout the Highlands and Islands have used the energy from burning peats for heating and cooking for centuries and this practice continues to this day. Most crofting traditional dwellings would not be complete without their peat stack at the back door. The black houses of a bygone age had an open hearth in the middle of the room and smoke from the fire permeated every corner of the dwelling, covering everything and every occupant with a residue of soot.

The ubiquitous peat resource comes from the more than one million hectares of bog that covers a large part of the Highlands and Islands. Over time, great accumulations of *Sphagnum* moss build up and the lower layers are transformed into peat. In many parts of Scotland, over 7 metres of peat have accumulated, in some cases representing an unbroken sequence since soon after the last glaciers melted.

Peat has been a source of domestic energy for centuries in the Highlands and Islands. It is carefully cut, stacked and dried, ready for use.

The shale oil industry in West Lothian

James 'Paraffin' Young discovered how to extract oil from the Carboniferous shales of West Lothian and he set up one of the world's first commercial oil refineries to process this local resource. He produced a range of prized products including naphtha, wax, lubricating oils and paraffin. In the 1860s, Scotland's shale oil business boomed. But, again, there was an environmental downside. Over a fifty-year period around 3 million tonnes of shale were mined and treated. The waste products from this process were piled into hills of red blaes that are a very visible reminder of the area's industrial past. As the wisdom of using recycled material for bulk-fill in construction projects is now recognised, some of these bings have been removed. But the flat-topped hills will still be characteristic of the West Lothian skyline for years to come since they are now recognised as a significant element of the local industrial heritage.

Blaes bings in West Lothian

Recycling for a more sustainable future

Mineral deposits are not renewable on a human timescale, so sustainable development in this context is difficult to achieve. Inevitably, the capital asset is diminished as more of the resource is worked. But the red blaes of West Lothian, and the many other industrial by-products, such as colliery spoil and furnace slag, that litter the Central Lowlands of Scotland give us the ideal opportunity to achieve a more sustainable future. Of the 220 million tonnes of various types of this secondary resource that have been identified, some has already been put to beneficial use. Recent developments, such as the Gyle shopping centre on the western approaches to Edinburgh, have used considerable quantities of red blaes in their construction. Many other developments have taken advantage of this discarded material. The benefit to the environment is considerable. The blaes substitutes for virgin deposits of rock, sand and gravel that can be left in the ground for another day. In some cases, the source of the material may be closer to the construction site, thereby reducing transportation costs and associated environmental disturbance. For many uses, secondary material and construction wastes are equally suitable as freshly quarried building materials. Exploiting this resource, albeit a finite one, could meet the demand for construction fill for the next twenty years. Making this change by exploiting this resource to the full would be a real contribution towards sustainable development in Scotland.

The Glensanda superquarry

As much of Scotland is made from rock that is suitable for quarrying, it is no surprise that the extractive industry thrives in this country. Over 30 million tonnes of rock are extracted

Lingerabay, South Harris. The mountain, Roineabhal, is made of anorthosite, a rock type similar in composition to granite. This area was the subject of one of the longest-running public inquiries ever held in Scotland.

throughout Scotland every year, in order to build and repair our infrastructure of roads, schools and hospitals. Aggregate is now an internationally traded commodity. However, for increased profits to be made, the holes in the ground to extract this material have become bigger and bigger, and superquarries are now in vogue in some parts of the world. Only one is in operation in Scotland, at Glensanda on Loch Linnhe. Here, since 1985, the Strontian granite has been quarried on an industrial scale that is unmatched by any other single quarry in the country. Its annual output of around 5 million tonnes is exclusively for the export market. Bulk container ships transport the crushed stone through the Firth of Lorne to the open sea and onwards to markets in Germany and the USA.

One of the longest-running public inquiries ever held in Scotland debated the economic benefits and potential environmental downside of a proposed second superquarry to be sited at Lingerabay in South Harris. After an inquiry lasting for six months, the application was finally turned down. The detrimental effect on the island's way of life and the natural environment was thought to outweigh the economic benefits that might accrue from quarrying this natural resource.

Sand and gravel

Scotland is also blessed with an abundant supply of sand and gravel deposits. Most owe their origin to the meltwaters from the last ice sheet that covered Scotland. Around 10 million tonnes are extracted each year and these materials are used for a wide range of end uses, such as aggregate for concrete, road making and the construction industry. The distribution of natural sand and gravel deposits is not evenly spread across the country, but there is no region that is entirely without its own indigenous supply of these materials. However, many of the best deposits in the most strategic locations have already been worked out and in some areas it may be difficult to find additional resources that can be removed without some lasting environmental damage.

STONE FOR BUILDING

The architectural styles of the towns and cities of Scotland are many and varied. Every part of the country has its own vernacular style that is in part influenced by the availability and nature of the local building stone. The bedrock of Scotland provides a huge variety of materials for architects and builders to work with. For over 5,000 years, this resource has been exploited to create a great variety of structures – fortified castles, dwellings of every type, civic buildings, bridges, monuments and places of worship and work. Sandstones and granites of every hue, quartzites, slates, marbles, serpentine, basalts and limestones are among the materials that have been exploited. From the early inhabitants onwards, Scots have built their settlements and other structures from this bounteous resource.

Edinburgh is known as the Athens of the North and its classical architecture is justifiably celebrated. There were many sources for the stone used to build the New Town and the older parts of the city, but Craigleith Quarry was perhaps the most productive. Situated on the western outskirts of Edinburgh, it was the principal source throughout the eighteenth and nineteenth centuries. First worked to provide stone for Edinburgh Castle in the seventeenth century, Craigleith sandstone was last used in the late 1890s during the construction of Leith Docks. The majestic Ionic columns at the front of the University of Edinburgh's Old College represent the largest single pieces of stone cut from the quarry. Other famous Edinburgh landmarks constructed from this material include the City Chambers (1761), the City Observatory (1792), West Register House (1814), the Scottish National Gallery of Modern Art (1825), the

Left. Old College, University of Edinburgh, was constructed from sandstone from Craigleith Quarry.

Right. This fossil tree trunk, which now stands in the grounds of the Natural History Museum in London, was recovered from Craigleith Quarry in 1854. The tree is around 330 million years old, from a time when reptiles first appeared and much of the land was covered by swampy forest.

Dean Bridge and Charlotte Square. Andrew McMillan, an authority on Scotland's building stones, recounts a story about the transport of these enormous blocks from the quarry to the west end of the city. "Sixteen horses were required to haul each stone, placed on a special carriage. Considerable doubt was expressed as to whether the old North Bridge would stand the strain, as each pillar weighed nine tons. In 1835, sixty carts were each making an average of four journeys per day into the city from Craigleith." This level of traffic gives a clear indication of the importance of this quarry in the construction of the New Town. The fossil tree trunks that sit outside the Natural History Museum in London were also quarried from Craigleith.

The former quarry is now fully infilled and the restored site occupied by a supermarket. With the exception of one small wall of rock that was left to demonstrate a representative section through the strata, the magnificent exposures from which the heart of the City of Edinburgh was hewn are lost for ever.

The stone industry has undergone something of a renaissance in recent years. Many new civic and commercial buildings have been erected or extended in Edinburgh in the 1990s and beyond. The most controversial by some distance is the Scottish Parliament. Plans for the building were unveiled in celebration, but the project was completed in 2004 amid controversy. The main concerns were the escalating cost and, to a lesser extent, the unusual design. However, now the dust has settled, the building endures as the focus of civic decision-making in Scotland and as a significant adornment to the historic Royal Mile. Kemnay granite cladding and Caithness flags were amongst the indigenous materials used in this building project. The project architect, Enric Miralles, chose this variety of granite to "empha-

Craigleith, Edinburgh, as a working quarry. It supplied the building stone that was used to construct much of Edinburgh's New Town. The quarry is now restored and is the site of a supermarket.

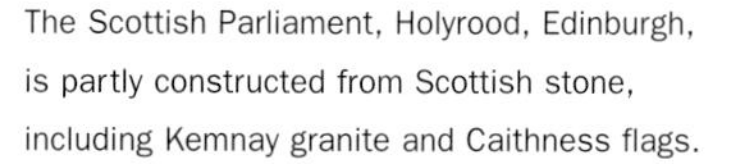

The Scottish Parliament, Holyrood, Edinburgh, is partly constructed from Scottish stone, including Kemnay granite and Caithness flags.

Carved in stone

Set into the concrete boundary wall of the Scottish Parliament on the Canongate in Edinburgh are a variety of stones from around the world and also some poetry etched in stone. In the 'Written in Stone' competition, organised in 2004 by Scottish Natural Heritage, youngsters from around Scotland were encouraged to compose a verse inspired by their appreciation of the rocks and landscapes of Scotland. Robert Adam's winning poem has now been carved in stone near the main entrance to the Scottish Parliament for all passers-by to read and enjoy.

Look, What can you see?
I see beauty in the lochs,
I see majesty in the mountains,
I see legend in the rocks,
And it is ours.

Mairead MacNeil from Castlebay, on the Isle of Barra, won the Gaelic language verse. Here is her offering, also carved in stone:

Beanntan Àrda
Fo cheò
Mar chaistealan glasa
A'fleòdradh sa mhuir.

The translation reads:

Towering mountains
Shrouded in mist
Like grey castles
Floating in the sea.

sise the complex's relationship to the landscape and for its sparkle when wet". Some 14,000 tonnes were extracted from Kemnay Quarry near Aberdeen.

The extension to the National Museum of Scotland in Chambers Street is built from Clashach sandstone of Permian age, quarried near Elgin. The darkest shade of sandstone was selected from the quarry to ensure that it quickly blended with the façade of the existing buildings. Precision workmanship was required to ensure that the building's corner tower achieved a regular curvature.

The glowering grey granites of Aberdeen are dreich in wet and windswept weather, but the mica flecks dance with reflected light in the sun. From the seventeenth century onwards, the city was built using the plentiful supplies of local granite. Rubislaw Quarry, now disused, was the source for much of that stone. It was one of the deepest quarries ever sunk and worked the granite to a depth of over 100 metres. Black powder was used to split the rock along natural planes to produce regular blocks. A complex system of powerful cables was used to bring the blocks of granite to the surface. Other granite quarries included the Kemnay and Corrennie quarries. Another iconic building, Marischal College, in the heart of Aberdeen, stands as a magnificent testament to the skills of the stone masons who erected this, the second biggest granite building in the world, after the Escorial Palace in Spain. The older parts of the city, including the splendid King's College, were built from imported sandstone, before techniques for working the much harder local granite had been developed.

Unlike Edinburgh and Aberdeen, Glasgow is not associated with one single building material. The civic buildings of Glasgow demonstrate a catholic taste in stone; there are red sandstones from Ayrshire and Dumfries, cream-coloured sandstones from Lanarkshire and granites from more exotic locations. The Clydesdale Bank building in George Street is built from granites quarried in Sierra Chico in Argentina. Next door is Dale House which is clad in a granite gneiss from Brazil. In George Square, the buildings comprise stones of various provenances – granites from Aberdeenshire and Finland, and sandstones from Stirlingshire and Linlithgow. In adjacent city-centre streets, limestones from Portugal and the Shap granite from Cumbria are also in evidence. But most characteristic of the architecture in Glasgow are the red sandstone tenements, halls, churches, monuments and bridges found throughout the suburbs and further afield into Paisley, Ayr and Kilmarnock. The stone for many of these buildings came largely from the red sandstone quarries to the south at Mauchline in Ayrshire and Locharbriggs near Dumfries.

Dumfries is an example of a town constructed predominantly in a single medium – the red sandstones from the nearby quarries of Locharbriggs and Corncockle. They give a pleasing uniformity and consistency to the townscape. These quarries have been worked for hundreds of years and the products exported to many parts of Scotland. The quarrying techniques have changed little during that time. The stone is worked by splitting it into manageable-sized blocks using wedges driven into natural weaknesses in the stone.

Dumfriesshire also boasted some fine granite quarries, such as Craignair Hill and Silver Grey quarries near Creetown. Although the materials are entirely different, the quarrymen worked the granite blocks in a similar way to the techniques used at Locharbriggs. Steel wedges were inserted

Jedburgh Abbey, in the Scottish Borders, is one of the architectural wonders of Scotland. Only part of the original structure now exists, but it still dominates the surrounding landscape.

in a line and a series of hammer blows delivered to each wedge in succession to create a clean straight-edged split. These blocks of more manageable size could then be dressed and transported to where they were required.

In addition to the building stones described above that are quarried in industrial quantities, there are a few specialist rock types that are more sparingly available. Marble, for example, has been quarried at a number of locations across Scotland. Perhaps best known is the marble quarry on Iona that has provided ornamental blocks and slabs for, amongst other buildings, the island's own abbey.

The quarries at Torrin on Skye are still producing white marble chips for the building industry. The marbles of Skye were formed when the Beinn an Dubhaich granite cut through older limestones. The heat and pressure caused the limestone to be transformed into marble. The rock is shot through with exotic minerals that stain the limestone a variety of different colours. The quarry lies close to a zone of alteration, known as a skarn, which is often formed when molten granite magma comes into contact with limestone.

The slate industry is more often associated with Wales than Scotland, but a limited number of quarries have been opened to win this material over the years. Ballachulish, at the mouth of Glencoe, is perhaps the best known, but the quarries on Easdale Island, in the Firth of Lorne, also produced prodigious quantities of this roofing material. One of the features of such workings is the quantity of waste material generated. The sprawling waste tips are still very much in evidence at Ballachulish, despite being largely reclaimed in the 1980s. These rocks started off as fine muds, but were squashed by deep burial within the Earth's crust. The pressure applied caused a grain or cleavage to be developed in the rocks, which allows them to be split into fine leaves of regular size. These quarries may be reopened at some time in the future to supply demand from the building trade for these traditional materials.

Caithness flagstone, another roofing material, has been extensively quarried in Caithness and activity has recently been resurgent. These fine-grained silts and sands date to Old Red Sandstone times and many slabs contain beautifully preserved fossil fish. The colour of the flags varies considerably across Caithness. This very versatile building material has been put to good use since people first arrived in the area, being used in the construction of numerous chambered cairns and brochs. The Maes Howe chambered cairn on Orkney, which dates from the Neolithic, is constructed from flags that are similar in all respects to those found in Caithness. Caithness flags still find favour with builders and architects and, as previously noted, were used extensively during the construction of the Scottish Parliament.

The Border abbeys

Local stone from the Scottish Borders helped to create another series of architectural wonders – the abbeys of Melrose, Jedburgh, Kelso and Dryburgh. These towering structures were built just under a thousand years ago. They were constructed on a truly epic scale. What remains after the pillaging raids by successive English kings, particularly Edward II and Richard II, is just a pale shadow of their former glory. The Premonstratensian Order, a religious order of the time, chose the site for Dryburgh Abbey close to a ready source of warm,

Minerals aplenty at Wanlockhead and Leadhills

Minerals have been exploited at Leadhills and Wanlockhead since Roman times. The variety of minerals hosted by the folded sandstones and shales of the Southern Uplands is unrivalled in Britain. Around seventy different mineral species, some extremely rare indeed, have been recovered from the nineteen separate veins that comprise this ore field. Leadhillite, Lanarkite, Caledonite, Macphersonite, Matthaddleite, Susannite and Scotlandite are all minerals that were first described from the Leadhills and Wanlockhead area. These discoveries have established a worldwide reputation for this area.

Lead and zinc ores were the primary reason that the area was mined. Galena or lead sulphide (PbS) was the basis for the lead-mining industry of the eighteenth and nineteenth centuries. Sphalerite or zinc sulphide (ZnS) is the principal ore of zinc and this mineral was exploited at Wanlocklead from the 1880s onwards. Native gold has also been discovered within the area, particularly during the fifteenth and sixteenth centuries. Native, in this context, means gold in a pure form, uncombined with other elements, such as sulphur or oxygen. Several nuggets have been found, mainly through panning. No native silver has been recovered at Wanlockhead itself, although Argentite, a sulphide of silver, has been recorded. When galena was smelted, small amounts of silver were isolated then sent to Edinburgh for making into coins and jewellery. The Museum of Lead Mining at Wanlockhead tells the story of mineral working through the ages and provides a display of some of the stunning specimens that have been recovered from the ore field.

In 1999, HM Queen Elizabeth presented a silver mace with gold embellishments to the new Scottish Parliament as a symbol of its authority. The gold was panned from the burns and streams around the Wanlockhead area.

These beautiful crystals of aragonite originate from Leadhills.

pinkish-coloured sandstone of Upper Old Red Sandstone age, according to I T Bunyan, the expert on the building stones of the Border abbeys. Ploughlands Quarry, located between St Boswells and Kelso, is thought to have met the building requirements for both Dryburgh and Melrose Abbeys. Limestones were also imported from a source near Kelso for use in Melrose and a volcanic rock, called agglomerate, was quarried from close to the site of the abbey. After the abbeys were sacked, the building stones were recycled to construct many of the town-houses in the settlements that sprang up around these religious centres. This is perhaps one of the earliest examples of building materials being successfully recycled.

The Scottish Stone Liaison Group has taken a keen interest in the architectural stone quarries of Scotland. It has established a register of working sites and their products to ensure that new buildings or repairs to historic structures can be carried out using the most appropriate materials. There are also a number of excellent guides to the geology of the building stones of our historic cities. For example, *Building Stones of Edinburgh* is a handsome volume by Andrew McMillan and others that describes the geology of the area and the sources of the stone used in the construction of many prominent buildings across the city.

GEOLOGY AND AGRICULTURE

In the same way that geology shapes the nature of our landscapes, the rocks, landforms and soils of Scotland have a fundamental influence on the way in which the countryside is farmed. The variety of styles of farming and estate management in Scotland is considerable. The

Top. The productive East Lothian plain, looking southwards towards the Lammermuir Hills, where agriculture is much less intensive.

Middle. Sporting pursuits are the main source of income from the land on many Highland estates.

Bottom. Machair lands in flower

crops grown and animals kept are largely determined by the capability of the land and the climate, which in turn are strongly tied to its geological character. Clearly external factors, such as economic subsidies, also play their part.

The fertile lands of the Merse around Kelso, underlain by base-rich basalt lava, sandstones and limestones, support some of the finest arable farming land in Scotland. Thick soils floor the Tweed floodplain in the south-east corner of the country and a full crop rotation of cereals and roots crops is possible. The higher ground to the south around Yetholm, once an active volcano some 400 million years ago, is famed for its sheep and the Cheviot breed takes its name from these hills. The fertile lands of the Borders are separated from the equally productive East Lothian plain by the Lammermuir Hills. These hills are built from unyielding sediments laid down on the floor of a long-disappeared ocean. The Lammermuirs are marginal agricultural land in comparison with the areas lying immediately to the north and south.

The Central Lowlands, particularly Strathearn and eastwards towards the Kingdom of Fife, is another area with a strong farming tradition. The deep, freely draining sandy soils create growing conditions that are ideally suited for soft fruit, such as strawberries and raspberries. A belt of rich red soils, overlying the brick-coloured Old Red Sandstone bedrock, runs from the west coast to Stonehaven. North of Strathmore, the land rises sharply and the Highland Boundary Fault separates the productive lands to the south from the Highlands. The combination of the higher ground, harsher climate and the underlying acid bedrock has left a legacy of marginal agriculture. Glenesk, Glen Clova and Glen Dye are south-east trending glens that are cut into the hard Dalradian rocks. This rough ground is only suitable for sheep grazing. These hills lead into the Cairngorms massif that dominates the area to the north. This higher hill land is beyond the limits of farming or sheep grazing activities and has been managed as sporting estates for centuries. Much of the income comes from sporting pursuits, such as stag and grouse shooting.

East of the Cairngorms towards Aberdeen, the land is lower and again more suitable for intensive agricultural use. This land-use pattern continues along the coast with mixed farming, pigs and poultry husbandry supported by the relatively richer soils developed over the Old Red Sandstones of the Moray Firth. Around Inverness and northwards to the Black Isle, cereals can be grown and general cropping can be sustained.

The only other area of Scotland that is intensively farmed is Ayrshire and the south-west. The sandstones and volcanic rocks of Carboniferous age that underlie the soils of Ayrshire and the Solway coast support lush pastures used to graze the dairy cattle for which this area is world renowned.

The rest of Scotland is described in farming terms as 'less favoured areas'. But in some places, nature has sweetened the pill by providing more fertile ground. The machair lands of the North-west Highlands and the Western and Northern Isles are like no other place in Scotland. The shell sands swept in by Atlantic rollers have created a land of rare biodiversity and cultural depth. Poet John MacCodrum captured the mood in this verse composed around 1750:

'Tis a beautiful land, the land of the machair,
the land of the smiling coloured flowers,
the land of the mares and stallions and kine,
the land of good fortune which shall never
be blighted.

GEOLOGY AND RENEWABLE ENERGY

Soils – not just dirt

Soils are often forgotten about or largely taken for granted. But it is the variety and complexity of this medium that gives the vegetation of Scotland much of its character, be it pastoral or semi-natural. Soils support agricultural and forestry production and have an important role in water filtration and storage – key aspects of the water cycle. They also absorb contaminants that are a potential threat to water supplies and agricultural land, and store large amounts of carbon.

Soils are a vital part of the natural environment and just as important as the plants and animals they support. Soils influence the distribution of habitats and provide a home for a wide range of organisms. An analogy can be made with a tropical rain-forest. Surprisingly, a common-or-garden soil profile is just as biodiverse as the most exotic tract of Equatorial habitat. Worms, beetles, caterpillars and an amazing number of microscopic bacteria and algae make up this underground biodiversity hot spot. Soils are not just dirt and 'out of sight and out of mind'. We farm them intensively for food; we pamper them in our gardens; we rely on them to absorb excess rainfall and we marvel at the wilder landscapes they support. They too are a key part of our geological heritage and a key link between the geosphere (the bedrock) and the biosphere above.

Hydropower – a hundred years of green energy

In an age when the burning of fossil fuels is known to accelerate global climate change, we already have a head start in developing our green credentials. The intense glaciation, to which Scotland's surface was subjected, created a large number of deeply incised glens. In one of the most far-sighted and intensive engineering initiatives, many of these glens were dammed in the thirty years after the Second World War to create a dual-purpose resource. Damming the glens of Highland Scotland created reservoirs to supply the population with water, and a huge capacity to generate hydro-electricity was also developed. The Laggan Dam, east of Fort William, is a fine example of this dual purpose. During this major engineering programme, the North of Scotland Hydro-Electric Board built fifty major dams and power stations, almost 200 miles of tunnels and 20,000 miles of power lines.

Wind power – energy of the future?

Wind power is presented by its supporters as the power of the future – clean, safe and infinitely renewable. Geology helps to determine at which locations the turbines will be most productive. Higher ground, usually underlain by rocks more resistant to erosion, is often favoured because wind speeds are normally higher. Coastal locations are also suitable places for wind turbines. There is concern, however, that the visual intrusion and associated environmental disturbance that they create are greater than the limited amounts of energy that they add to the national grid. Building these structures offshore is another option that is currently being exploited in some parts of the world.

Right. The dam at Loch Laggan provides the dual resource of water and electricity.

Below. Wind farms are now widespread across Scotland as we strive to meet renewable energy targets.

NATURAL ROUTEWAYS

The glaciers sliced great swathes through the mountain landscapes of Scotland during the Ice Age, creating natural communication routes that run uninterupted for miles through the most difficult terrains. Perhaps the finest examples are the road and rail routes that follow the path of the ice through the Grampians, linking Perth with Inverness, the capital of the north.

Gleneagles and Glen Devon meander through the Ochil Hills, linking Kinross with Crieff to the north. They too were carved by ice. An ancient drove road that was used to take cattle to market runs parallel to the modern tarmac road. This scenic route was carved in andesite, a volcanic rock that is as hard as flint.

The A86 from Fort William to Kingussie is another route that exploits the passage of the ice. It runs alongside Loch Laggan, following a route excavated by a powerful ice stream cutting through the tough Dalradian rocks.

One of Scotland's newest motorways, the M74, also follows where the ice led. This arterial road leading south from Glasgow to the border would have been nigh on impossible to construct if the ice had not carved a pathway. The six-lane highway runs through the Southern Uplands, one of Scotland's least populated areas.

The sea has also done its bit to help our road builders, both ancient and modern. Raised beaches, abandoned as the land rose, almost ring the island of Arran. No surprise then that a road circumnavigates the island, never leaving the flat coastal lands for long. The same is true for the coastal routes of Ayrshire.

One of the most impressive natural routes, although in this case not exploited by the road or rail builders, is the Lairig Ghru in the Cairngorms. Glaciers gouged this perfectly proportioned landform,

Opposite. Glen Devon and Gleneagles were created as the ice pushed south-eastwards into the Ochil Hills, cutting a swathe through the tough andesite rock.

Below. A Speyside distillery produces the 'water of life' from simple ingredients. The water is 'flavoured' by the nature of the rocks over which it flows.

creating a breach through the Cairngorm plateau. The glen is now used by long-distance walkers, but no more environmentally damaging traffic has been routed this way.

The Caledonian Canal, which runs from Fort William in the south-west to Inverness in the north-east, also follows a geological lead. The line of loch basins that link the sections of canal was gouged out by the ice. But they, in turn, exploited an earlier weakness. The lochs lie along the line of the Great Glen Fault, which was active around 470 million years ago and has been a line of weakness in the Earth's crust ever since. That the ice preferentially eroded this pathway is therefore no surprise. The canal was started in 1803, using a ground survey prepared by James Watt some years earlier. This ambitious civil engineering project was an early form of job creation scheme, designed to staunch the flow of people from the Highlands. As with many large civil engineering projects, it was completed ten years late and cost over twice the anticipated budget. But the route is still fully navigable and runs for 60 miles, linking Lochs Ness, Oich and Lochy with Loch Linnhe at its southern end and the Moray Firth to the north. Some twenty-nine locks were constructed to deal with the natural rises and falls in the land. The Great Glen Way, a walker's delight, largely follows the banks of the canal.

OUR NATIONAL DRINKS

The spirit of Scotland – uisge beathe

Scotland's association with whisky, or, as it is otherwise known, 'the water of life', is almost as old as the hills. And the hills from which the water is drawn to make the golden spirit have a key role to play in its manufacture. The ingredients for this iconic drink are simple – malted barley and water, with a pinch of yeast. There are a large number of malt whisky distilleries across Scotland and the product from each has its own particular flavour, colour, 'nose' and other characteristics. Many take their malted barley from the same suppliers, so the key variable in this magical process is water. Distilleries that sit side by side in Strathspey produce spirits with different characteristics simply because they draw their water from river catchments that have different geological signatures. This ability of rocks, glacial deposits and peat to 'flavour' the water is well known.

Water used in whisky making comes from the most natural of all sources – the sky. Scotland 'enjoys' a high rainfall across the country, with the west being wetter than the east. But regardless of location, water is available in adequate quantity at many locations for the distilling process. After the rain falls, the resulting water percolates through the soil and any glacial deposits to meet the bedrock. Sometimes that contact is fleeting, as the water courses quickly towards the point at which it is extracted for what is essentially an industrial process. Under these circumstances, the interaction with the bedrock is limited and the water remains acidic. However, where the host rocks are more permeable, such as sandstones, or where the rock is a jointed igneous rock, the water has more time to absorb minerals from them and may even become slightly alkaline.

The River Spey rises to the south-west of the Monadhliath Mountains, and flows north-eastwards towards the sea at Spey Bay. This river and its tributaries support a number of the hundred or so distilleries that produce malt whisky across Scotland. Glenlivet, Glenfarclas and Glenfiddich are some of the internationally famous names that draw their water from this catchment of Dalradian rocks, cut through by granites. Glenlivet even has a granite named after it – the Glenlivet Granite.

Other parts of the country, particularly the islands, also have a thriving whisky industry. The Inner Hebridean island of Islay is a case in point. Place names from the south coast of the island, such as Laphroaig, Ardbeg and Lagavulin, are redolent of the tangy seaweed and peat flavours that infuse these island malts. The Talisker Distillery on Skye draws on spring waters from the basalt lavas of Palaeogene age that built much of the northern part of the island. The Talisker malt is described as 'peaty, full-bodied and aromatic'. So the more recent geological deposits of peat also make a significant contribution to the flavour.

The existence of whisky distilleries in Scotland since the fifteenth century is well documented. James IV was the first royal personage to be associated with the taking of this strong drink, and later, a mention of 'uiskie' was made in the funeral account of a Highland laird in 1618. So whisky has been emblematic of Scotland for centuries and the industry continues to make a huge contribution to the economy.

Our brewing legacy

Beer has been brewed in Edinburgh for over 200 years. Many small breweries prospered in the city, attracted by the unpolluted underground springs issuing from strata of Lower Carboniferous age. The first location exploited by the brewers stood on the current site of the Scottish Parliament at the foot of the Royal Mile. It is reported that the Abbey Brewery was first built in the grounds of the Abbey of Holyrood-house in 1777. These independent brewing operations were gradually acquired by William Younger and Sons, the brewing multinational. The brewery and bottling plants have now moved to another part of the city, although the springs that once fed the brewing vats continue to flow. These springs are now utilised as 'grey water' to flush the toilets, amongst other uses, in the Parliament building.

Let Robert Burns have the last word on this subject:

Let other poets raise a fracas
'Bout vines, an' wines, an' drucken Bacchus,
An' crabbit names an' stories wrack us,
An' grate our lug:
I sing the juice Scotch bear can mak us,
In glass or jug.

GEOLOGY AND CULTURE

Landscapes celebrated by artists, writers and composers

The Scottish countryside is now universally celebrated and admired. The diversity of rocks and landforms across Scotland are at the root of this scenic resource. The grandeur of the Highlands was only made possible by the presence of the resistant bedrock of granites and altered rocks, known as schists, that were later carved by erosion into towering peaks and incised glens. This scenery has proved to be an inspiration for artists, writers and composers in more recent times.

But for centuries, the glens were the sole preserve of the common people and their function was entirely utilitarian. They were a source of wood, a place to quarry stone and a space to grow crops and keep animals. Other than on hunts and other organised forays into the countryside, the more genteel Scots folk stuck to their towns and cities. It was the Romantic poets and painters, in particular, who helped change attitudes.

Between 1760 and 1763, James Macpherson published a collection of poems that he claimed were translations into English of ancient Gaelic works dating back to the third century. Known as *Macpherson's Ossian*, these works described the noble deeds of ancient heroes set against the backdrop of an elemental landscape. This is one of the first celebrations of the landscape in Scottish literature. But Macpherson is remembered as a literary hoaxer, as, when challenged, he was unable to produce the original accounts he claimed to have translated. Whatever his motivation was for this deceit, his work gave the landscape of Scotland a literary presence that it had not previously enjoyed, a presence that was

Loch Katrine by Horatio McCulloch (1778–1840)

readily built upon by the generations of writers who came after him.

Thomas Gray in 1765 published an account of his trip to the Highlands that celebrated the unique qualities of these wild and untamed lands. "I am returned from Scotland, charmed with my expedition; it is of the Highlands that I speak: the Lowlands are worth seeing once, but the mountains are ecstatic and ought to be visited in pilgrimage once a year . . . A fig for your poets, painters, gardeners and clergymen that have not been amongst them, their imagination can be made up of nothing more than bowling greens, flowering shrubs, horse-ponds, Fleet ditches, shell grottoes and Chinese rails. Then I had so beautiful an autumn. Italy could hardly produce a nobler scene, and this is so sweetly contrasted with that perfection of nastiness and total want of accommodation that only Scotland can supply."

Dr Samuel Johnson, writer, wit and critic, is well known for his Dictionary that remained the authoritative reference work on the English language for over a century. He is perhaps equally well known north of the border for his dislike of Scotland and all things Scottish. He made a visit with Scots writer and his own biographer James Boswell to the Western Isles in 1773 and the account, *A Journey to the Western Isles of Scotland*, was published two years later. He is one of the most oft-quoted writers in the English language. He famously wrote that "when a man is tired of London, he is tired of life". He also said that the only good thing to come out of Scotland was the road to England. Although many Scots regard his views on their homeland with a degree of derision, the account of his trip with Boswell to the Scottish islands was significant as one of the first popular and widely read descriptions of the people and countryside of Scotland.

Sir Edward Landseer RA (1802–73) was amongst the most popular romantic painters of the nineteenth century and also a royal favourite. Arguably his most famous work is *Monarch of the Glen*, painted around 1850. It depicts a stag at bay, placed in a dramatic Highland setting. It is now seen as a caricature of reality and has fallen somewhat out of favour. But with Queen Victoria and Prince Albert as patrons, his representation of Highland landscapes and lifestyles was popular at the time. Equally significant was the painter Horatio McCulloch. His epic canvas, which hangs in Perth Museum, depicts Loch Katrine in dark and melancholic shades under a stormy sky.

Working in the early twentieth century, the Scottish colourists Francis Cadell, John Fergusson, George Hunter and Samuel Peploe represented the Scottish countryside in a new and exciting way. Cadell had a particular fascination for the land and seascapes of Iona and returned there many times to capture its changing moods.

The Romantic poets were also working to similar ends. Robert Burns helped turn the wild lands of the Highlands into a place of belonging in his poem 'My Heart's in the Highlands' –

My heart's in the Highlands, my heart is not here;
My heart's in the Highlands a-chasing the deer;
A-chasing the wild deer, and following the roe,
My heart's in the Highlands, wherever I go.
Farewell to the mountains high-cover'd with snow;
Farewell to the straths and green valleys below;
Farewell to the forests and wild-hanging woods;
Farewell to the torrents and loud-pouring floods.

Sir Walter Scott also captured the changing mood with this verse entitled 'My Own, my Native Land!'

O Caledonia! Stern and wild,
Meet a nurse for a poetic child!
Land of brown heath and shaggy wood,
Land of the mountain and the flood,
Land of my sires! What mortal hand
Can e'er untie the filial band
That knits me to thy rugged strand?
Still, as I view each well-known scene,
Think of what is now, and what hath been,
Seems as, to me, of all bereft,
Sole friends thy wood and streams were left;
And this I love them better still,
Even in extremity of ill.

In 1810, Sir Walter Scott's epic poem *Lady of the Lake* was published. Set in the Trossachs, it celebrates the beauty and natural heritage of the area and became a best-seller. The effect on visitor numbers was equally dramatic and some would say that this publication kick-started the tourism industry in the Highlands of Scotland.

Music also helped to celebrate some of our most dramatic geological landscapes. Mendelssohn visited Staffa in 1829 and was so inspired by the cathedral-like qualities of Fingal's Cave that he penned the *Hebrides Overture* or *Die Fingalshohle*. The poet Wordsworth also paid a visit to this place over ten years later, but he could not attain the solitude that he sought because of a large

Fingal's Cave, Staffa, inspired Mendelssohn, who wrote his *Hebrides Overture* after a visit to the island in 1829.

gaggle of fellow travellers accompanying him. He wrote of his trip:

> We saw, but surely in the motley crowd
> Not one of us has felt, the far-famed sight;
> How could we feel it? Each the others blight,
> Hurried and hurrying volatile and loud.

The Scottish landscape continues to capture the attention of the media and contemporary commentators. The recent BBC series *A Picture of Britain*, fronted by David Dimbleby, was a celebration of the British countryside that has been "a source of deep pleasure, intense national and regional pride and public debate and controversy for hundreds of years". So the fascination with the countryside continues, even in our popular culture that is crowded by the cult of celebrity and passing fads and fashions. Another popular TV series, *Monarch of the Glen*, an everyday tale of landed gentry folk, was set in the Scottish Highlands. This tableau of 'whisky and whimsy' was filmed against the stunning backdrop of the Monadhliath Mountains and took a romanticised view of the Scottish landscape to an international audience.

GEOLOGY CREATES A PLAY-GROUND FOR MANY PURSUITS

Golf is part of the fabric of life in Scotland. The development of the game of golf stands alongside the invention of the television and the discovery of penicillin as one of Scotland's greatest gifts to the world. The game has been played in the 'home of golf' since the fifteenth century and increases in popularity with every passing year. The first courses were built on the links land of the east coast. The undulations of the sand dunes are perfectly suited to the game. The turf is hardwearing and the wild grasses are tolerant of drought, but unsuitable for arable farming. The links land had finally found a beneficial use. Despite the efforts of King James II to ban the game by Act of Parliament, golf continued to thrive. The Kirk was also hostile to this new Sunday pastime, as more and more churchgoers preferred the fairways to the pews. Repeat offenders even risked excommunication!

Every year, golfers from around the world compete for the claret jug, the famous Open Championship trophy. The Open is always played on a links course and Scotland is the venue in alternate years. A variety of courses host this most prestigious of tournaments, such as Carnoustie near Dundee, Muirfield in East Lothian and Royal Troon on the Ayrshire coast. But the Old Course at St Andrews has most cachet of all. Arguably the most famous course in the world, this sand spit has been created over thousands of years by the action of the sea. The process of longshore drift carries sand grains along the coastline, building the shoreline seawards as layer upon layer of sediment accumulates. Sand spits have undulating surfaces, as the sand dumped by the sea is whipped up into dunes by the strong onshore winds. The pioneer marram grass then stabilises the new landform. Marram thrives in exposed conditions and is remarkably resistant to drought conditions. Its

The site of the Old Course, St Andrews, has been created over thousands of years by the interaction between land and sea, waves, tides and wind. Nature has provided the perfect playground and sporting venue of world renown.

tussocks and deep roots encourage more sand to build up and help stabilise the dunes. The dunes nearest to the sea are often unstable with too much sand blowing about to be used as a golf course. Further inland, the dunes become more stable with grasses, herbs and mosses binding the surface and with only limited amounts of fresh sand being blown in. They are often called 'grey dunes' because of the greyish colouration of the lichens that cover the ground. Beyond these stable dunes are the links themselves – undulating plains of vegetation-covered sand. They are very stable and are ideal playing surfaces for the game of golf. The undulations of the land that are so characteristic of the links courses reflect the fact that successive ridges of sand have built seawards over the last few thousand years, driven by surging tides when relative sea levels were higher than today. This relatively simple natural process has thus created one of the most celebrated and recognised sports venues in the world.

Gleneagles, set amongst the rural splendour of lowland Perthshire, is another famous golfing venue. The King's Course, designed and built by James Braid, the doyen of golf course architects, was created in 1919 around a series of linear natural ridges, known as eskers. These prominent features were formed as the ice melted and a series of meltwater rivers flowed in tunnels underneath the ice. The ridges comprise sand and gravel that were carried and eventually dumped by the fast-flowing glacial meltwater rivers. This intricate lacework of ridges and hollows is perfect golfing country. The fairways follow the lower ground that lies between the sinuous esker ridges and the ridges themselves provide the perfect viewing

The golf course at Gleneagles, where nature has provided perfect sites to view the action – a network of glacial ridges, known as eskers.

platform for the spectators.

The broad floodplains created by rivers also provide an ideal venue for other sports. The Jed Water, a tributary of the River Tweed, has carved a wide floodplain that runs through the Borders town of Jedburgh. Although recent experience has demonstrated that building on river floodplains has inherent dangers, Jed-Forest Rugby Club has occupied this site for over a hundred years. The river has also carved a vertical cliff in the Old Red Sandstone deposits that underlie the site. It provides an arresting backdrop to this sporting venue.

Skiing

Scotland's winter sporting venues have been on the edge of economic viability since they were built in the 1960s. But they have survived many lean years of sparse snow. Aonach Mór, Glencoe, Cairn Gorm, the Lecht and Glenshee are the main focus of winter sports activity in Scotland. All five centres are located on some of the higher ground in the country and benefit from the higher snowfall and colder conditions that are characteristic of these altitudes.

The Cairn Gorm centre is largely located in two of the glacial scoops carved out of the north wall of the granite massif, known as the Northern Corries. Corrie Cas and Corrie na Ciste support runs for beginners and more experienced skiers alike. By contrast, the ski runs at the Lecht are on more open hill slopes and at lower altitude. The Nevis Range resort sits on the northern flanks of Aonach Mór. Glencoe is close-by, in a corrie on the northern slopes of Meall a'Bhuiridh on the edge of Rannoch Moor. Finally, the Glenshee ski centre is located just south of Braemar on rocks of the Dalradian.

The Old Man of Hoy – a climber's paradise

One of the most familiar views from the starboard side of the ferry as it heads from Scrabster northwards towards Stromness harbour is the Old Man of Hoy. It stands 137 metres high and dwarfs the tower of Big Ben by an amazing 40 metres. Originally, this feature had two legs, but lost one during a severe storm in the early years of the nineteenth century. Since that time, this natural wonder has stood largely unchanged, although it will eventually collapse into the sea. It is not the only stack along this stretch of coast, but it certainly is the most famous. It was thrust into the limelight when the leading figures of the climbing world scaled it in 1967. This dramatic event was broadcast live on television and the nation held its breath as Joe Brown and his team inched their way towards the summit.

Paddy's Milestone – the home of the curling industry

Ailsa Craig, an island in the Firth of Clyde, is built from a granite that is ideal for the manufacture of curling stones. Rock has been quarried from the island since the nineteenth century and the stones that have been fashioned from this fine-grained material have been sold around the world. This micro-granite, containing a distinctive fleck from the blue mineral called riebeckite, is ideally suited for the intense buffeting received during the average curling match! This small island is also known as Paddy's Milestone, as it located part-way between Ireland and Scotland.

Top. Ailsa Craig. This pillar of granite emerges from the Firth of Clyde. It has been the source of curling stones for centuries.

Middle. Toe Head and Tràigh Scarasta, Western Isles, is one of the most beautiful places in Scotland.

Bottom. Sandwood Bay, Sutherland. This thin strip of sand is regularly lashed by Atlantic storms, but on quieter days it is the epitome of tranquillity.

The first curling club was established in Kilsyth in 1510 and since that time, the sport has been 'exported' to many countries around the world. The early venues were the many frozen lochs and ponds found across the country during the winter months. The game continued its development at a time when Scotland was gripped by the 'Little Ice Age'. Winter temperatures plummeted during the late 1600s until around 1850, and harsh weather conditions set in. Rivers and lochs developed thick crusts of ice that easily supported the heavy curling stones, so ready-made curling venues were widely available.

Beautiful beaches

Scotland has some of the finest beaches to be found anywhere. Once again, it is nature alone that we have to thank for this dazzling recreational resource. The wide expanse of sand and the view from the shore at Tràigh Scarasta, which links Toe Head with the rest of Harris on the Western Isles, is mercifully free from any of man's interventions – roads, houses, ice-cream vans and, best of all, any other human beings. It is truly a wilderness experience.

Sandwood Bay on the Sutherland coast offers an even wilder experience. The coastal walk from Cape Wrath to Sandwood Bay is one of the most spectacular in the British Isles. There is no cliff-top path to define a route, so the walking is hard and unrelenting. The hike culminates in the desolation of Sandwood Bay, a linear strip of sand and dunes that is regularly lashed by Atlantic storms. It is a Site of Special Scientific Interest since this beach system is perhaps the least affected by human intervention in the whole of Britain. The dynamic system of the beach and the spectacular sand hills migrate and change naturally. It is a place of rare isolation from the modern world and well worth the 8-kilometre walk from Sheigra to the south.

Bird islands

Scotland's coastal geology also provides an invaluable resource for birdwatching. There are over 800 islands around the coastline of Scotland and that is just counting those with some sort of vegetation cover. Many are uninhabited and left largely to nature. The majority are located in the west and north, built in the main from rocks resistant to erosion. Their steep-cliffed coastlines provide ideal nesting locations for seabirds and Scotland is blessed with some of the largest colonies found anywhere in the world. The populations supported by these remote locations are of international conservation significance.

Handa has been a nature reserve since 1961 and it supports over 170 species of bird. Razorbills, fulmars, puffins and great skuas are amongst the many species that inhabit this remote outpost. These cliffs are cut in red Torridonian sandstone that is widely distributed across the North-west Highlands. The natural layering of the rock provides ledges aplenty for the nesting birds. The Bass Rock in the Firth of Forth is another birders' delight. This eroded volcanic plug of Lower Carboniferous age stands proud from the leaden waters of the Forth, stained white by a coating of guano deposited by its huge colony of gannets.

Left. The cliffs of Torridonian sandstone at Handa, Sutherland, are home to large colonies of seabirds.

Right. Beinn Laoigh (Ben Lui) has been described as the 'Queen of Scottish Mountains'. This peak in the Tyndrum Hills is 1,130 metres high, carved from the schists of the Grampian Highlands.

Scotland's highest peaks – the Munros

The mountains of Scotland are remarkably varied in their appearance and characteristics. Compare, for example, the steep and jagged ridges of the Cuillin, the isolated monoliths of the Torridon and Assynt ranges, and the vast stony plateaux of the Cairngorms. This diversity of mountain forms reflects the geological history of the landscape. Many different rock types, brought together by continental collisions and volcanic activity at different times in the past, provide the foundations of our mountain scenery. More recent processes of uplift, weathering and erosion have shaped the present mountain forms from these geological foundations, particularly during the period before the Ice Age and throughout the Ice Age itself. To this geological inheritance, we owe one of our most valuable recreational and scenic assets.

In 1891, Sir Hugh Munro of Lindertis published a table of all the peaks in Scotland above 3,000 feet. The Munros, as they became known, today enjoy unprecedented popularity and outdoor enthusiasts from across the country rush to climb them. So the popular pursuit of 'Munro-bagging' became established. In total, 284 peaks qualify for this accolade and all are to be found north of the Highland Boundary Fault, built largely from the Moine, Dalradian and Torridonian and igneous rocks of various geological ages. The glacially steepened crags on many of these mountains, including Ben Nevis, the Glencoe hills and the Cairngorm corries, also provide technical challenges to even the most skilled rock and ice climbers. But some Munros, such as Ben Chonzie, have gentler profiles.

The Mar Lodge Estate in the southern Cairngorms, has one of the biggest concentrations of Munros – fifteen in all. They are arguably part of Scotland's most important mountainscape, comprising a granite massif with high plateaux, corries and glens that support some of our rarest wildlife. In contrast, Ben Hope in Sutherland, the most northerly of all the Munros, sits in splendid isolation from any other peaks of substance. Perhaps the most spectacular are the Munros on the gabbro of the Black Cuillin on Skye, carved by frost and ice into our most 'alpine' mountains. The most technically difficult summit is the Inaccessible Pinnacle, which lies on the Cuillin Ridge. It is said that this is one of only two peaks on his original list that Sir Hugh Munro failed to climb, despite having attempted the ascent on a number of occasions.

The Corbetts

After the Munros came the Corbetts. John Rooke Corbett, a member of the Scottish Mountaineering Club, catalogued the hills with heights that fall between 2,500 and 3,000 feet. These often become the next challenge for hillwalkers who have completed the Munros. The list was published after Corbett's death under the heading 'Other Tables of Lesser Heights'. But there is no doubt that the Corbetts can be every bit as challenging as the Munros. Ponder Ben Loyal in Sutherland, Askival and Ainshval on Rum and Beinn Bhan in the Applecross range, for example. These peaks are hardly a gentle stroll. Corbetts are to be found in the south of Scotland, albeit in small numbers, although there is no doubt that the harder rocks of the north and west of Scotland build by far the largest collection of these celebrated hills.

Arkle, in the far north of Sutherland, is formed of Cambrian quartzite resting on a platform of Lewisian gneiss. The white screes of this Corbett and its neighbour, Foinaven (behind), are particularly distictive.

MANAGING THE ENVIRONMENT

Earth scientists have an important contribution to make to contemporary issues. Environmental management, such as land use planning and resource exploitation, requires their considerable technical expertise to be applied. No profession is better placed to advise on which areas of the country are too hazardous to build on because of the risk of flooding or coastal erosion, which areas may be subject to subsidence and where the best supplies of building materials are to be found. All of these considerations and many more besides help influence the way in which we determine land use planning policies and take resource management decisions. The fiercely competing pressures on our restricted land surface in Scotland make it ever more important that we 'get it right'.

Important decisions, such as the siting of a superquarry, waste disposal facility or new motorway, are always preceded by an intensive period of planning and study. Establishing the likely consequences of a development on the immediate environment is a key part of that process. Environmental assessment, as this exercise is known, is now a well-entrenched part of the development control process in this country. Geologists and geographers have a key role in helping planners, developers and local communities, who are likely to be affected by a particular project, to understand the predicted environmental consequences. It is therefore crucial that we are able to explain any complex issues in non-technical language so that all parties are able to participate in the debate. This recognition of the need for effective communication is in itself a major step forward. No longer can the specialists take cover under their polysyllabic technical vocabulary.

Coastal protection, Montrose. Management of the coastline will become an increasingly important and contentious topic as sea levels continue to rise.

There is also a requirement to undertake a Strategic Environmental Assessment (SEA) of development plans that concern potentially contentious topics, such as mineral development, waste disposal and transport. The purpose of this exercise is to find the best-fit compromise between essential developments and any potential damage to the natural and built environments. A waste disposal facility, for example, is unlikely to be welcomed wherever it is sited. The SEA process is designed to make sure that all the potential locations are considered and that the best option is more likely to be chosen.

There is also a growing realisation of the need to work with nature, rather than stand, like King Canute, in the path of powerful natural processes. In managing dynamic systems, such as rivers and coasts, the forces that nature can deploy easily outgun our capabilities. So instead of engineers routinely building sea walls even higher than before to protect the coastline, we now work with the power of the sea to find an effective solution. These concerns are likely to grow in the future as sea levels continue to rise.

The most appropriate solution, in some situations, may be to adapt to the natural processes. This may include a strategy known as 'managed realignment', which has been adopted in East Anglia. Sea walls have been breached to allow the land behind to flood and salt marsh to be re-established. These fringing habitats act as a natural buffer to the sea and are a cost-effective and natural form of defence against marine erosion in situations where it is uneconomic or more damaging to the wider coast to maintain hard defences. An understanding of how the natural processes will respond to these kinds of change is an essential part of the planning process and requires expert geological knowledge.

Environmental geology maps are a fairly recent innovation and represent a partial solution to the problems of communication. They present information on the nature and distribution of natural resources, such as coal, sand and gravel. Natural hazards, such as landslides, areas prone to flooding and poor foundation conditions, are also important factors that help guide the location of housing and other infrastructural developments. Environmental information is presented in thematic maps which translate complex technical data into a format that lay readers such as planners, elected councillors and members of the public can consult and use.

Disposal of waste products

Geologists can also help find places where society's waste products can be safely disposed of. Perhaps not the most glamorous of roles, but without it life would soon become intolerable. Many disused quarries find an afterlife as receptacles for our flotsam and jetsam. But not any old hole in the ground will do. The bedrock has to have special characteristics. Water management is the vital consideration. After either rainwater or groundwater comes in contact with rotting matter, which constitutes a fairly high proportion of all household refuse, it becomes polluted. If this liquid runs from the waste disposal site, either as a surface stream or groundwater, it will taint water supplies and generally create an environmental nuisance. Some rocks are better at containing this hazard than others. Clay is best, as it is impervious to water – in other words water cannot pass through these layers. The nuisance is effectively sealed within the site and can be most effectively managed with least environmental impact.

Since the beginning of the nuclear age some sixty years ago, we have produced significant quantities of highly toxic and radioactive waste in Scotland. Dealing with this waste is a very high-profile issue and the public have a keen interest in, but only a partial understanding of, the technical complexities involved. The role of the geologist is to identify the most suitable locations for disposal and to assist the competent authorities in implementing the most appropriate solutions. However, despite intensive research over many decades, no purpose-built deep repository for this material has been constructed and all of the most toxic and radioactive wastes are currently in temporary storage awaiting a decision on how to permanently dispose of them safely. If a new generation of nuclear power stations is considered to be the best option for electricity generation in the future, the issue of radioactive waste disposal will have to be tackled with greater urgency. Geologists will be at the forefront of the search for a suitable disposal site – be it at home or further afield.

CONSERVING THE BEST OF OUR EARTH HERITAGE

Most people are aware of the need to protect what remains of the planet's vital biological systems and other gems of the natural world. Ecosystems, those complex interdependencies of plants and animals, rely on rainfall and other natural gifts, for survival. But the rocks and soils that support these natural systems are also vital. And, as we will see in the following chapters, these rocks, fossils and landforms, our Earth heritage, are also of historical, cultural and scientific value in their own right.

There has been a legal requirement for over the last fifty years to protect a network of the key geological and landform sites throughout Scotland. The need to conserve our diverse Earth heritage is already widely recognised by academics, planning professionals and other land managers. However, to the wider public the need to conserve geological and landform features is less obvious than, for example, protecting a rare species of butterfly or an endangered wildlife habitat. But geological sites are often just as vulnerable to changes in land use and therefore are equally deserving of protection against inappropriate development or damage through misuse.

Selecting the key sites for conservation

Although the conservation agencies, principally Scottish Natural Heritage (SNH), take a holistic and 'wider countryside' approach to nature conservation, the designation of nationally and internationally important sites remains the principal statutory instrument in conserving our Earth heritage. Over 500 Sites of Special Scientific Interest (SSSIs) are currently notified throughout Scotland primarily for their Earth heritage importance. They range from the internationally important historical sites first

Opposite. The distribution of Sites of Special Scientific Interest (SSSIs) across Scotland that are notified for geology and landform interests.

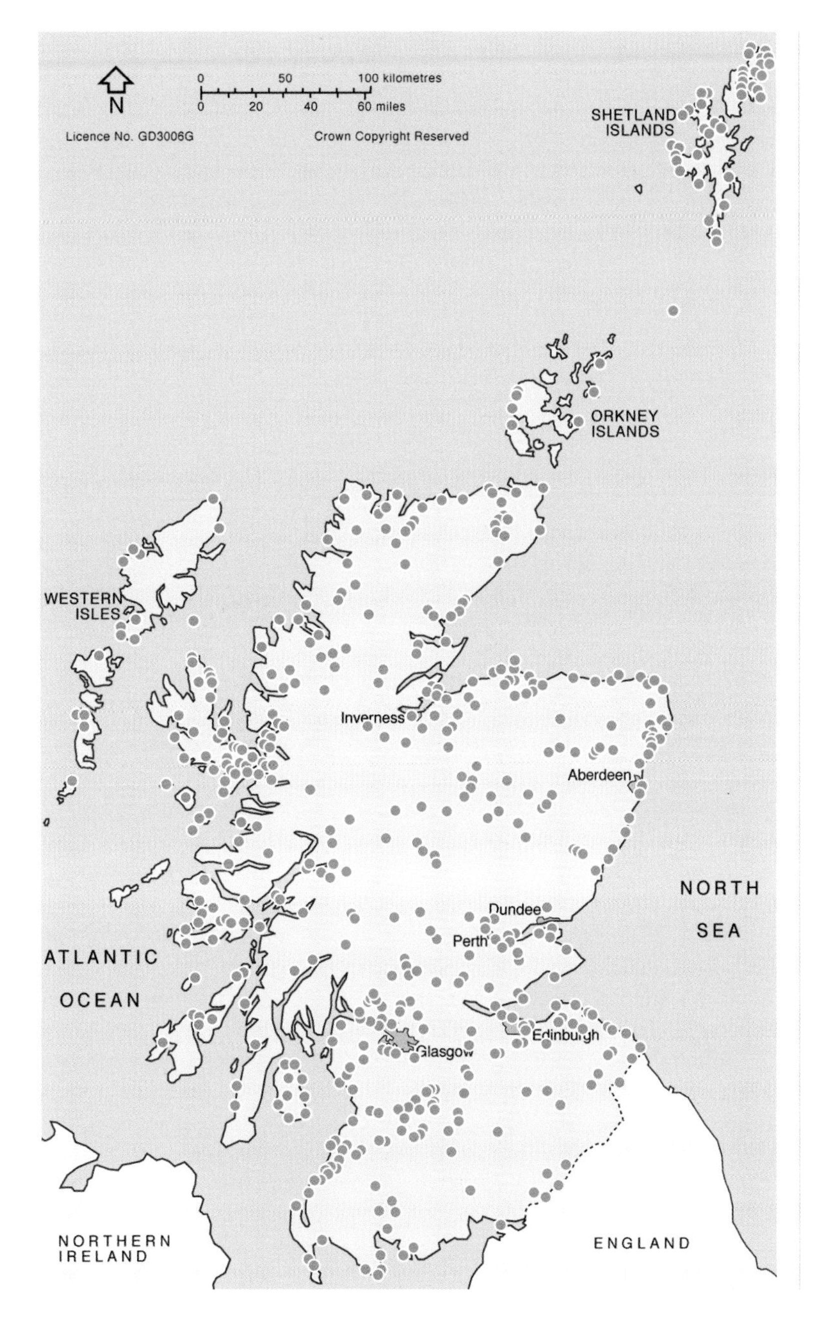

described by James Hutton at Siccar Point and Glen Tilt, to the representative and sometimes esoteric features of Scotland's geology. The suite of representative sites comprises by far the largest number of SSSIs and, taken collectively, demonstrates the geology and landforms of Scotland in all their splendid complexity and variety.

Identification and protection of key localities for research and education lie at the core of geological conservation. This approach is based on selection of special or representative sites using scientific criteria. Historically, this system dates back to the mid-nineteenth century. Early examples of Earth heritage initiatives include the listing of erratic boulders in Scotland in the 1870s and the enclosure of the stumps of a former forest of Carboniferous trees at Fossil Grove in Glasgow in 1887.

Formal identification of key sites began in the 1940s with a series of site lists that were then added to in an ad hoc way. This approach was superseded by the Geological Conservation Review (GCR), a major programme of systematic assessment of the conservation value of geological and geomorphological sites throughout Great Britain. Site assessment was undertaken between 1977 and 1990 and the GCR is the most comprehensive review of sites in any country. It was designed to reflect the full diversity of Earth heritage in Great Britain, spanning all the major time periods from the Precambrian to the Quaternary. Publication of the results in a series of forty-two scientific volumes is now nearing completion. Individual site descriptions of the best and most representative locations throughout the country can be accessed online (www.jncc.gov.uk).

Many of the sites identified by the GCR process are now designated as SSSIs and are accorded a measure of legal protection, including a requirement for consultation with SNH over developments requiring planning consent and other notifiable activities. The Geological Conservation Review resulted in the de-designation of sites that no longer met the stringent criteria for site selection and the identification of new sites that merited inclusion on the basis of recent research or reappraisal.

The criteria for site selection are threefold. Sites should be: internationally important; or exhibit features of exceptional importance; or be representative of a particular period of time or geological process. Providing that they meet the criteria, sites may be added to the GCR register as the results of new research or discoveries are reported.

A key teaching and research resource

Rocks, fossils, landforms and soils are a valuable resource for scientific research and education. Geology and its allied disciplines are fundamentally field based, so exposures of rocks and landforms are essential for continued research and education. As part of the process of scientific advancement, there is a continuing need for these sites to be available for further research to test and develop new theories and for demonstration and reference purposes. Scotland's varied landscape of uplands, river basins and active coastlines can therefore be regarded as an outdoor laboratory where a huge range of natural phenomena can be observed and described. Unsurprisingly, Scotland is a popular destination for academics from universities and institutes throughout Britain researching different aspects of the natural world. Journals such as the *Scottish Journal of Geology* and the *Scottish Geographical Journal,* publish a varied selection of articles on the rocks and landforms of Scotland. For example, the Rhynie Cherts of Aberdeenshire are the subject of an active research programme to find out more about the exotic life forms and precious metal content of these singular deposits. This is just one of hundreds of Earth science research projects that are currently active in Scotland.

Like any field of specialist research, the subjects under discussion can seem esoteric, almost irrelevant to a lay audience. But it is vital that this work continues and that the results of this research are communicated to as wide an audience as is prepared to listen. Most academic institutions are supported by money from the public purse by one route or another, so effective communication to that wider audience about the value and relevance of this research work is something that the profession neglects at its peril.

On economic and social grounds, geological and landform SSSIs are essential for the training of new generations of geologists to help locate and develop the oil, gas, mineral and aggregate resources that modern society ultimately depends upon. The applied value of geoscience is also apparent in the use of land for agriculture, forestry, mining, quarrying, building and infrastructure, all of which are closely related to underlying geology, landforms and soils. Rocks, landforms and sediments also record climate change and the dynamic history of the landscape. These environmental records allow the present situation to be set in the context of past natural variations. This is important since past landscape evolution provides a basis for understanding present changes and predicting future events and risks such as those associated with flooding, earthquakes, volcanic activity and coastal erosion.

Voluntary conservation

The voluntary wildlife movement is strong in Britain. The Royal Society for the Protection of Birds (RSPB) currently has more members than any one of the political parties. This high level of popular participation gives non-statutory wildlife organisations a strong voice. The RSPB, Scottish Wildlife Trust and The National Trust for Scotland also manage significant tracts of land throughout Scotland for wildlife conservation.

In comparison, the movement to conserve local geological and landform sites is in its infancy. Regionally Important Geological/Geomorphological Sites (RIGS) groups are mainly located in the Central Lowlands. Their primary purpose is to undertake practical conservation work, like site management and raising awareness through the publication of trail leaflets and other publicity materials. Their efforts are largely, although not exclusively, concentrated on sites that are not designated as SSSIs. Some of these RIGS have been notified to the local planners as key non-statutory sites. A similar tier of local nature conservation sites exists for wildlife interests.

Sites of local geodiversity importance are selected on the basis of their scientific and educational importance, historic interest

Landform sites, such as Carstairs Kames in Lanarkshire, are often threatened by mineral extraction. Quarrying on a large scale can completely remove the features of scientific interest.

and aesthetic and cultural significance, reflecting local rather than national values. Although these sites do not have statutory protection, many local authorities now have conservation policies for RIGS as well as other local nature conservation sites. An important recent initiative is the preparation of a Local Geodiversity Audit for West Lothian. This will provide the basis for a Local Geodiversity Action Plan that should help ensure greater protection for rocks, landforms and soils of local interest. It should also encourage local awareness and involvement and provides a model for the development of similar plans in other areas.

Threats to geodiversity

Pressures on geodiversity are many and varied and arise principally from engineering schemes, such as road developments or quarry infill and other changes in land use. Geological sites typically occur as natural and man-made exposures. While there is a view that rocks and landforms are robust, significant damage and loss of key sites have occurred in the past and remain an ongoing concern. The main types of impact include physical damage, destruction or removal of the interest, loss of visibility and access to exposures through burial by landfill or concealment by vegetation, damage to site integrity through fragmentation of the interest, and disruption of natural processes. For example, exposures in disused quarries can be lost through landfill, prime glacial landforms destroyed by quarrying for sand and gravel, and key exposures sealed and natural processes disrupted by coast protection measures, river bank protection and flood defences.

Mineral extraction can have positive and negative impacts. Quarries and gravel pits are a significant geological resource, particularly in areas where natural exposures are poor or scarce. Quarrying may reveal new sections of value, and many important sites are in former quarries where the geological interest would not otherwise have been exposed. In some cases, quarrying may pose a direct threat to particular landforms; for example, sand and gravel quarrying may destroy glacial landforms. In other cases, the loss of a landform through quarrying, for example an esker system, may need to be balanced against the potential value of new sections revealing the internal structures of these deposits. A range of agriculture and forestry operations may also impact on geological sites and landforms. Landforms are often concealed beneath blankets of commercial afforestation and damaged by extraction haul roads. Soils are under pressure from land use practices and contamination, intensification of agriculture, afforestation, waste disposal, acid deposition and urban expansion.

Traditional approaches to coast and river protection from erosion and flooding typically involve large-scale heavy engineering, which seals key exposures behind concrete seawalls, rock armour or gabions. Natural processes of sediment supply and movement are disrupted, usually displacing the problem elsewhere. Engineers have learned that working with nature is often more effective and cheaper in the long run than attempting to hold back change. The value of such alternative approaches is considered further in chapter 5.

There is a consensus that responsible fossil collecting can promote the science of palaeontology, providing that a code of good practice is followed. However, irresponsible collecting of rare fossil and mineral specimens, often through excavation of key sites for commercial gain, represents a significant loss to science as the context to the finds is not recorded. Irresponsible collecting can also cause damage to exposures and loss of other specimens. Mechanical excavators, explosives, crowbars and rock saws have all been used to remove fossil material in a search for high quality, commercially saleable specimens. Practical conservation therefore requires a combination of statutory protection, management of key sites and a heightened awareness of the value of geodiversity.

Engineering for conservation

Development and effective conservation of geological features can often coexist. Indeed, many of the key conservation sites across the country were created by engineering activities, such as quarrying or road construction. Some disused quarries are of considerable value as a research and teaching resource. But conflicts of interest arise where sites are scheduled for restoration, as infill inevitably leads to a loss of geological exposure. However, techniques have been developed to allow the majority of such sites to be infilled with waste whilst preserving a conservation section that is selected to be representative of the scientific interest.

Many geological sections are located on eroding coastlines and a point is usually reached when local pressure is exerted to protect the coastline from further retreat. But where such protection measures are installed, there is an immediate loss of access to exposures of scientific interest. Again, techniques have been developed to ensure an effective safeguard to the features of scientific interest whilst slowing down erosion to an acceptable level. In England, installation of offshore structures, such as breakwaters constructed from blocks of stone, has been effective in attenuating the force of the waves attacking the soft cliffs, so the rate of erosion falls considerably. Access to the key exposures is safeguarded whilst reassuring landowners and residents that the threat from coastal retreat is under control. Another technique that has been used in Scotland is known as 'beach feeding'. Quantities of sand are added to the beach in an attempt to slow down rates of erosion.

Geodiversity and sustainable management of the natural environment

In a wider context, conservation management of Scotland's geodiversity involves much more than site protection for scientific study and education. Rocks, landforms and soils are an integral part of the natural heritage. They provide the basis for the diversity of Scotland's habitats and species. The active processes that shape our mountains, rivers and coasts also maintain our ecosystems. Scotland's biodiversity depends on the continued operation of these processes. For example, the Atlantic salmon relies on the availability of river channel features such as pools, riffles and glides, which in turn depend on the underlying geology and processes such as floods, erosion and deposition. It is now increasingly recognised that conservation and sustainable management of the natural world depend on understanding the links between geodiversity and biodiversity.

Geology is not just something that happened in the distant past – it is relevant today. The geological record clearly demonstrates how the Earth's natural systems have evolved in the past and can show how they might behave in the future. Natural processes frequently impact on human activity (e.g. through flooding, coastal erosion and landslides), resulting in economic and social costs. Management responses often involve locally engineered solutions, such as riverbank and coast protection measures, that are inappropriate because they transfer the problem elsewhere. Typically, also, management is based on the timescale of human experience and is not informed sufficiently by the longer-term geological perspective. However, this perspective is vital in assessing natural hazards, such as flood risk and sea-level changes, and implementing sustainable management of the natural environment.

Human activity is now an important 'geological force', reshaping the surface of the Earth through movement of rock and soil, building cities, motorways and dams, fixing the coast through concrete barriers, deforestation and soil erosion, and causing extinctions of species. Not only are many resources being used at a rate greater than their rate of replenishment, but there are also many uncertainties about the long-term environmental effects of the disposal of waste from human activities. Human activity is also having a potentially significant impact on global climate. A major challenge is to apply our understanding of the Earth's processes to mitigate future human impacts, to contribute to the restoration of areas already damaged by human activities, and to help develop sustainable use of the Earth's resources.

A clear message is that sustainable management of the natural world and use of natural resources depends on the effective application of Earth science knowledge and conservation of geodiversity. This involves working with natural processes, rather than against them, and developing sustainable approaches to the use and management of our natural resources based on understanding the properties of rocks, landforms and soils, how they interact with the living world and how natural systems behave.

CELEBRATING OUR GEOLOGICAL HERITAGE

Raising wider awareness about the importance of our heritage of rocks, fossils and landforms is vital for geological conservation. As part of an integrated approach to the management of our natural heritage, public bodies, industry, land and water managers, planners and policy makers should be aware of the value of geodiversity and its conservation, so that informed decisions can be made. There is a need to raise awareness of geodiversity with the general public and in schools to help form the basis for a wider constituency of support for geological conservation. If communities value and take pride in their local geodiversity, they are more likely to support its stewardship and conservation. Effective geological conservation will ultimately depend on better public awareness, understanding and support.

Children make a geological map of Scotland as part of the Scottish Geology Festival

Scottish Geology Festival

Geology has been the preserve of scholars for too long. The stories of our geological past have been known for decades, but they have been written by scientists for the benefit of other experts. Yet the story of moving continents, wide oceans forming and disappearing like seasonal puddles and the formation of great mountain ranges is for everyone to read and understand. SNH, with many enthusiastic partners including Our Dynamic Earth, the British Geological Survey (BGS), the National Museums of Scotland and the Hunterian Museum, initiated the first Scottish Geology Festival in 1995 to take the subject to a new and wider audience. Guided walks, talks and demonstrations were organised across Scotland to explain the special geological significance of each part of the country. Over a hundred events now take place during the biennial Scottish Geology Festival in every corner of Scotland with the intention of engaging everyone with an interest in the countryside. Whether it is finding out about the geology of Wester Ross, meteorites from outer space, or going on a fossil foray, all interests are catered for. Local ranger services, staff from national and regional museums, geological societies, universities, industry, science centres, SNH and BGS continue to support these events. For many, knowledge of our past stops at the Stone Age and any understanding of our geological heritage is a completely closed book. But the Scottish Geology Festival seeks to raise the level of understanding by offering events that will be of interest to those who have yet to discover the subject.

This very effective outreach programme has become a fixture in the diary for the foreseeable future.

It is also important that geological information is available to the public in written form. In conjunction with the BGS, SNH publishes a series of booklets that describe the geology and landforms of Scotland. When complete, the series will comprise twenty titles covering the country from the Shetland Islands to the Scottish Borders. Each describes the rocks, fossils and landforms of the area for the benefit of local people and tourists alike.

Geotourism – a contribution to the rural economy

Interpretation of geodiversity and geology-based tourism are not new, as demonstrated by the worldwide appeal of show caves, glaciers and other natural wonders. Traditional geological interpretation, however, was often based on interpretation boards using too detailed and overly technical language. The visitor may be bombarded with what it is felt they should know, rather than the words and pictures required to enhance what is, in most cases, a leisure time activity. Recent developments have seen more effective communication, resulting in the production of more appropriate materials, presented in stimulating ways using a range of media, and based on the best interpretive practices and sound educational principles. As well as helping to raise awareness of Earth heritage, these activities have an economic dimension. For example, the interpretive centre at Knockan Crag in North-west Scotland (see chapter 6) is already helping to hold visitors in the area for longer, making it more likely that additional bed nights will be spent in the locality. This makes a considerable contribution to the local economy, where tourism is the mainstay.

Walkers enjoying the view over Loch Maree, an ancient geological landscape.

Scotland has a great deal to offer the visitor. The landscapes, particularly those of the Scottish Highlands, are world-renowned. VisitScotland, the government agency that promotes the Scottish tourism brand at home and abroad, has established that the majority of visitors come to Scotland to visit the wilder, more scenic areas of the country. Most are also aware of the historical heritage that the country has to offer, in terms of castles and other sites of historical importance. So that mix of natural and historic heritage is an irresistible draw for many visitors from abroad and also other parts of the Britain. Tourism is now one of Scotland's biggest industries, earning the country almost £5 billion every year and supporting 200,000 jobs. Some of those jobs are in the remotest parts of the country and are an essential adjunct to farming, fishing, crofting and other indigenous occupations where jobs in manufacturing and other forms of service

The Polar Gallery from Our Dynamic Earth, Edinburgh

industry are thin on the ground. Eco-tourism is already well established in Scotland, with marine wildlife tourism particularly well developed. Nationally, wildlife tourism earns the country some £57 million directly and supports an estimated 2,200 jobs.

Perhaps the most ambitious project undertaken to date in Scotland that connects with the geotourism theme is the Millennium project, Our Dynamic Earth, constructed in the heart of Edinburgh in the shadow of Salisbury Crags. The futuristic building was opened by HM Queen Elizabeth on 2 July 1999. A number of other science centres have been built across Scotland with the express purpose of taking science to the public, but Our Dynamic Earth is the only one that explores aspects of the Earth sciences in any detail. It is a tourist destination of some significance and attracted over half a million visitors in its first full year. Since opening, it has presented a holistic view of Planet Earth: focusing on its formation, early development, stewardship of our finite planetary resources and, generally, the way in which the human species has interacted with the natural environment. It differs in its presentational style from a museum or other static displays by telling these stories in an interactive and involving way. In the words of its first scientific director, Stuart Monro, "the story is communicated as an immersive experience, taking people on a trip through the Universe, to the edge of a volcano, across glaciers, and into the oceans, polar areas and the tropical rainforest. It is a story that is gradually built up, first exploring the physical formation of the planet and its landscapes, then of its life and diverse environments."

Geotourism is not entirely underwritten by government agencies and big-budget Millennium projects. Independent operators also have an important place in the market. Geowalks has been operating out of Edinburgh for over five years. Thousands of visitors and locals have joined Angus Miller, the company's proprietor and rock enthusiast, for his perambulations around the city's geological high spots and many other locations throughout Scotland. He uses his extensive knowledge of the local area to inform and entertain those who join him on his regular walks. Edinburgh is richly endowed with heritage sites, both natural and cultural. James Hutton developed the ideas that informed his seminal book *Theory of the Earth*, published in 1795, by studying the local geology. Although Hutton travelled more widely throughout northern Europe, it was always to his beloved Edinburgh that he returned for inspiration. Hutton's Section in the Royal Park of Holyrood is famous in the annals of geology and is a regular part of Angus's trips. It was here, or so the story goes, that Hutton saw the evidence for intrusion of dolerite in a molten state into the host rocks of Carboniferous age. So, in the company of an

Tourists visiting the North-west Highlands enjoy the dazzling scenery of the area. Interpretive panels help the visitors to 'read the landscape' of predominantly Lewisian gneiss. This viewpoint overlooks Loch Glencoul, with the Stack of Glencoul in the far distance.

engaging tour guide telling fascinating historical tales and interpreting the excellent geology, with a backdrop of dazzling views over one of the most beautiful cities of the world, what more could a geotourist want!

A geologist's paradise

The colourful geological history of the North-west Highlands has now been recognised internationally through the award of European Geopark status, endorsed by UNESCO. It is one of thirty such areas throughout Europe and a first for Scotland. The award was made on the basis of the outstanding landscapes, existing interpretation of the area's geological gems and the sustainable approach to development taken by the local authorities. Geopark status, unlike some of the preceding conservation designations, has been warmly welcomed by the local people. It is seen as an accolade for the area and potentially of great benefit to the local economy in terms of its likely boost to tourism. The North-west Highlands now sits alongside the Petrified Forest of Lesvos in Greece, the meteorite crater of Rochechouart-Chassenon in France and the TERRA.vita Naturepark in Germany as one of the key geological landscapes of Europe. The aims of the European Geopark initiative are to protect the geological heritage, raise awareness of geology with the public, and promote the use of natural and cultural heritage to achieve sustainable economic growth. Let us hope that this very welcome initiative does exactly that.

CHAPTER 2
HOW THE EARTH WORKS

Millions, millions – did I say millions?
Billions and trillions are more like the fact.
Millions, billions, trillions, quadrillions,
Make the long sum of creation exact.
J. S. Blackie, *Song of Geology*

EARLY BEGINNINGS

The Earth is the third planet from the Sun. Along with seven other major planets, it spins around the Sun and has done so since its formation as an independent entity, some 4,500 million years ago. But scientists can map out our history even further back than that, to the Big Bang almost 13,000 million years ago. At that point, all matter which now builds the Universe began to expand rapidly and that process continues to this day. It is a mind-blowing concept to think that everything we see around us, and are ever likely to see in our lifetimes, was concentrated into a single point prior to the Big Bang. From these early beginnings, the Universe has expanded into the multiplicity of stars, planets and other heavenly bodies that now decorate the night sky. How small is our home planet and how vast is the timescale over which these processes have been played out.

Scientists and philosophers have debated the origins of the Universe for millennia. The Greek philosophers Aristotle and Ptolemy favoured the view that the Earth was the centre of the Universe, with all of the constituent stars, suns and moons revolving around it. This view was in line with contemporary theological teachings in which the creation of the Universe was seen to be for the express purpose of supporting humankind. This view prevailed until Nicolaus Copernicus, the Polish mathematician and astronomer, revolutionised humanity's conception of itself and its place in the Universe. Copernicus revived an earlier idea that the Earth and the other planets revolved around the Sun, whilst the Sun remained in a fixed position. His work, and that of astronomers who came after him, such as Galileo, was condemned by the Church as heresy and the work of the Devil. In fact, Galileo threatened the establishment with his development of Copernicus' ideas to such an extent that he was held under house arrest for many years. However, Johannes Kepler, a German mathematician writing in the early seventeenth century, provided arithmetical and observational 'proof' of Copernicus' ideas. He also developed the idea that the planets did not orbit the Sun in perfect circles, but in a series of ellipses. And so he established the basis for our modern understanding of the solar system and all the constituent elements in it.

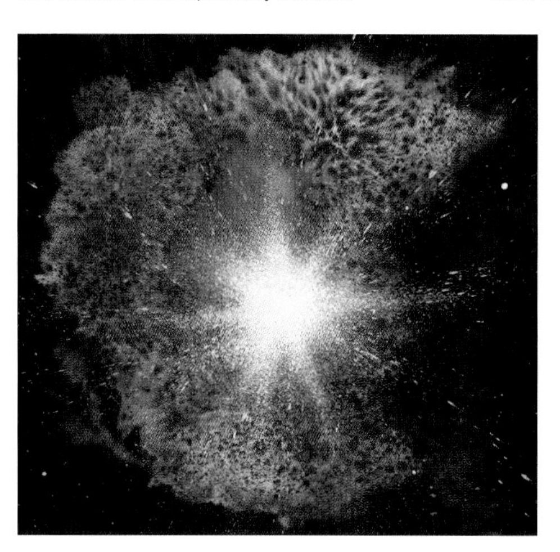

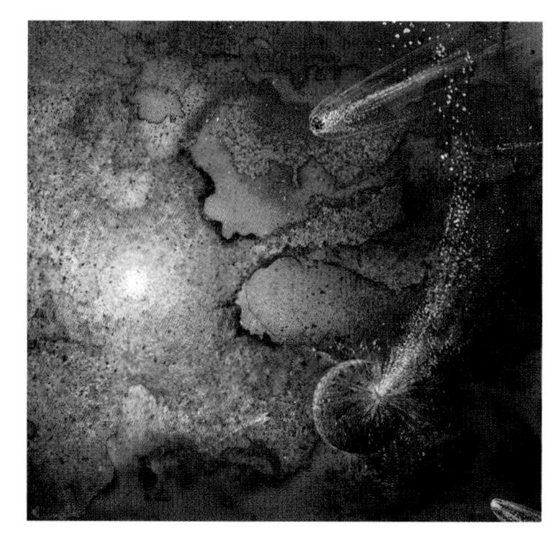

Left. The Big Bang took place around 13,000 million years ago.

Middle. Discs of matter coalesced to form our solar system. Over time, eight major planets and many associated smaller bodies emerged from the stellar dust.

Right. Planet Earth formed around 4,500 million years ago. Our moon was subsequently created as a result of an interplanetary collision.

EARTH'S STORY

We take up the story of the Earth about 5,000 million years ago. A dying star collapsed and exploded. Hot gases formed into rotating discs, swirling around a newly ignited star at its centre – our sun. Over time, the gases condensed, forming great lumps of matter that orbited the Sun, colliding and coalescing to form ever-bigger and more coherent bodies. As these bodies increased in size, so they developed their own gravitational fields and started to attract more matter towards them. By this process of accretion, these proto-planetary bodies grew in size. In these early times, the Earth had already secured its ability to support life; its overall mass created a gravitational pull that was later able to retain an atmosphere of sufficient density for water to condense and ultimately to assist the evolution and sustenance of terrestrial life.

A full Moon

As the primordial Earth circled the Sun, it was showered by inter-planetary debris. The Moon was formed as a result of one of these random collisions. A stray object, thought to be about the size of Mars, collided with the Earth. A cloud of matter was spat back out from the Earth after the collision and this gaseous material condensed, over time, to form the Moon. We still live with the consequences of this cataclysmic collision. The axis around which the Earth spins once a day, creating periods of light and dark, was tilted by a few degrees as a direct result of the impact. The planet's orbit around the Sun was also made more elliptical. These irregularities of spin and orbit are fundamentally important to the way in which we live our lives. They affect how much solar radiation reaches the Earth's surface and at what time of year, and hence determine the seasons of the year and their duration. The tilt away from and then towards the Sun as the Earth completes its annual orbit around the Sun accounts for the changing seasons. As the Northern Hemisphere is titled towards the Sun, we experience summer conditions and the opposite applies for the winter.

Both the Earth and Moon were heated to melting point as a result of the collision. As its temperature dropped, the Earth 'fractionated' or developed a series of layers as it cooled to a solid state. The heavier elements migrated to the centre and the lighter fraction formed the crust. By around 4,200 million years ago, the surface of the Earth had begun to congeal and form a solid outer skin or crust. Shortly after that time, the Earth was a sufficiently stable environment for life to develop and thrive.

The Earth as we know it

From that time onwards, the overall structure of the planet has changed little. Its inner core is solid, comprising of a mixture of iron and nickel. Temperatures in the inner core exceed 4,000°C. The outer core is a white-hot semi-liquid, consisting mainly of iron, with other elements such as nickel and sulphur. The next two layers of the onion-skin structure are the inner and outer mantle. These are solid layers, rich in magnesium and iron. The outermost layer is known as the crust which, in proportion to the Earth as a whole, is as thin as a postage stamp stuck to the side of a football. But this is the realm that we know best.

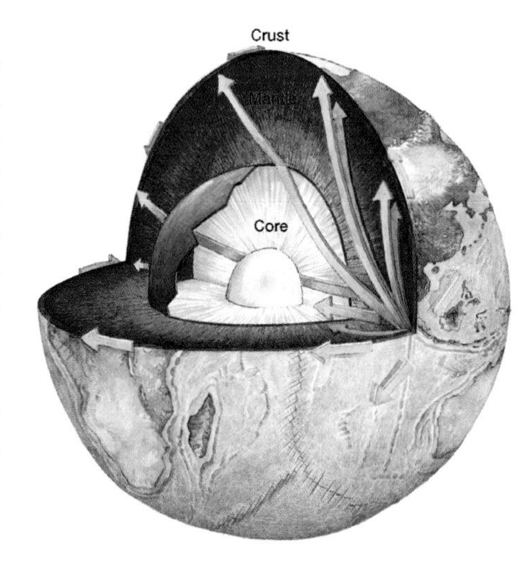

A section through the Earth showing its 'onion-skin' structure and the passage of shockwaves generated by seismic activity. Some waves, known as body waves (shown in blue), travel through the Earth and the time they take to make this journey provides information about the nature of the material they have passed through. A second group of waves (shown in yellow) are known as surface waves. These travel around the Earth just beneath the surface. These waves cause extensive damage in built-up areas, as they push the rocks sideways at right angles to the overall direction in which the waves are moving.

A slice through the Earth

Our knowledge of the interior of the Earth is gained by inference and indirect observation. Even the deepest borehole is a pinprick that gets nowhere near the lower layers of the Earth's structure. Seismology is the study of the passage of shockwaves generated by earthquakes. These shockwaves can reveal a great deal about the inner secrets of the Earth's structure; for example, the liquid nature of the Earth's outer core is inferred from seismological observations. Severe earthquakes generate a range of different types of shockwave that travel variously across the Earth's surface and deep into its interior. These waves are detected by arrays of instruments called geophones that have been placed in the ground for the specific purpose of monitoring earthquakes. Eskdalemuir, near Dumfries, is the main seismic monitoring station in Scotland. It regularly picks up the distant rumblings of earth movements from sites around the world. Most days of the year, around forty-five earthquakes are recorded across the globe. Such occurrences only hit the headlines where the main 'quake site' or epicentre coincides with populated areas. Most take place in remote locations, largely coinciding with the natural boundaries in the Earth's crust.

Comrie leads the way in earthquake detection

Earthquake House, near Comrie in Perthshire, was one of the first purpose-built seismic stations to be constructed in the western world. Built in 1874, it subsequently fell into disrepair, but has now been fully restored. A particularly large earthquake, thought to be around 4.9 on the Richter scale, shook the Comrie area in 1839. Local resident David Milne took a keen interest in the event. He described the experience of a local church minister, who "felt himself being lifted up and heeled over first to the East, then let down to the West, and lastly made to lean over again to the East. He felt three distinct oscillations, which could only have been caused by an irregularity or curvature in the surface of the ground." Milne went on to sketch a wave with a height "that may not have exceeded an inch" and a wavelength "that may have exceeded a mile. This would be enough to give the sensation felt." As a precursor to the construction of the Earthquake House, crude pendulum devices were installed at three separate locations around the Comrie area to monitor further shocks.

The village of Comrie lies close to the Highland Boundary Fault and earthquakes are regularly recorded in the vicinity. In January 2005, residents were 'rocked' by the biggest earthquake to hit the area in recent times. It measured 2.7 on the Richter scale and was felt over a wide area.

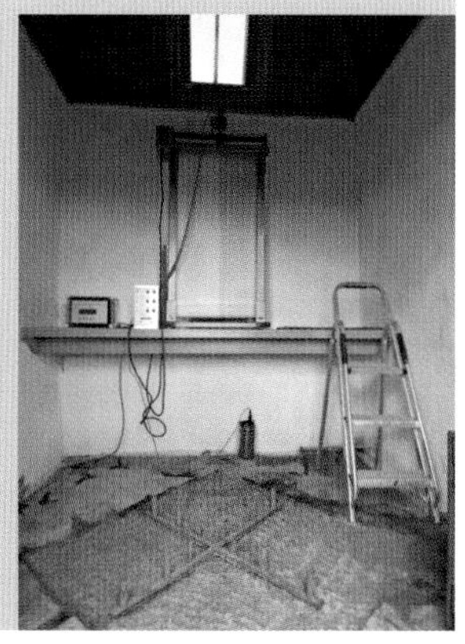

Earthquake House, Comrie

The Earth moves

As well as direct consequences for human populations, earthquakes can also have devastating indirect impacts where they occur under the sea floor. This was strikingly demonstrated when an earthquake-induced tidal wave, known as a tsunami, caused extensive loss of life and damage to property around the shores of the Indian Ocean on Boxing Day 2004. An earthquake of magnitude 9.3 on the Richter scale was generated as the Earth's tectonic plates ground past each other off the coast of Sumatra. A sudden movement of the plates caused the sea floor to rise and a huge body of water was displaced and set in motion. At the ocean surface, powerful waves were created that moved away from the epicentre of the earthquake. The waves travelled at many hundreds of kilometres per hour. It was not long before they made landfall. As they approached the shorelines, the wave heights increased dramatically and unleashed their full force on the coastal settlements and infrastructure of Indonesia, India, Sri Lanka and many other countries. The precise toll in human terms will probably never be known, but there is no doubt that the Boxing Day Tsunami is the worst natural disaster to be visited on this region in living memory.

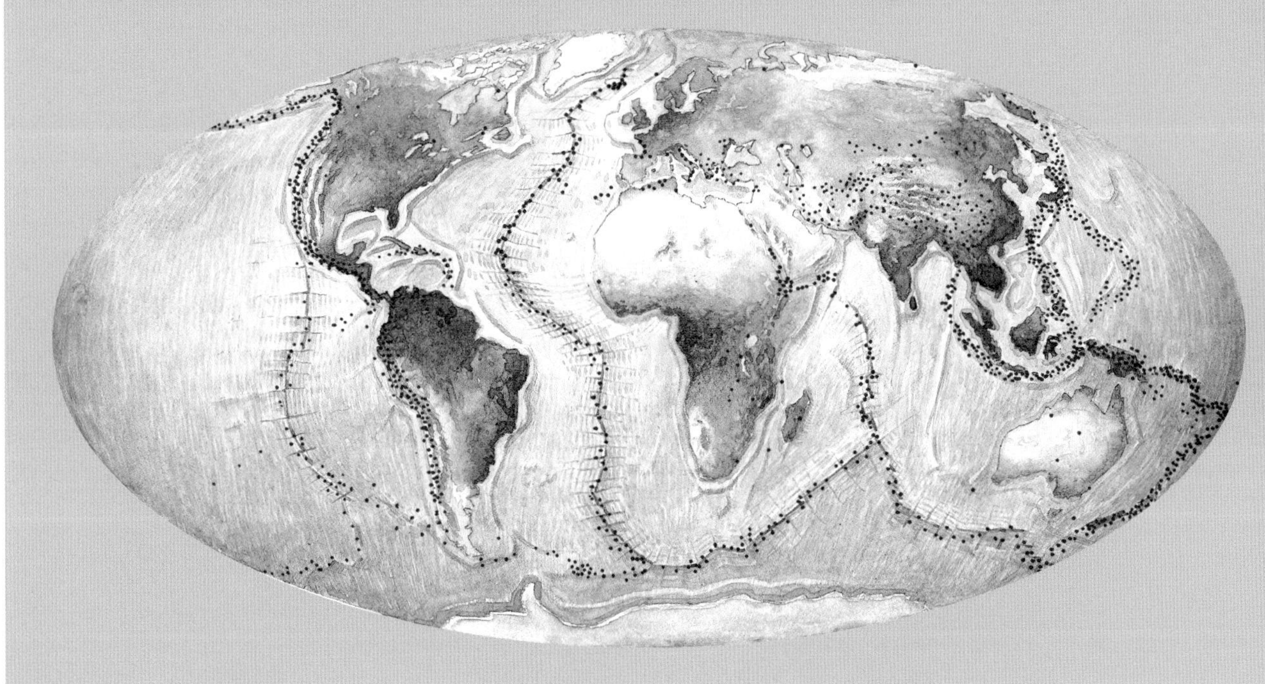

The surface of the Earth is fashioned from thirteen plates. Most earthquake activity (shown by the red dots) is concentrated at the margins of these plates. These are the most active areas on the Earth's surface. They are the focus for volcanic activity and are also the places where mountains form as continents collide.

The Earth as a magnet

As the Earth spins, it acts as a dynamo, generating a pulsating magnetic field. The North and South Poles seem reassuringly fixed reference points – unchanging poles at the top and bottom of our world that have drawn intrepid explorers like magnets. But nothing could be further from the truth. The magnetic polarity of our world – north at the top and south at the bottom – has flipped on countless occasions during the geological past. The polarity of the Earth is determined by the ebb and flow of currents deep within the planetary core. The magnetosphere, or area influenced by this force field, reaches out some 60,000 kilometres into space. And a good thing too, as it protects all life on Earth from the solar winds – deadly streams of charged particles emanating from the Sun. The Aurora Borealis, or Northern Lights, that enlivens our night skies, are a manifestation of these solar winds interacting with the Earth's atmosphere. When the polarity changes, the magnetosphere temporarily collapses, leaving the Earth unprotected against the Sun's lethal breath. Study of the geological record shows that such events have taken place at approximately half-million-year intervals throughout the history of the planet. The most recent change of polarity took place about 790,000 years ago, so another reversal is overdue.

The Aurora Borealis lights up the night sky. This light show is created by the interaction of solar winds with the Earth's magnetic field.

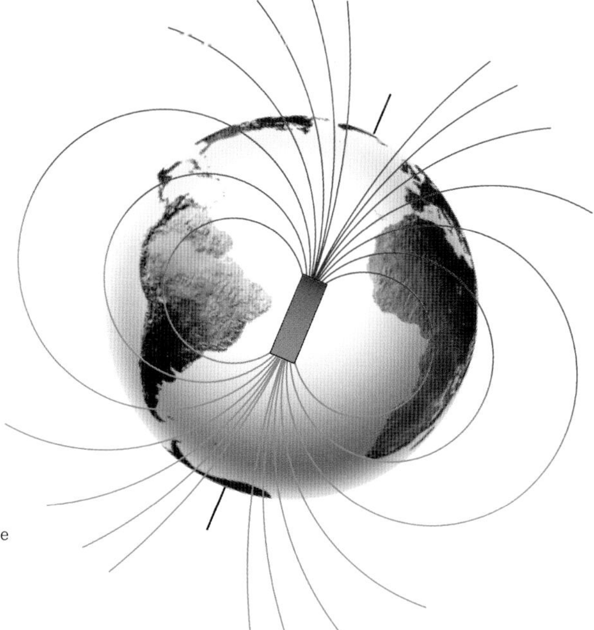

The Earth acts as a powerful magnet with a north and south pole. This magnetic force field is created by movements that take place in the dense liquid core of the Earth. The diagram shows the lines of magnetic force that emanate from the Earth's core.

WANDERING CONTINENTS

There is no doubt that the continents have wandered the globe. But to the lay reader, the certainty of statements such as 'Scotland started life near the South Pole' or 'Scotland and England were at one time separated by an ocean wider than the Atlantic' can be rather puzzling. What is the evidence for such outlandish claims? One of the reasons for the certainty is the insight that the study of palaeomagnetism provides. This study has proved vital in understanding how the Earth works and also how it functioned in the geolog-ical past. The basic premise is that Planet Earth functions as a magnet with north and south poles. The Earth's magnetic field is a powerful force that leaves a magnetic signature, like a tape recording, on any newly formed rock rich in iron or other material susceptible to magnetisation. This signature is retained unchanged from the time of its creation. Volcanic rocks, for example, acquire this magnetic signature on cooling from a molten state. Measurement of the direction of this remanent magnetic signature allows the investigator to plot where a particular rock formation was formed relative to the magnetic poles as they existed at that time. So the latitude at which a particular sample of rock was formed can be established with a high degree of accuracy. If this procedure is repeated for a number of rocks of different ages from the same piece of crust, the movements of that crustal plate can be tracked across the globe over long periods of geological time.

The Earth moves

Turbulence at the centre of the Earth has had a profound influence on the surface of the planet. Movement in the semi-liquid outer core creates convection cells in the plastic mantle

Gravitational forces and circulation in the mantle move the plates across the surface of the Earth slowly but inexorably. These circulation cells are driven by turbulence in the Earth's core.

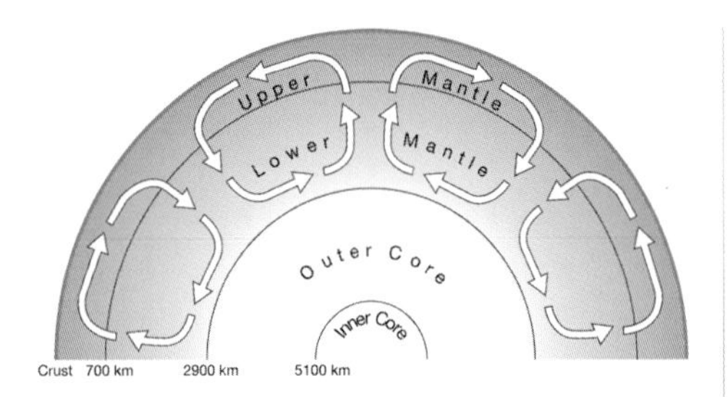

above, which, in turn, have affected the outermost layer – the crust. This crust first developed around 4,200 million years ago, like the skin on a cooling rice pudding. The crust is now differentiated into two separate types – an oceanic crust comprised of basalt and a thicker, less dense continental crust that formed the land. The crust then formed into distinct plates that moved around the surface of the Earth, largely powered by the movement of the mantle below. So from the very earliest times, the surface of the Earth was dynamic and constantly shifting in response to these deeper forces. This movement is not rapid, but it is constant, driving continents across the globe at rates of up to 20 centimetres every year with an average rate of around 8 centimetres per year. Gravitational forces and circulation in the mantle move the plates across the surface of the Earth. Slowly, but inexorably, these circulation cells are driven by turbulence within the Earth's core.

Plate tectonics

The whole process of shifting continents is known as plate tectonics. This concept is now the cornerstone of our understanding of the way in which the Earth works. It was in the 1960s and 1970s that our understanding of plate tectonics was formally articulated, although the idea of continental drift had been around for some time. Sir Francis Bacon, writing in the seventeenth century, noted that the coastlines on either side of the Atlantic Ocean seemed to fit each other, as though in the past they had been one. Alfred Wegner published his account of drifting continents almost one hundred years ago and scientists who came after him, such as Arthur Holmes, supported the idea that the distribution of land and sea has altered radically over geological time. As our knowledge of the inner workings of the planet has grown, so the details of the plate tectonic model may have been further refined to fit the emerging facts.

As the Earth is not growing in size overall, for every site where new crust is added, such as the mid-ocean ridges, there

Arthur Holmes – a man ahead of his time

Arthur Holmes (1890–1965) was born in the industrial North-east of England and educated at Imperial College in London, but it was as Regius Professor of Geology at the University of Edinburgh that he reached the peak of his career. His contributions to the science were many and varied, but two stand out. His initial and persistent interest was in geochronology – the science of dating rocks – and, later, he became interested in the theory, as it was then, of continental drift. Professor Holmes was the first to propose that very slow moving convection currents in the Earth's mantle caused continents to move, oceans to open and mountains to form. In 1944, his book entitled *Principles of Physical Geology* was published, a work that was to make him known to every serious student of the subject of geology for generations to come. Holmes lived long enough to see the theory of sea-floor spreading and plate tectonics being formally proposed, although his earlier contributions had been largely forgotten. The second edition of *Principles of Physical Geology* noted, with perhaps a whiff of irony, that "mantle currents are no longer regarded as inadmissible". Arthur Holmes' work is regarded as amongst the most influential of the twentieth century.

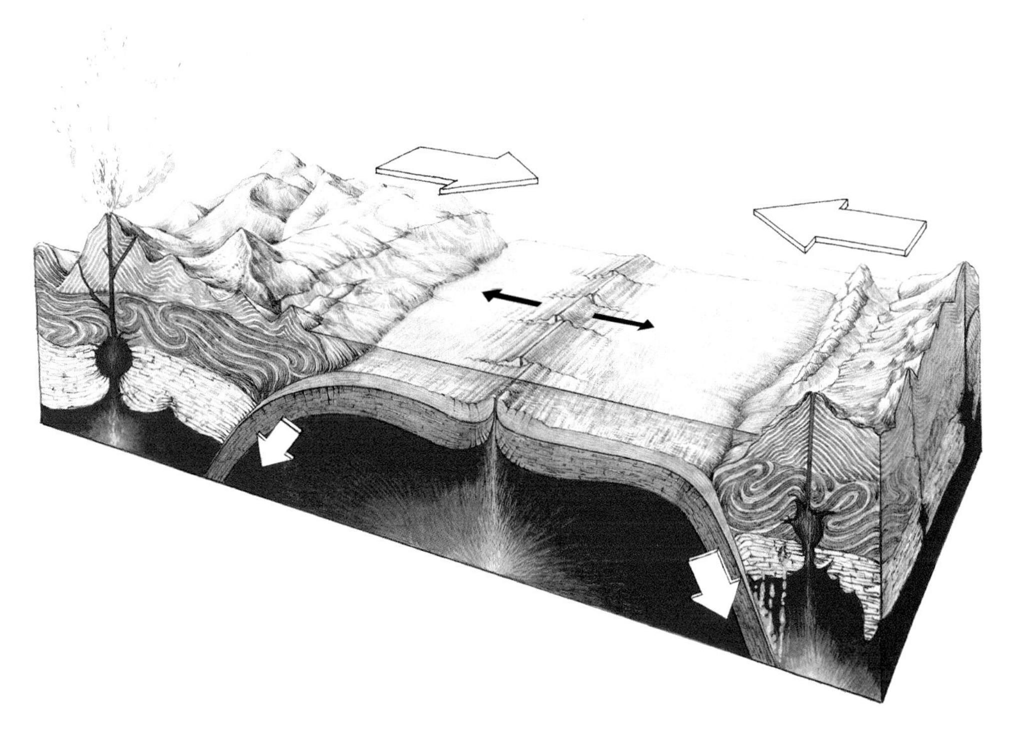

Plate margins are the most 'active' places on Earth. Earthquakes shake the ground and volcanoes erupt. This action takes place at the edges of the plates, which are either being added to or destroyed. This view of the world shows both 'constructive' and 'destructive' plate margins. In the centre of the ocean, at the mid-ocean ridge, material is being added as magma ascends from the upper mantle. Beneath the shorelines, subduction zones have formed where the ocean plates dive under the lighter continental rocks. Rates of spreading at the mid-ocean ridge and rates of subduction are broadly in balance, as the overall area of the Earth's crust must stay the same. In this case, the net effect is that the ocean narrows over time.

must be places where the crust is being destroyed. This process takes place at what are known as destructive plate margins. At these locations, ocean crust is reabsorbed into the mantle and mountains form as plates collide. These processes are going on today and we can observe their surface manifestations. But it is also crucial to recognise that they have been operating throughout most of our geological history. Over the last 3,000 million years, the movements of plates have caused every mountain on the face of the Earth to rise and have been responsible for most of the world's earthquakes, tsumanis and many other destructive events. This process, more than any other, has created the world as we know it today.

Driven by the movement of the Earth's plates, continents can travel from one side of the globe to the other, given enough time. Since the planet's early beginnings, the map of the world has constantly changed as continental landmasses have come together and parted in a multitude of different configurations. Rock masses, hundreds of square kilometres in size, have broken away from the main continental areas and become welded to adjacent landmasses, linking areas that had no previous association. This process of continental drift created a series of patchwork landmasses. Continents were created that comprise fragments of rock that had previously been on opposite sides of an ocean. For example, northern Scotland once formed a small part of a major continent, known as Laurentia, that included North America and Greenland. For many millions of years, northern Scotland was separated from England and Wales by an ocean wider than the present-day Atlantic Ocean. Known as the Iapetus Ocean, it closed around 420 million years ago in a collision that united northern Scotland and continental Europe in a new continental land-mass. Over time, this land-mass moved through many of the Earth's climatic zones. The point of collision for the land that was to become Scotland was around 20°S of the Equator and we now sit at a latitude of 57°N, so the journey across the globe within this relatively limited period of time has been prodigious.

Similar continental rearrangements have affected every single landmass around the globe. And it is certain that this process will continue into the future. So, a map of world drawn in 100 million years' time will reveal a very different arrangement of land and sea to that of the present day.

Stripes on the sea floor

The crucial evidence confirming the shifting continents idea was provided by two British scientists, Fred Vine and Drummond Matthews. They suggested that the crust underlying the world's ocean comprise basalt lavas. These rocks are mostly younger than the rocks that build the continents. Volcanic activity also takes place under the sea along zones known as mid-ocean ridges. At these locations, new strips of rock are periodically added to the ocean floor. With the addition of new material, the continents on either side of the mid-ocean ridges are continually being pushed further apart.

As new sea floor is erupted at the mid-ocean ridges, the iron-rich basalts are magnetised by the Earth's magnetic field. As previously described, the polarity of the Earth's magnetic field regularly flips, so that the North and South Poles reverse. Vine and Matthews demonstrated that the sea-floor basalts are magnetised in a series of parallel strips that are alternately of normal and reversed polarity. There is a similar pattern of magnetic stripes either side of the mid-ocean ridges. Each stripe can then be accurately dated using advanced analytical techniques. So the history of sea-floor spreading for each of the major oceans can be established. The movements of the associated continental landmasses can also be plotted as the opening of great oceans can be 'rewound' to their earlier positions.

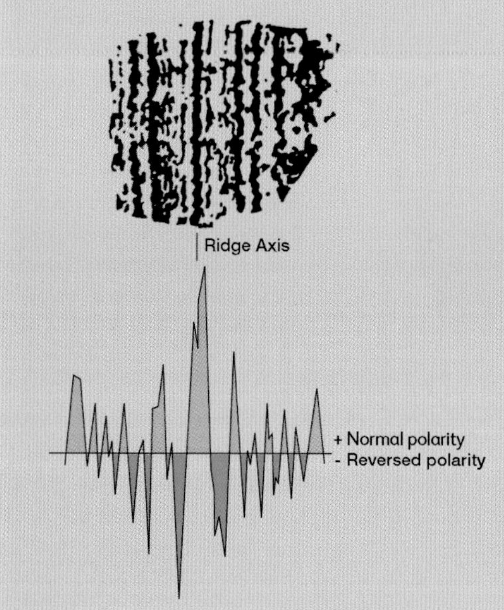

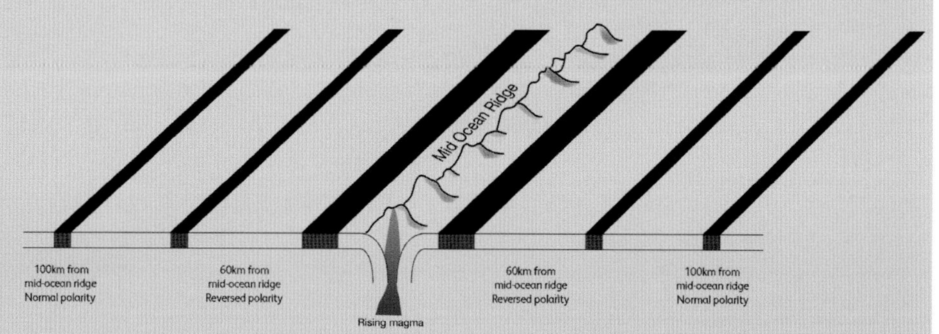

Left. The peaks and troughs of the graph show the normal and reversed polarity of the magnetic stripes on the sea floor. As new material is added at the mid-ocean ridge, it is magnetised according to the nature of the magnetic field at that particular time. This is also represented in the diagram above where the black lines represent normal polarity and the white, reversed polarity. The lines are symmetrical across the mid-ocean ridge. These magnetic reversals, where the poles flip over, take place approximately every 500,000 years.

Above. With the addition of new material along the central spine of the ocean, the rocks gradually 'migrate' away from the mid-ocean ridge. This is an important part of the 'conveyor belt' mechanism that drives continents across the globe. The stripes are alternately magnetised with normal and reversed polarity.

Building mountains

Continents wander the globe, so, from time to time, they are bound to collide. There is incontrovertible evidence for this process in the geological record and also from present-day observations. When continents collide, great stress is created in the Earth's crust and the character of the rocks involved is altered – or metamorphosed – as they respond to the changes in pressure and temperature imposed by burial deep in the crust or by exposure to intense heat. As continents approach each other, so the sediments that were laid down on the ocean deeps that once separated them also become caught up in the collision. Mountain ranges from around the world all mark the site of major plate collisions. The Himalayas were formed when India collided with Asia, and the Andes developed in response to the South American plate being under-ridden by the oceanic plate of the western Pacific. The many islands of Japan only

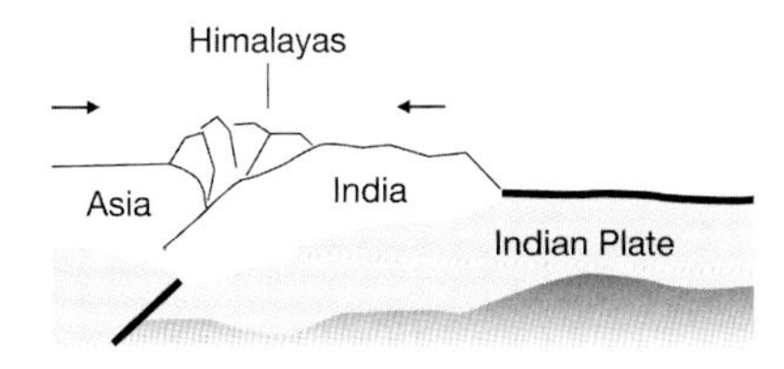

As continents collide, the sediments laid down in the ocean that formerly separated them are squeezed and uplifted to form fold mountains. After collision, the crust is considerably thicker than before.

exist because of plate subduction – where one plate dips beneath another and is consumed in the process. Where plates collide, the crust thickens and mountains rise.

But, for every metre that these mountains reach to the sky, so their roots also plunge downwards into the Earth's crust. Some of the world's highest mountains are thought to have roots over 70 kilometres in depth. In comparison, the average thickness of the crust is around 35 kilometres. In this way, a newly emergent mountain chain is able to remain supported. But such environments are inherently unstable. Their height and lack of substantial cover of soil or vegetation means the mountains are exposed to the full force of weathering and erosion. Rivers and streams transport the weathered rock away from the mountains and deposit the material as huge accumulations of sediment in the lowlands, while the finer material is carried to the sea.

Scotland bears its fair share of scars from past continental collisions. The spectacular Highland scenery derives directly from such events. As the continents rearranged themselves around 420 million years ago, so Laurentia collided with neighbouring continental fragments to create a mountain belt that extended from the north of Norway southwards through Scotland to what we now recognise as the Appalachian Mountains of the eastern USA.

Rocks cooked and squashed

While studying a rock sequence in Glen Clova in Angus in the 1890s, George Barrow discovered that rocks, originally laid down as sands and muds, had later been changed as they were progressively cooked and squashed deep within the Earth's crust. The circumstances that led to their deep burial are explored in the next chapter. This process is more accurately described as metamorphism. Barrow noted progressive changes in the physical appearance and mineral content of the rocks, which he postulated were proportionate to the conditions of alteration. The appearance of minerals such as garnet, silliminite and kyanite indicated a scale of progressive severity of these conditions. Deeper burial within the Earth's crust gives rise to higher temperatures and pressures, so causing a higher degree of alteration. The precise temperature and pressure conditions under which these minerals are stable can be replicated in the laboratory, so an accurate assessment of the conditions required to create a particular mineral assemblage can be made.

These temperature and pressure stability fields can in turn be related to the depth of burial in the Earth's crust, so geologists can accurately model the environment in which individual rocks were formed. This approach is used today throughout the world by geologists to assess the degree to which rocks have been altered by metamorphic processes. Regions of different alteration are now called Barrovian zones, after George Barrow.

The science of experimental petrology, or the creation of rock-forming minerals in the laboratory, was also pioneered by a Scotsman, Sir James Hall (1761–1832). Working in his foundry in East Lothian, he took charges of powdered limestone, sealed them in a gun barrel and fired them to high temperatures. He created marble from the original sediment, the first time that such a transformation had been achieved outside the realms of nature. Today, it is commonplace to investigate the temperatures and pressures at which a wide range of minerals are stable and in so doing, reveal their origins.

Sir James Hall

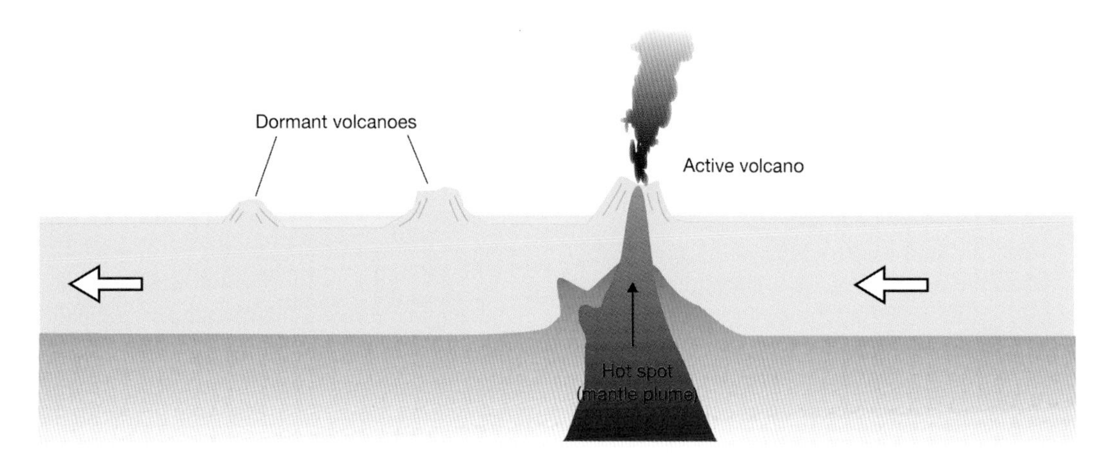

Most volcanic activity is concentrated at the plate margins. But volcanoes do occur in the middle of plates as well. Mantle plumes originate deep within the mantle, around 3,000 km below the Earth's surface. Like blow-torches acting on the under-surface of the plate, these streams of super-heat eventually pierce a hole and the resultant magma then flows from a deep source to the surface. Mantle plumes remain in the same place whilst the plate moves across the surface of the Earth, so a chain of volcanic islands is formed.

Volcanoes

There are over 1,500 active volcanoes distributed across the Earth's surface. This figure is greatly exceeded by volcanoes that are now entirely extinct. Some of the world's great cities, Edinburgh and Rio de Janeiro for example, are built on the eroded roots of previously active volcanoes. The locations of currently active volcanoes around the globe are closely linked to tectonic plate boundaries, which are natural weaknesses in the Earth's crust. Along the Mid-Atlantic Ridge, a string of sub-sea volcanoes pump out new slivers of ocean-floor material, forcing the continents of North America and Europe to move in opposite directions. These sub-sea volcanoes are amongst the highest natural structures on Earth and define a spine that runs down the centre of the North and South Atlantic Ocean and other major oceans of the world. They are an important and integral part of the plate tectonic story.

Volcanoes also occur at continental margins, known as subduction zones, where the crust is being destroyed. For example, the chain of islands that makes up Japan comprises a series of active volcanoes that result from one layer of crust dipping beneath another, causing the steeply descending plate to de-water and the overlying mantle rocks to melt as a direct result of this addition of water. The molten rock generated at depths of anything up to 120 kilometres then rises to the surface as explosive molten rock, forming a chain of active volcanoes.

Volcanoes can also appear away from plate boundaries, where their origins are related to areas of super-intense heat, known as mantle plumes. The sequence of events that gives rise to these volcanoes is described in the diagram above. The Hawaiian Islands were created in this manner. Over many millions of years, the Pacific Plate moved over a stationary mantle plume and, over time, a volcanic island built up from the ocean floor to emerge above the waves. The first volcano was formed and built, layer upon layer, a lava cone. In due course, the plate moved on, exposing a second point of the crust to intense heat, and a new volcano developed in response. This new volcano then emerged from beneath the Pacific Ocean and the first fell dormant, and so on. In this way, an island chain developed.

Volcanoes can also affect the weather of the whole planet. Dust and ash ejected from a volcano as it erupts are thrown high into the Earth's atmosphere and can partially blot out the Sun, sometimes leading to substantial drops in temperature for several years in the case of the largest eruptions.

Volcanoes have been discovered on other planets, so the processes that create these outpourings of ash and lava are not unique to our world. Olympus Mons on Mars, for example, stands 25 kilometres high, dwarfing any volcano to be found on our planet.

Defining rock types

Rocks can be assigned to one of three types – sedimentary, igneous or metamorphic. Sedimentary rocks cover the greater part of the continental landmasses across the globe and for most residents in Scotland, the likelihood is that your house is supported by sedimentary rocks of some type – sandstones, limestones, shales, breccias and conglomerates. These sedimentary rocks all have one thing in common: they all comprise debris derived from older rocks and laid down afresh, usually in layers, at the bottom of seas, oceans, streams and rivers or on land.

Igneous rocks – literally rocks from the fire – take many forms – from jet-black basalts to decorative pink-flecked granites. All were molten at some time in their history and cooled at some point in the Earth's crust; on the surface as a lava or at depth where the cooling process takes longer, which results in the rock-forming crystals growing to a larger size. Basalts, extruded as lava onto the surface, and gabbros, formed at depth in the Earth's crust, have similar mineral content and overall composition, but they look very different in fist-sized specimens. This difference is accounted for by the different grain size.

Metamorphic rocks have been changed from a sedimentary or igneous parent material by heat, pressure or both to become a new rock. This new rock is different in appearance and mineral composition from the original parent material. In many instances, this change in appearance and mineral content is dramatic; other metamorphic episodes result in more subtle changes to the original parent material. Deep burial in the Earth's crust is the commonest cause of metamorphic change. Intricate folding and banding often result from this intense alteration. Fold mountain belts around the world largely comprise metamorphic rocks that have resulted from continental collisions.

Top. Conglomerate – sedimentary strata are formed as pre-existing rocks disintegrate. Quartz pebbles sit in a matrix of finer grained material.

Middle. Granite and related igneous rocks are widespread throughout Scotland.

Bottom left. Section of igneous rock as seen under a microscope.

Bottom right. Folded metamorphic rocks – schists and gneisses form much of the bedrock of the Highlands of Scotland.

THE ROCK CYCLE

Rocks, like rubbish, can be recycled. In fact, rock recycling is one of the most important ways in which the Earth replenishes itself. Rocks are neither created nor destroyed but, under certain conditions, can simply change from one type to another. Sediments laid down as sandstones, shales or limestones can be transformed through deep burial to metamorphic rocks such as quartzite and schist. Under extreme conditions, metamorphic rocks can melt and become igneous rocks. Where igneous, metamorphic or sedimentary rocks become exposed to the elements on the Earth's surface, the process goes full circle as the rock decays, erosion takes place and sediments build up.

It was Dr James Hutton, Scotland's foremost geological pioneer, who first described this phenomenon in his seminal book *Theory of the Earth*, published in 1795. He wrote, "We are led to conclude that all strata have their origin at the bottom of the sea, by the collection of sand and gravel, and of earths and clays . . . The strata which now compose our continents are all formed out of strata more ancient than themselves." He further observed that "The strata of the globe are also found composed of bodies which are fragments of former strata, which had already been consolidated, and afterwards were broken and worn away by attrition, so as to be made gravel." For a man who had nothing more than his observational skills to rely upon at a time when the first book of the Bible was a literal and guiding truth, these were remarkable insights.

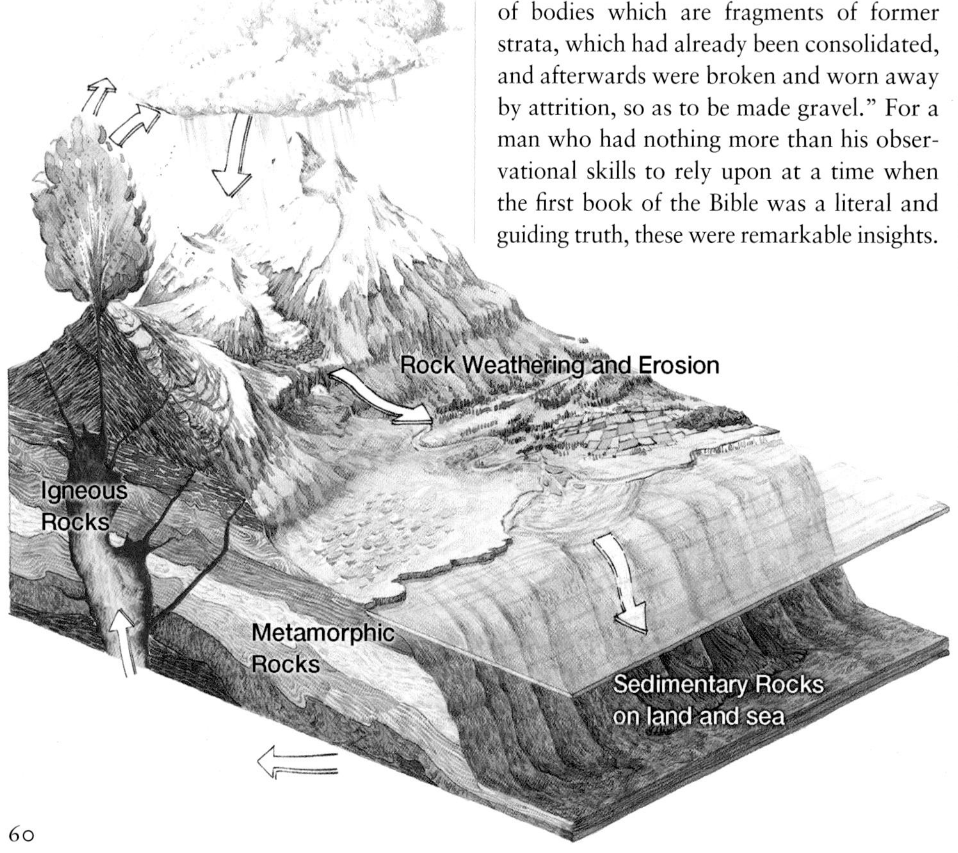

The rock cycle illustrates the relationship between the three main types of rock – sedimentary, igneous and metamorphic. Sediments may be metamorphosed if sufficient heat and pressure are applied. If this process is taken to extremes, then metamorphic rocks can melt and 'change' to become igneous rocks. If erupted onto the Earth's surface, these igneous rocks may at some stage be eroded and form sedimentary layers. And the process is ready to start again.

BUILDING THE FOUNDATIONS OF MODERN GEOLOGY

The abyss of time

We now accept that the Earth is more than 4,500 million years old, but earlier estimates were slightly more cautious. For example, an Irish cleric called Archbishop James Ussher, writing in *The Annals of the World* in 1658, made the startlingly precise estimate that the Earth was formed on the "twenty-third day of Octob. in the year 4004" BC, using the simple device of counting the number of generations in the first book of the Bible. Ussher also said that the world would end after 6,000 years – that was October 2004. Sir Isaac Newton used a more scientific methodology in reaching the conclusion that the Earth had taken around 50,000 years to cool from a molten state, but this estimate was ignored as incompatible with the teachings of the Church.

It was Dr James Hutton, a star of the Edinburgh Enlightenment, who came closer to the truth than anyone had previously done. James Hutton was born in Edinburgh on 3 June 1726. He studied medicine at Edinburgh University, completing his studies at Paris and Leyden. Throughout his illustrious career, he maintained an interest in many aspects of the world around him,

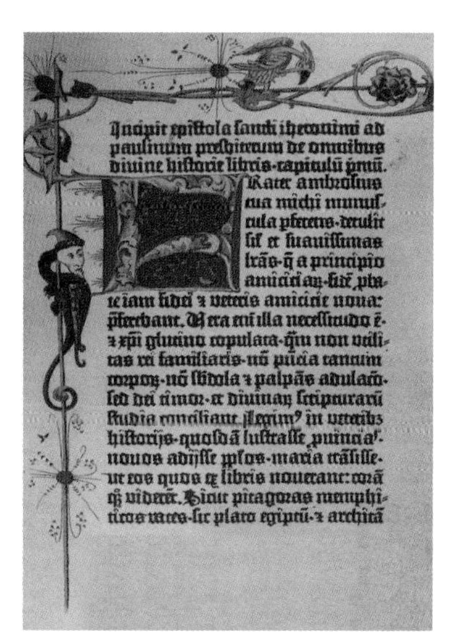

Above. Archbishop Ussher's manuscript that estimated the Earth was formed on the "twenty-third day of Octob. in the year 4004" BC.

Below. Portrait of James Hutton by Henry Raeburn (1756–1823)

Edinburgh, a hotbed of genius

The Scottish Enlightenment was a pivotal period in Scotland's history. For sixty years towards the latter part of the eighteenth century, advances were made in just about every field of endeavour, including science, economics, medicine and philosophy. Between 1730 and 1790, Scotland became the cultural leader of Europe. The philosopher David Hume, the economist Adam Smith, the chemist Joseph Black, the engineer James Watt, the poet Robert Burns and the geologist James Hutton revolutionised thinking and achievements in their own particular fields. An English gentleman who lived in Edinburgh at that time observed, "Here I stand at what is called the Cross of Edinburgh, and can, in a few minutes, take fifty men of genius and learning by the hand." Not surprisingly, this golden age in Scottish history was known as 'a hotbed of genius.' During this period, Scotland progressively ceased to be a nation at war with itself and its neighbours to become, in its upper echelons at least, a cultured and intellectual society, with a strong focus on innovation and discovery. The Royal Society of Edinburgh was founded during this golden age, the *Wealth of Nations* was published by Adam Smith and the city of Edinburgh was enhanced by the construction of the elegant New Town. The intelligentsia met regularly at dining clubs throughout Edinburgh to engage in discourse and develop their ideas. Adam Smith, Joseph Black and James Hutton met frequently as dining companions in a group known as the Oyster Club. According to an account of their meetings, "the conversation was always free, often scientific, but never didactic or disputatious". The golden age of free-wheeling intellectual development and growth was the background against which James Hutton made his seminal contributions to the science of geology.

including agriculture, chemistry, meteorology, philosophy and most bizarrely the study of the Chinese language. But it is for his writings on geology that he will be best remembered. He is universally acknowledged as the founder of modern geology and his insights are still the basis for many aspects of the science.

James Hutton's early interests were in farming. He inherited the family farm near Duns in Berwickshire where he undertook the study and observation of the natural world around him. His ideas on farming were later collected in an unpublished manual called *The Elements of Agriculture* which ran to 1,100 pages. But soon, the nascent science of geology had his undivided attention. He travelled his native land and further afield into England largely on foot, undertaking investigations and observations as he went. Over a period of twenty years, he built up a series of innovative and challenging ideas that were to change the way in which we were to view the working of the planet. Quite simply, he did for geology what Charles Darwin was to do sixty years later for the science of biology.

In *Theory of the Earth*, Hutton painted on a big canvas. The notion that the Earth was formed on a timescale in line with biblical teaching was challenged. Far from being a construct created in six days of labour and one of rest, Hutton could see "no vestige of a beginning and no prospect of an end" in terms of the timescale for the formation of the planet. John Playfair, Hutton's biographer, visited some key geological localities with him and wrote of "the palpable evidence presented to us, of the most extraordinary and important facts in the natural history of the world

The map that changed the world

William Smith, geological map-maker extraordinaire and canal builder who masterminded the construction of the Kennet and Avon Canal in the early 1800s, was the first to understand the significance of fossils in dating rocks. His employment as a canal builder and mineral surveyor took him to many parts of the country, including southern Scotland. He noted that rocks from different areas had a similar content of fossils, and so was able to link them in broad sweeps across the country in terms of their similar age. He went on to prepare the first accurate geological map of England and southern Scotland, based on the growing appreciation of the emerging science of stratigraphy that his work had championed. His ideas were largely based on fieldwork undertaken in the south of England, initially near his home in Bath and later further afield. But he realised that his work had universal application and that the principles established on the Jurassic limestones of the west of England, were equally applicable on the other side of the world. So the science of stratigraphy was born. Smith's map hangs to this day in Burlington House in London, home of the Geological Society. This recognition is just reward for earlier setbacks when membership of this most prestigious society was initially denied to him because of his lack of social standing. This wrong was eventually put right when, in old age, William Smith was honoured by the Geological Society and presented with the Wollaston Medal.

. . ." What Hutton demonstrated on their visit to Siccar Point in Berwickshire (see chapter 6) was that the rocks they observed must have taken thousands, perhaps even millions, of years to form, well outside the timescales previously accepted. Playfair further wrote of this trip with Hutton that "the mind seemed to grow giddy by looking so far into the abyss of time". And so, with these simple observations carried out without the aid of sophisticated equipment or the benefits of computer simulations, Hutton and his fellow travellers started to unlock one of the enduring conundrums of the universe – an appreciation of the enormity of the geological timescale.

Subsequent investigators have built on this key observation. Sir Charles Lyell, born at Kinnordy House, near Kirriemuir, in the year that Hutton died, further developed ideas on a geological timescale for the Earth in his book *Principles of Geology*. In it, Lyell propounded his theory of uniformitarianism, a theory that still holds good to this day. This rather indigestible word describes a simple concept: that 'the present is the key to the past'. Generations of geology students have been brought up on the idea that the geological record can be better understood and explained if we have a full understanding of the processes that are currently shaping the surface of the Earth. Plate tectonics and the associated processes of volcanic eruptions, erosion and deposition are operating today and we can observe their effects. And these observations can be used to explain what we see in the geological record. If, for example, a geologist observes a sequence of rocks made up of sands, water-worn rounded pebbles and occasional layers of mud, it is reasonable to assume that these diverse deposits were laid down by an ancient, but long-disappeared river. This assumption is valid, as river courses of today are choked with similar deposits. Similarly, layers of pure sand containing the shells of burrowing animals could reasonably be interpreted as ancient beach deposits, and limestones are likely to have been formed as nearshore deposits laid down on the continental shelf. Moreover, ancient lavas are in many cases identical to contemporary volcanic flows. So by understanding how the present-day geological world works, we can indeed unlock the secrets of the past. Lyell also began the process of dividing the geological record into time periods, largely

Portrait of Sir Charles Lyell

using the occurrence of fossils contained within the rock strata as the basis for subdivision.

As with Hutton before him, Lyell's legacy to his science is inestimable. Lyell's ideas also had a strong influence on Charles Darwin. The two scientists had a warm friendship and Darwin read the first edition of Lyell's *Principles of Geology* whilst on board HMS *Beagle* in the southern oceans and the second version of this classic text was waiting for him when he docked in Montevideo in 1832.

Dating the geological past

A pile of newspapers, stacked as they are read and discarded one on top of another, has the oldest paper at the bottom of the pile and the youngest at the top. And so it is with rocks. When studying a rock sequence, the assumption can normally be made that the oldest rocks are at the base and the youngest ones at the top. As each page in our pile of newsprint tells a different tale, so every layer of rock records a unique fragment of the planetary story. The geologist's task is to read and interpret these individual pages of the Earth's history and to place them in the correct order, building up a picture from small scraps of evidence of the way in which these rocks have formed.

Dating the geological past can be accomplished in two ways – by comparative and absolute methods. Using the former, rocks from one area can be compared with those of another and their relative age determined by the fossil assemblage they contain. Study of the fossil record over the last two centuries has established the relative order in which ancient life forms made their appearance in the geological record. By carefully cataloguing the fossil content of rock layers, its age, relative to strata in another part of the country, can be determined. 'Absolute' dating of rocks can only be accomplished using advanced analytical techniques. This process requires that the minute quantities of radioactive material that some rocks, particularly granites, contained at their time of formation, are accurately measured. Radioactive substances decay at a fixed and characteristic rate to more stable products. Measuring the relative quantities of the original to decay products against the rate at which radioactive decay takes place allows investigators to make a numerical assessment of the age of any rocks that contain radioactive material. As measurement techniques have improved, so the reliability and value of these dating techniques has increased in recent years.

Geological time

The geological timescale has been considerably refined since Lyell's early attempts. The abyss of time, as Playfair so eloquently put it, is now divided up into periods and then stages that have universal application across the globe. Each stage has its 'type area', which is regarded as the best and fullest development of strata of that particular age to be found anywhere in the world. As Britain was in the forefront of much of this early pioneering geological survey, many of these geological periods and the component stages were defined in this country, specifically the Cambrian, Ordovician, Silurian and Devonian. In terms of the further subdivision of periods into stages, the Jurassic Period, for example, is divided into a number of stages including the Kimmeridgian, Oxfordian and Bathonian, which take their name from well-known locations in the south of England.

A geological timescale for Planet Earth. From the earliest beginnings 4,500 million years ago, the Earth has evolved from an inhospitable place to a world of incredible beauty and biodiversity. This graphic represents the various changes that have taken place through geological time, illustrating the significant stages of development with scenes and life forms from the land that was to become Scotland.

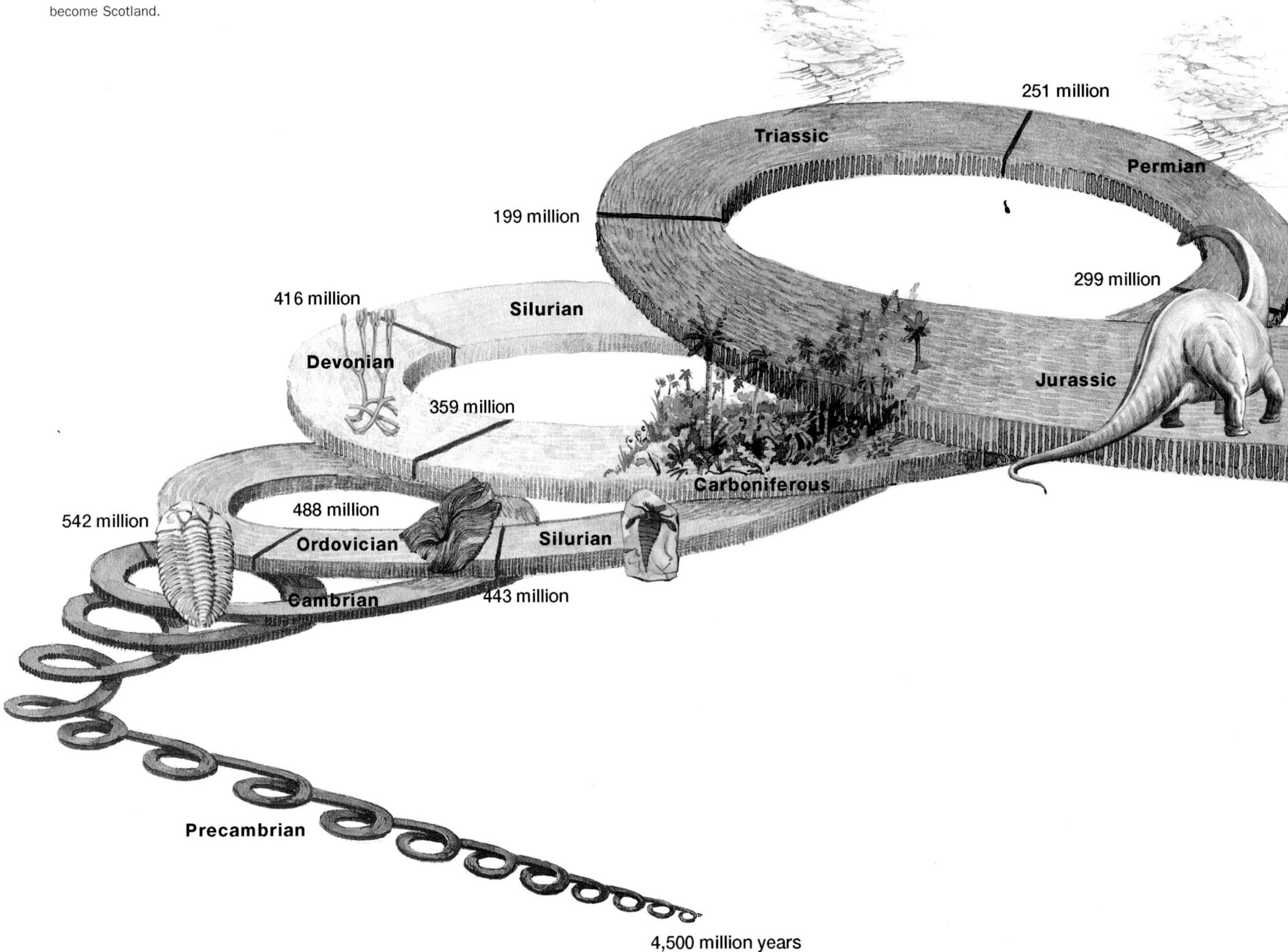

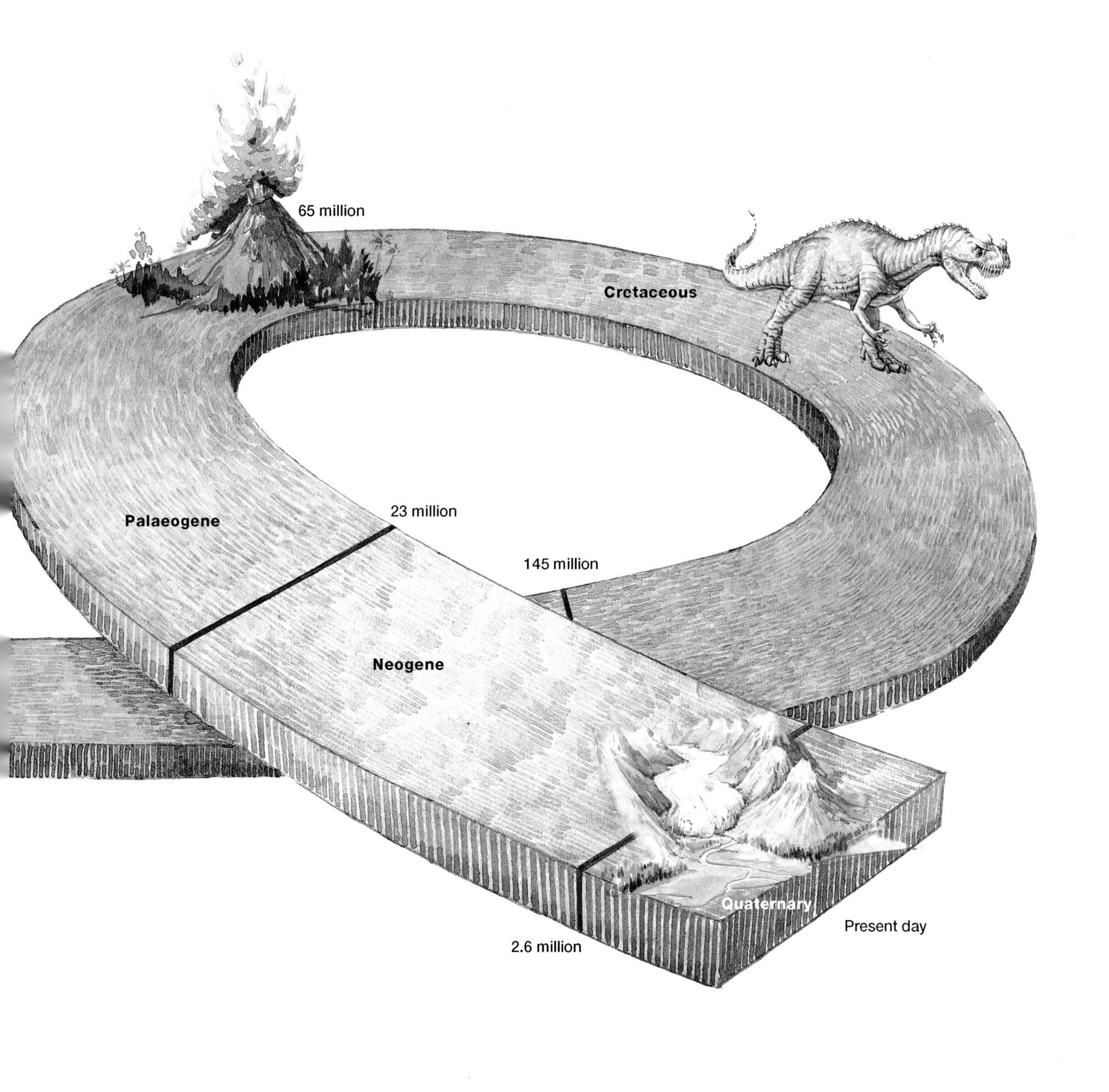
65 million
Cretaceous
Palaeogene
23 million
145 million
Neogene
Quaternary
Present day
2.6 million

A Geological Timescale for Planet Earth

The "abyss of time" has been divided into geological periods that are universally recognised. The table provides a commentary on the most significant events worldwide and also in Scotland.

GEOLOGICAL PERIOD	GLOBAL EVENTS	EVENTS IN SCOTLAND
Quaternary 2.6 million years ago to present day	For much of the Quaternary, the climate has been cold. Popularly known as the 'Ice Age', this period of time has been characterised by many episodes of glacier advance and retreat. During the coldest periods, ice sheets advanced across large parts of North America and Northern Europe, as well as in polar areas.	This period, more than any other, has shaped the landscape. Ice sheets and mountain glaciers carved out glens and corries in the uplands and formed moraines and other deposits on lower ground. The relative levels of the land and sea also varied as the ice sheets advanced and retreated.
Neogene 23–2.6 million years ago	Mammals continued to extend their range and became one of the dominant life forms on the planet. Grasslands spread across the southern plains of Africa forming extensive habitats. India collided with mainland Asia forming the Himalayas.	The climate was warm temperate for much of this period, and further weathering and erosion of the rocks occurred. Temperatures fell progressively and cooling intensified as the Ice Age approached.
Palaeogene 65–23 million years ago	Mammals extended their range to fill most available ecological niches. The Atlantic, North and South, continued to widen as the Americas drifted further away from Europe and Africa. Tropical climates prevailed across the globe and a wide variety of trees developed, including beech, chestnut and sycamore. The distribution of land and sea across the globe took on a familiar look.	A chain of volcanoes developed, including St Kilda, Skye, Rum, Mull, Ardnamurchan, Arran and Ailsa Craig, and many lava flows were erupted. During the early part of this period, the climate was tropical and some lavas on Skye show evidence of deep tropical weathering. Periods of uplift and erosion of the Scottish landmass occurred and great thicknesses of sandstones and muds were laid down in the offshore basins of the North Sea.
Cretaceous 145–65 million years ago	The fifth mass extinction took place at the end of the Cretaceous Period. Dinosaurs were the main casualty. A meteorite strike in Mexico, causing widespread disruption to the world's weather systems, is thought to be the primary cause of the dinosaurs' demise. The first flowering plants appeared and birds evolved. The North Atlantic began to develop as North America split apart from Europe.	For much of this period, Scotland, with the exception of the highest land, lay under a tropical sea. Thick layers of chalk were laid down on the floor of this tropical sea, most of which was later removed from present land areas by erosion. Great thicknesses of chalk were formed on the floor of the North Sea.

Jurassic 199–145 million years ago	This was the age of the dinosaurs worldwide. *Archaeopteryx*, a link between dinosaurs and birds, evolved. Pangaea, the supercontinent comprising all of the world's landmasses, began to split asunder.	Rapid sea-level rise marked the beginning of this period, flooding much of Scotland. Meat- and plant-eating dinosaurs roamed the coastal fringes and an abundance of sea-life existed, including ammonites, corals, belemnites, ichthyosaurs and marine turtles.
Triassic 251–199 million years ago	The first mammals and dinosaurs evolved. The fourth and most extensive mass extinction took place at the end of the Triassic Period. The Earth's ecosystems were changed for ever as most marine life was wiped out in a geological instant.	Scotland was located in near-equatorial latitudes, in a similar position to sub-Saharan Africa today, and desert conditions largely prevailed. Red sandstones of the Elgin area, formed under these conditions, preserve the footprints of long-extinct reptiles.
Permian 299–251 million years ago	Plants such as cycads, conifers and ginkgos evolved. The third mass extinction took place at the end of the Permian Period, perhaps triggered by extensive volcanic eruptions in Siberia. These large-scale eruptions would have filled the atmosphere with dust, which in turn would have changed climates worldwide.	Scotland lay around ten degrees north of the Equator. Desert conditions were widespread across the land. The red sandstones of the Dumfries area and the Hopeman sandstones of Elgin are characteristic of the period.
Carboniferous 359–299 million years ago	The first reptiles evolved and amphibians made their first appearance. Tropical forests were the dominant ecosystem across the planet. Dry land was divided into two continental landmasses – Gondwana remained astride the South Pole. Coal forests were widespread on land and sharks and spiny fish colonised the sea.	Tropical rain forests covered much of the Central Belt, as Scotland moved northwards to reach the Equator. Coral reefs flourished in azure-blue tropical seas. Numerous volcanoes were active during this period, such as Arthur's Seat, Edinburgh Castle Rock, Traprain Law and the Garleton Hills. Thick sequences of lava were erupted to form the Campsie Fells.
Devonian 416–359 million years ago	Four-limbed vertebrates evolved from finned fish and spread to all corners of the world. The climate was warm and mild. The second mass extinction took place at the end of the Devonian Period. Rapid development of terrestrial vegetation occurred.	The mountain range created by the colliding continents was rapidly eroded during the Devonian Period. The rock was unprotected by vegetation cover and was quickly broken down into mud, sand and pebbles that were carried to lower ground by rivers and under the force of gravity. Between the high mountain peaks, great lakes developed, which teemed with primitive fishes.

Silurian 443–416 million years ago	The plants and animals that survived the end-Ordovician extinction continued to thrive. Brachiopods, molluscs, trilobites and graptolites increased in diversity. The first animals, including arthropods, colonised the land. The first vascular plants, all below 20 centimetres in height, also made an appearance on dry land. Sea levels rose as the Ordovician ice sheets melted.	The Laurentian continent collided with Avalonia, uniting Scotland with England for the first time in their geological history. Movements took place along major faults to assemble Scotland from four previously separate continental fragments. Many granites and related igneous rocks date from this period.
Ordovician 488–443 million years ago	Corals and fish first appeared in the fossil record. Another burst of evolutionary activity gave rise to many new animals and also the first tiny land plants. Mosses and liverworts colonised the land, which until this time had been bereft of life. The Iapetus Ocean was at its widest, separating Laurentia from Avalonia. The first mass extinction took place at the end of the Ordovician Period. An ice age gripped the Southern Hemisphere.	The Caledonian Mountain Belt, stretching from Norway to the Appalachians, started to form. Much of the land that was to become Scotland was created during the cataclysmic episode. The Iapetus Ocean, which separated northern from southern Britain, was at its widest, although by the end of the period the ocean had narrowed considerably.
Cambrian 542–488 million years ago	An explosion of different life forms appeared in the fossil record, including molluscs, brachiopods and trilobites. Most of the world's landmasses were united as part of the supercontinent, Gondwana.	The land that was to become Scotland was located near the South Pole, at the margins of the Laurentian super-continent. Beach deposits of this age now occur in the North-west Highlands and preserve evidence of early life in the form of worm burrows and trilobites.
Precambrian including the Hadean 4,500–542 million years ago The lower age limit of the Precambrian is not formally defined.	Increased accumulation of oxygen in the atmosphere facilitated the evolution of simple organisms, such as worms, jellyfish and sponges. A primitive crust developed as the Earth cooled and water covered the lower-lying land to divide the planet into land and sea. Tectonic plates started to move around the globe, colliding, spreading, sliding and generally driving activity on the Earth's surface thereafter. The creation of Planet Earth from interstellar dust marked the start of the Precambrian era when conditions would have literally been 'Hell on Earth'.	Rocks formed during the Precambrian underlie about half the surface area of Scotland. The oldest rocks, known as Lewisian gneisses, build the Western Isles, Coll, Tiree and parts of the north-western seaboard. They were forged deep within the Earth's crust up to some 3,000 million years ago. They are overlain by a thick sequence of Torridonian sandstone that has remained unaltered since its formation. The earliest traces of life to be found anywhere in Scotland have been described from the Torridonian of the North-west Highlands. Great thicknesses of sandstone, limestones, muds and lavas accumulated during later Precambrian times that were later altered by metamorphism to form Dalradian rocks. Similarly, Moine rocks started off as layer upon layer of sandstone, only to be altered by deep burial in the Earth's crust.

IN THE BEGINNING – LIFE APPEARS

The early Earth had an atmosphere almost bereft of oxygen. Hydrogen sulphide, methane, nitrogen and carbon dioxide dominated. But as oxygen-producing bacteria developed and thrived in the primitive oceans, so the atmosphere became more benign. Over time, thick layers of rocks rich in iron oxide, known as banded ironstones, developed in the oceans, indicating an increasing abundance of oxygen. Between 3,000 million and 2,000 million years ago, great accumulations of these red- and black-striped rocks built up in the world's oceans as a direct result of the increased amounts of oxygen. Earth was rapidly developing a planetary life support system.

The precise point at which life began on the planet will, in all likelihood, remain an enduring mystery. But searching for clues in some of the most inaccessible parts of the Earth's crust, the ocean deeps, has provided some ideas about where it might have begun. The black smokers of the mid-Atlantic are volcanic pipes that pump super-heated water, heavy with sulphurous compounds, into the ocean deep. It is in an inhospitable environment similar to this that the earliest, most primitive bacterial life forms are thought to have evolved. The origins of these bacteria are still uncertain. Probably they evolved through chemical reactions, but some scientists have speculated that organic compounds, the building blocks of life, were transported from extra-terrestrial sources, carried in a bombardment of icy comets from outer space.

Stromatolites – the givers of life

The atmosphere of the primordial world was largely inimical to life. But with the advent of the stromatolite, things changed dramatically for the better, paving the way for all life that followed. Stromatolites were one of the first organic structures to be recorded in the fossil record. These rather unprepossessing structures were formed by countless colonies of single-celled bacteria, formerly known as blue-green algae and now known as cyanobacteria. They characteristically formed as a series of mats, giving the rock a layered form in section. Their great gift to the early world was the ability to convert carbon dioxide to oxygen. These bacterial reefs are found worldwide in the fossil record, from Australia to Scotland, in some of the very earliest rocks. Over time, these single-celled organisms changed the nature of the Earth's atmosphere and encouraged the later explosion of life that happened during the Cambrian Period. But the success of these organisms led ultimately to their undoing. As the atmosphere became richer in oxygen, so the bacteria that built the stromatolite structures were put under increasing stress as the composition of the atmosphere gradually changed. The adaptation of micro-organisms to an oxygen-rich atmosphere was a key step in the evolutionary process that led to the development of the more advanced cell structures and ultimately to more advanced organisms. Live stromatolites are found today in Australia, so some clearly coped with the changing atmospheric composition. Their gift of oxygen to the early world changed the course of evolutionary history. Oxygen was poisonous to many other early organisms that existed in Precambrian times on the Earth. So a whole new range of life forms, including our antecedents, were favoured at the expense of many existing organisms. Such was the lottery of life in those early days.

The Cambrian explosion of life

Around 550 million years ago, something quite remarkable happened on Planet Earth. The simple forms of life that had previously inhabited the Precambrian seas rapidly gave rise to a multiplicity of more complex creatures. A huge variety of invertebrates, or animals without backbones, suddenly appeared in a previously relatively barren fossil record. The Cambrian explosion of life had happened. In Earth history, this was the greatest period of evolutionary innovation, with the forerunners of many of the animals that exist today making their debuts. At this time, all life was restricted to the sea and trilobites were one of the commonest forms. They look like the woodlice of today, having a head shield, segmented body and tail. Algae, sponges, brachiopods, echinoids and enigmatic colonial planktonic organisms called graptolites also left a voluminous fossil record. These primitive life forms that made their appearance during the Cambrian explosion have subsequently evolved into a myriad of different forms of life that occupy every available ecological niche on the planet.

Top. A crinoid or sea lily that lived in the tropical seas during the Carboniferous Period.

Middle. Trilobites inhabited the Iapetus Ocean in the Ordovician Period.

Bottom. Fossil fish from the Devonian Period.

The crinoid, trilobite and fish are examples of different life forms from Scotland's rich fossil record.

Life evolves

From that explosive beginning, biodiversity on Earth began to burgeon. Charles Darwin provided a model in his seminal book, *Origin of Species,* for the way in which plants and animals adapt over successive generations to fit changing environments and only the fittest survive. So over many generations, life forms change and adapt, giving rise to new species. Scientists have arranged species in the order in which they evolved, derived largely from a detailed study of

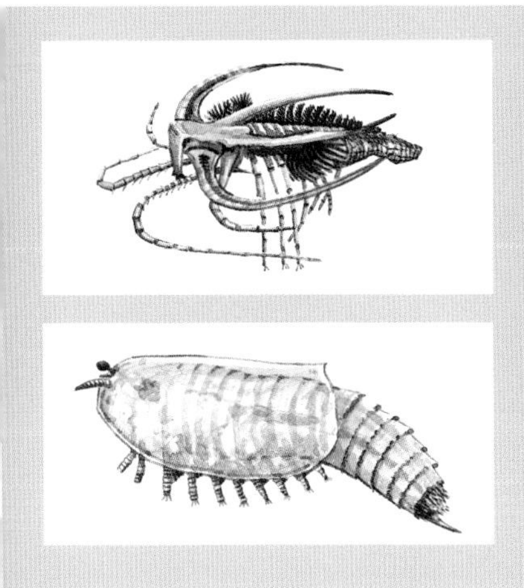

An evolutionary experiment

A most remarkable find was made in the Rocky Mountains of British Columbia, where Charles D Walcott, secretary of the Smithsonian Institute, first excavated the Burgess Shales in 1909. A dazzling array of primitive forms of seaweed, sponges, crustaceans, trilobites and sea urchins were discovered in these Cambrian rocks. The finds were of particular interest as the soft parts, such as gills and eyes, of many of the specimens were beautifully preserved. There were many creatures with bizarre anatomical features, such as an animal called *Opabinia* that had five eyes and a snout like a vacuum cleaner. It was as if evolutionary forces were experimenting with what combination of body parts would work best in the benign environment of the Cambrian seas. The Burgess Shales have now been declared a UNESCO World Heritage Site and are preserved as part of Yoho National Park. Similar deposits occur in China, where the Maotianshan Shales contain a similar collection of weird life forms.

Many of the life forms from the Cambrian were 'experimental' and few survived for long. These creatures are from the Burgess Shales in Canada.

the fossil record worldwide. Of all the species that have ever existed on the Earth, some 99 per cent of them are now extinct. Individual species seem to have a 'shelf life', beyond which viable populations do not survive. Some species remain viable for hundreds of millions of years, changing little over that time, whilst others have less staying power and disappear just as quickly as they chanced upon the scene.

There are many factors that determine success, but perhaps the most significant is the ability to adapt to new environments. The geological record attests to many and rapid environmental changes as the continents moved around the globe, travelling through different climatic belts. The fossil record also shows periods of catastrophic change where most of the life on Earth was wiped out in a geological instant. There were five major mass extinctions, as they are known, from the Cambrian onwards. Each episode had a natural cause, such as dramatic changes in sea level or widespread glaciations, but the results were the same: huge decreases in biodiversity. The most severe of these mass extinction episodes happened at the end of the Permian, some 251 million years ago, when it has been estimated that 96 per cent of all marine species were lost and 75 per cent of all vertebrate – animals with backbones – groups simply died out. Many causes have been proposed to explain this most devastating of moments in the Earth's history, including fluctuations in sea level, changes in the salt content of the sea and an increase in volcanic activity. It may indeed have been a combination of all of these factors.

The death of the dinosaurs 65 million years ago is the last mass extinction event known. But dinosaurs were not the only casualties of this end-Cretaceous/early-Palaeogene event. Around 75 per cent of all species that existed at that time, including most large land animals, marine reptiles, fish, brachiopods, some species of plankton and even the ammonoids, that had survived previous mass extinction events, finally disappeared off the face of the Earth. Crocodiles, however, survived this catastrophic event and its aftermath and continued to prosper.

The cause of this final mass extinction has

been the subject of intensive study and the most often quoted culprit is a giant meteorite strike off the coast of Mexico. The theory runs that this caused the oceans to boil and the atmosphere to be choked with dust. These events caused a winter of such intensity and longevity that ecosystems collapsed and only the hardiest and most adaptive of creatures survived. Whatever the cause, this episode ushered in the age of mammals.

In the early Palaeogene, some 65 million years ago, immediately after the demise of the dinosaurs, the world was entirely bereft of large creatures of any type. But soon, the ecosystems were replenished by the burgeoning ranks of the mammals. The earliest representatives of the mammal lineage have been around since Jurassic times, but it was in later times that this group of animals diversified. They radiated from small, simple shrew-like creatures, which probably fed on eggs, fruit and invertebrates, to fill almost every available ecological niche. At the beginning of the Palaeogene, mammals had divided into a number of groups, including carnivores, insect-eaters, primates, rabbits and bats. But it was during the subsequent part of that period that mammals evolved into many and highly specialised forms that exploited the riches offered by the varied habitats across the face of the Earth. Early modern man or *Homo sapiens* evolved from these early beginnings to become the most predatory and dominant force in the world. There are now over 6.5 billion individuals of this species, occupying nearly every habitat on Earth.

Crocodiles, lizards, turtles and frogs also thrived in the benign tropical paradise of the Palaeogene. Rainforests were the dominant ecosystem across the world; and not just confined to equatorial latitudes. Fossil evidence indicates that tropical forests also occurred in the polar regions during this period. Palms and horsetails formed the upper canopy, whilst magnolia, chestnut and sequoias grew closer to the ground.

Earth's changing climate: from 'hot-house' to 'ice-house'

The Earth's climate has fluctuated naturally between extremes of global warming and global cooling – between a 'hot-house' state and an 'ice-house' or ice age state. During the latter, ice sheets and glaciers developed in high latitudes and frequently expanded into middle latitudes. Ice ages have occurred throughout the history of the Earth, during the Archaean (about 2,900 million years ago), the Early Proterozoic (between about 2,250 and 2,400 million years ago), the Late Proterozoic (between about 950 and 580 million years ago), the Ordovician and Silurian (between about 445 and 430 million years ago), the Carboniferous and Permian (between about 340 and 260 million years ago) and the Quaternary (the last 2.6 million years). Between 750 and 580 million years ago, ice may even have covered much of the globe several times, turning the Earth into a giant 'snowball'. In Scotland, there is evidence of these early ice ages in the rocks on Islay, the Garvellach Islands and Schiehallion near Loch Tummel. In contrast, the climate was exceptionally warm during the Cretaceous (145 to 65 million years ago) and the Early Eocene (about 52 million years ago), when temperatures far exceeded those of today. Several factors may account for these dramatic fluctuations in climate: changes in the amount of solar radiation received by the Earth, changes in the amount of carbon dioxide in the atmosphere and changes in the global distribution of the continents arising from the plate movements described earlier in this chapter.

Much of the present landscape of Scotland was shaped during the Quaternary Ice Age. Today we are still in this ice age, but in a warmer interlude known as an interglacial. This interglacial is called the Holocene. It spans the last 11,500 years and follows a period of time when much of Scotland was covered by glaciers.

CLIMATE CHANGE AND THE ICE AGE

Falling temperatures after the 'hot-house' conditions of the Early Eocene marked the onset of a long period of stepwise cooling of the global climate that eventually culminated in the Quaternary Ice Age. Widespread glaciation first occurred in Antarctica about 34 million years ago. Later, as the cooling intensified about 14 million years ago, the Antarctic ice sheet became a permanent feature. Glaciers began to expand in the Arctic, in Greenland and Iceland, some time after about 14 million years ago. Further cooling saw the wider expansion of ice sheets on northern land areas, particularly after about 3 million years ago in North America, Greenland and Eurasia. This latter cooling is usually considered to mark the beginning of the Quaternary Ice Age. Glacial sediments dropped to the bottom of the sea from melting icebergs are first recorded after about 2.6 million years ago in cores from the floor of the North Atlantic Ocean, indicating that continental ice sheets had reached the coast.

Several explanations have been suggested for this global cooling and the onset and intensity of the Quaternary Ice Age. As the continents split apart during the break-up of

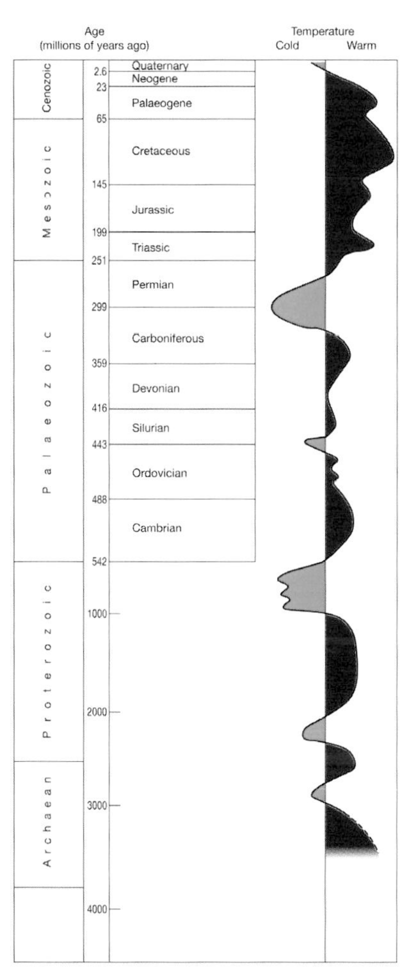

The broad-scale pattern of global climate change over the last 3,500 million years. The climate of the Earth has varied throughout this time between periods of 'hot-house' and 'ice-house' conditions. During the former, glaciers were generally absent; during the latter, ice sheets and glaciers expanded at high latitudes and sometimes extended into mid-latitudes and even low-latitudes. The climate of Scotland has varied according to these global patterns and to its changing position on the surface of the globe.

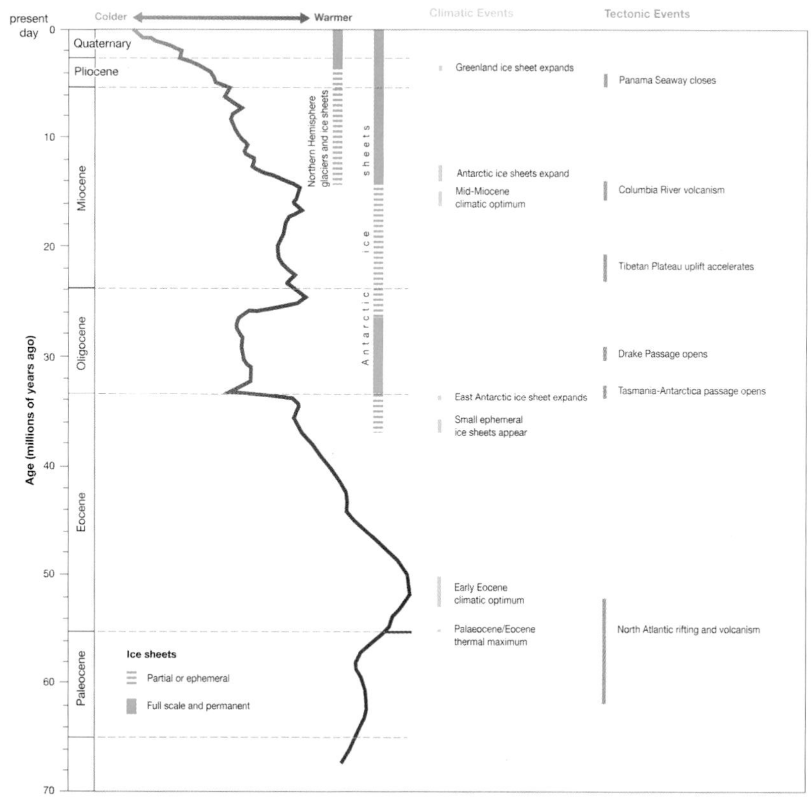

Climate change over the last 70 million years, showing stepwise cooling and the growth of polar ice sheets. The present Ice Age followed intensified cooling after about 3 million years ago. Selected geological events which probably affected the global climate are also shown.

Pangaea (see chapter 3), significant changes occurred in global ocean circulation patterns. In the Southern Hemisphere, Australia parted from Antarctica, and the Drake Passage opened between South America and the Antarctic Peninsula. This allowed the establishment of the Antarctic Circumpolar Current and the belt of westerly winds that encircle Antarctica today. In turn, they acted as a barrier to heat exchange, allowing cooling of the Antarctic continent and the eventual growth of ice sheets there. Later cooling and expansion of glaciers in the Northern Hemisphere has been attributed to closure of the Panama Seaway between North America and South America, closure of the Indonesian Seaway, stronger currents in the North Atlantic and more snowfall at high latitudes. Mountain building and uplift of extensive plateaux in western North America and in Tibet are also thought to have modified the circulation of the upper atmosphere and aided the formation of glaciers in the mid-latitudes, by diverting polar air southwards. At the same time, increased chemical weathering of freshly exposed rock in the uplifted mountains, particularly in the Himalaya and Tibetan Plateau, led to significantly less carbon dioxide in the atmosphere since carbon dioxide reacts with water and silicate minerals to form carbonates which are removed in solution by rivers and deposited in the oceans. It is unlikely that there was one single cause of the global cooling and all these factors may have contributed. Undoubtedly a key driver, however, would have been the decreasing amount of carbon dioxide in the atmosphere.

The last few decades have seen a revolution in the understanding of climate change during the Quaternary Ice Age. Advances in science and technology have allowed access to the remarkable climate archives preserved in the sediments on the floors of the world's oceans and in the layers of the Greenland and Antarctic ice sheets. The results from ocean drilling and ice coring research programmes have greatly changed our understanding of how often and how quickly the climate has changed. The Quaternary Ice Age was not one continuous episode of refrigeration. In fact, the climate has regularly swung back and forth in cyclic fashion from colder to warmer conditions. There have been many long, cold episodes (glacials), broken by shorter, warmer intervals (interglacials) when the climate was similar to that of the present day. During the period between about 2.6 and 0.75 million years ago, the major climate cycles lasted approximately 41,000 years. In the last 750,000 years, the glacial periods have lasted longer, for about 100,000 years, and increased in intensity. Even the glacial periods were characterised by marked fluctuations in climate, and the most extreme conditions when the ice sheets were at their maximum extent generally lasted only 10,000 to 15,000 years. Not only have the climate shifts been frequent, but they have also been extremely rapid. For example at the end of the last glaciation, 11,500 years ago, the climate in Greenland warmed by as much as 10°C from glacial to interglacial conditions in the space of a few decades.

Although we are still in the Ice Age today, we are fortunate to be living in one of the interglacials. However, the climate records show that such periods, geologically speaking, are relatively short-lived, lasting for only about 10,000 to 15,000 years. Therefore, the interglacial conditions in which we now live are exceptional when seen in the context of the longer-term climate record of the Ice Age and its strongly glacial signature. Our present Holocene interglacial began 11,500 years ago. So does this mean that we can expect the next glaciation to begin in the

Climate records from the ocean floor

As sediment accumulates over time on the floors of the world's oceans, shells of microscopic marine organisms called foraminifera become buried in the layers of mud. The proportion of heavy and light isotopes of oxygen in their carbonate shells reflects the temperature and composition of the seawater at the time when they lived, which vary with changes in global ice volume. During cold periods, relatively fewer water molecules with heavy oxygen are evaporated, so the ratio of light to heavy oxygen decreases in the oceans but increases in the ice sheets as the precipitation which accumulates as snow and ice contains more light oxygen molecules; these light oxygen molecules are stored in the ice sheets and are not returned to the oceans until the ice melts. During warm periods, the ice sheets melt, releasing water molecules with light oxygen back into the sea, and moisture with more heavy oxygen is evaporated in the warmer temperatures. Hence the ratio of light to heavy oxygen increases in the oceans and decreases in the ice sheets.

The layers of mud accumulating on the sea floor slowly build up an archive of past environmental conditions. By drilling cores into the floors of the oceans, scientists are able to sample the different layers of sediment back through time. Analysis of the ratio of the light and heavy isotopes of oxygen in the calcium carbonate of the foraminifera shells in different layers is used to measure changes in ice volume and hence, indirectly, in temperature.

The diagram below shows that the overall trend during the last 3 million years has been towards a colder planet. During the more intense glacial periods of the last 750,000 years, ice sheets periodically expanded and covered Scotland; during the many less cold episodes, smaller icefields and corrie glaciers probably existed in the Highlands. Interglacial periods, when the climate was similar to that of today or slightly warmer, have been relatively brief. In the context of the recent geological past, the present warm period in which we live therefore represents a very short interval in a largely glacial world.

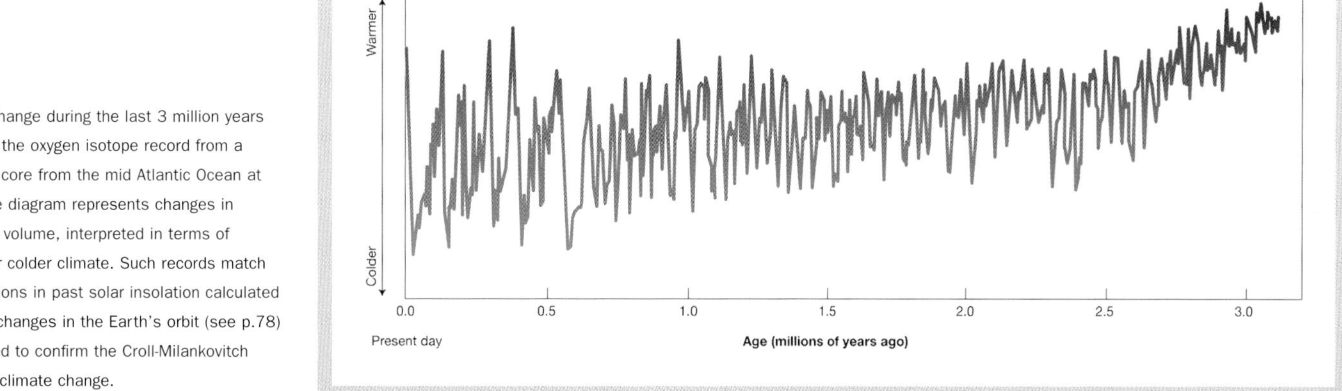

Climate change during the last 3 million years based on the oxygen isotope record from a sediment core from the mid Atlantic Ocean at 41°N. The diagram represents changes in global ice volume, interpreted in terms of warmer or colder climate. Such records match the variations in past solar insolation calculated from the changes in the Earth's orbit (see p.78) and helped to confirm the Croll-Milankovitch theory of climate change.

Climate records from the ice sheets

As snow and ice accumulate on the world's glaciers, tiny bubbles of air are trapped between the snow and ice crystals. These contain miniature samples of the gases in the atmosphere at the time the bubbles were sealed. Since the snow and ice build up in annual layers, it is possible to reconstruct past changes in the composition of the atmosphere in very great detail by counting the layers back through time, just like tree rings but in a vertical dimension. The longest records have been obtained from ice cores drilled down through the Greenland and Antarctic ice sheets. In Greenland, cores from the base of the 3-kilometre-thick ice sheet extend back about 150,000 years. In Antarctica, a core drilled in 1998 at Vostok Station, located in one of the coldest parts of the interior of the continent, provided a detailed record of climate change and variations in the concentrations of gases in the atmosphere over the last 420,000 years. More recently, in 2004, a core from a site at Dome Concordia (Dome C) reached some 3 kilometres down into the ice sheet and includes an even more remarkable record of environmental change over the last 740,000 years, spanning eight separate glacial periods. Full analysis of this core will add greatly to our detailed understanding of past climate changes.

The ratios of different isotopes of oxygen and hydrogen in the ice-core layers provide a natural thermometer, allowing the reconstruction of past temperatures. Analysis of the gases trapped in the air bubbles in the ice has also revealed past changes in the amounts of carbon dioxide and methane in the atmosphere and hence the levels of natural variation in these important greenhouse gases.

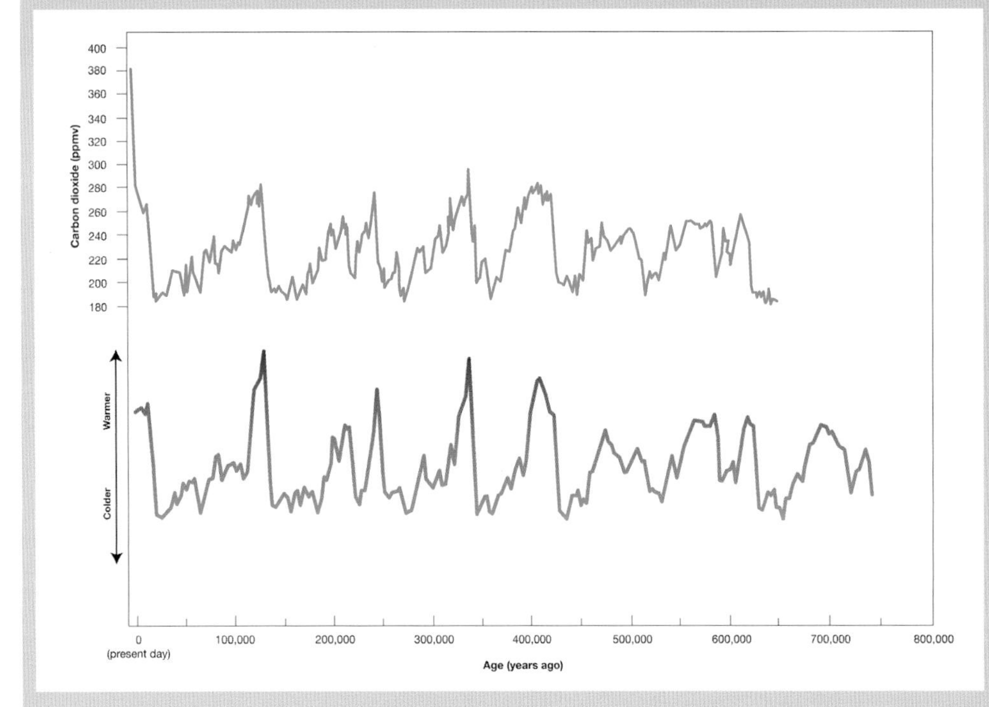

A 3-kilometre-long core drilled through the Antarctic ice sheet at Dome C provides a unique record of climate change spanning the last 740,000 years. The record is very similar to those obtained from ocean-floor cores. Within individual glacial periods, the overall pattern was one of progressive decline in temperatures, followed by very rapid warming into the next interglacial. However, there were also marked fluctuations in climate, with many short periods of rapid warming followed by more gradual cooling. The ice-core records provide a compelling demonstration that past temperatures and carbon dioxide concentrations in the atmosphere are very strongly correlated. Carbon dioxide concentrations were about 180–190 parts per million by volume during glacial periods and 280–300 parts per million by volume during interglacial periods. As a result of greenhouse gas emissions from human activities, the level of carbon dioxide in the atmosphere reached 381 parts per million by volume in early 2006.

not too distant future? Chapter 5 returns to this question.

The way the climate has fluctuated regularly during the Ice Age is related to cyclic variations in how the Earth orbits the Sun. The major climate cycles appear to be driven by a combination of changes in the shape (eccentricity) of the Earth's orbit and the variations in the tilt (obliquity) of the Earth's axis and the wobble (precession) of its axis. Together these orbital variations influence how much incoming solar radiation (insolation) reaches the surface of the Earth in different places and in different seasons, leading to changes in temperature. A Scotsman, James Croll, played a

James Croll

James Croll (1821–1890) is little known, but his work laid the foundation for our modern understanding of the links between climate cycles and the Earth's orbital variations. Born at Little Whitefield Farm, near Wolfhill, Perthshire, he was largely self-educated but read widely on philosophy and science. He worked variously as a millwright in Banchory, a carpenter, a shopkeeper in Elgin, a temperance hotel manager in Blairgowrie and an insurance salesman. In 1859 he was appointed janitor in the museum at the Andersonian Institute in Glasgow. The duties were not onerous and Croll had peace and quiet for study and, most importantly, access to a library. In 1864 he published a paper, 'On the Physical Cause of the Changes of Climate during Geological Epochs', in the *Philosophical Magazine*. In it, he examined the underlying principles of climate change and concluded that the regular succession of cold and warm climates indicated the existence of "some great, fixed, and continuously operating cosmical law". Croll's work attracted the attention of Archibald Geikie, Director of the Geological Survey in Scotland, and a leading exponent of the glacial theory. On the instigation of Geikie, Croll was appointed to a clerical position at the Survey in 1867 and encouraged to pursue his studies in his free time.

Although others had earlier proposed astronomical causes of climate change, Croll was the first to present a reasoned scientific case for linking ice ages with the Earth's orbital variations, in particular the changes in precession and eccentricity. He also recognised that the small changes in insolation arising from the orbital variations must be amplified by feedback mechanisms in the climate system. Croll's seminal ideas on climate change were brought together in his book, *Climate and Time in their Geological Relations: a Theory of Secular Changes of the Earth's Climate*, published in 1875. Croll's work added powerful support for the glacier theory and tied in well with growing evidence at the time for multiple glaciation provided by Archibald Geikie and his brother, James, among others. It also established a basis for the development of the modern theory that astronomical variations act as a 'pacemaker' for Ice Age climate fluctuations and that these fluctuations are cyclic.

Although a modest and retiring figure, Croll became renowned in the scientific world, a remarkable achievement for a self-taught man of humble origins. As fitting recognition of his achievements, he was elected a Fellow of the Royal Society and awarded an honorary LLD degree by the University of St Andrews. But sadly, like so many eminent Scots, he is now poorly known outside scientific circles. Croll was a lifelong teetotaller, but towards the end of his life he is reputed to have asked for a glass of whisky, saying, "I don't think there's much fear of me learning to drink now."

James Croll developed a theory of ice ages based on the variations in the orbit of the Earth and their effects on the amount of incoming solar radiation.

Orbital oscillations and climate change

Three orbital cycles determine the geographical distribution and seasonal variation of incoming solar radiation (insolation) received at the surface of the Earth. **A**. The shape (eccentricity) of the Earth's orbit around the Sun varies from nearly circular to more elliptical over cycles of approximately 400,000 and 100,000 years. This causes tiny variations in the amount of radiation reaching the surface of the Earth during different seasons; the more elliptical the orbit, the greater the seasonal variation in radiation received. Today, eccentricity is nearly at the minimum of its cycle.
B. The tilt of the Earth's axis causes the seasons as the Earth orbits around the Sun. The angle of tilt (obliquity) varies from 21.5° to 24.5° over a cycle of about 41,000 years and affects the strength of the seasons; higher tilt means a greater difference between the amount of radiation received in winter and that in summer (i.e. colder winters and hotter summers) and greater amounts of radiation in high latitudes. Today, the axial tilt is in the upper third of its range.
C. The axis of rotation of the Earth also wobbles, like a spinning top as it slows, describing a circle in space with one revolution every 26,000 years (precession). The wobble and other astronomical movements shift the dates on which the equinoxes occur and cause the seasons to move around the orbit of the Earth over a cycle of roughly 22,000 years. This precession of the equinoxes changes the time of year at which the Earth is closest to, or farthest from, the Sun, which strengthens or weakens the seasons. At present, the Earth is closest to the Sun in January, and farthest away in July, reducing seasonal contrasts. The eccentricity of the orbit determines how close or far the Earth is from the Sun at particular times of the year and hence how much the precession affects the strength of the seasons.

Together these changes act as the pacemaker for the Ice Age climate cycles (see box on p. 75) through their effects on summer insolation in the crucial area of the northern high latitudes. Here, times of short, cool summers allow ice sheets to grow and expand and vice versa. The conditions for glacier growth are most favourable when low obliquity coincides with Northern Hemisphere summers farthest from the Sun due to precession and greatest eccentricity. Other factors are required to amplify the radiation changes to produce the observed changes in climate, but these are not fully understood.

A. Eccentricity

B. Obliquity

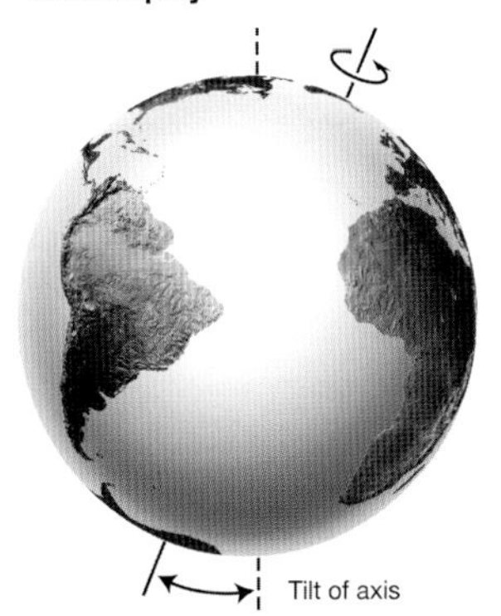

Periodicity: ~ 41,000 years

C. Precession

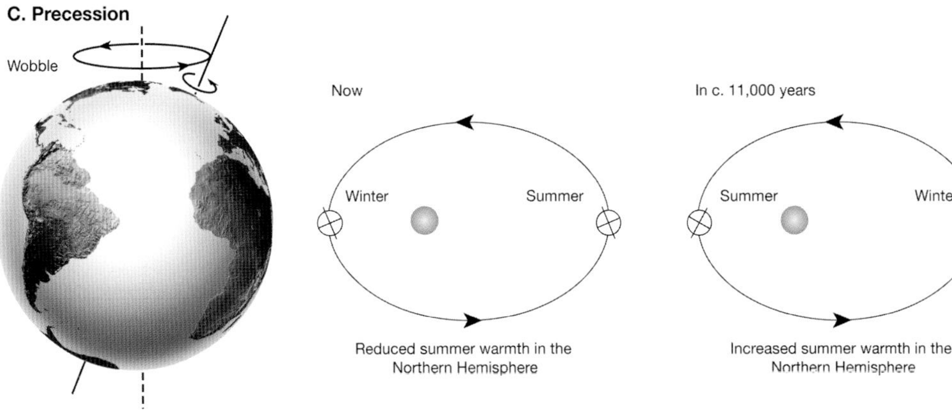

Periodicity: ~ 22,000 years

crucial role in developing this idea in the second half of the nineteenth century and it was later refined by the Serbian astronomer Milutin Milankovitch during the 1920s and 1930s. Milankovitch improved the methods of calculating past variations in the Earth's eccentricity, tilt and precession and developed a theory of climate change based on the changes in the incoming solar radiation in the Northern Hemisphere summer arising from these orbital variations. It was only in the 1970s, when long records from ocean floor cores revealed the full extent of past climate fluctuations, that the Croll–Milankovitch theory could be tested and confirmed. These records showed regular climate cycles with periodicities of about 100,000, 41,000 and 22,000 years, which closely matched the variations in insolation calculated from the orbital variations.

The changes in insolation produced by the orbital variations act as a 'pacemaker' for the Ice Age climate fluctuations. However, they are relatively small and alone are not enough to produce changes of the size seen in the climate records; additional factors are required to amplify the effects of the subtle variations in insolation. We still do not know exactly how this amplification works. However, changes in carbon dioxide appear to play a key part. Past atmospheric concentrations of this gas have been obtained from measurements on the air bubbles in ice cores. Changes in these concentrations over time closely match the temperature records, falling during glacial periods and rising again during interglacials. It appears that as global temperatures cool, upwelling of deep water in the oceans increases plankton growth: this draws down carbon dioxide from the atmosphere, which, in turn, cools further. When global temperatures increase, this process is reversed: carbon dioxide is released from the oceans, which increases the greenhouse effect and the temperatures rise further. This process has switched backwards and forwards during the Ice Age. In addition, as ice and snow cover expands across the surface of the Earth, more radiation is reflected back into the atmosphere, favouring further cooling and glacier growth, and vice versa.

Over shorter timescales of a few thousands of years, major reorganisations in ocean circulation patterns may have profound effects on the climate. A vast system of ocean currents moves around the globe (see p.80). It is driven by differences in seawater density, which depend on temperature and salinity, and is known as the thermohaline circulation, or 'ocean conveyor'. An upper arm carries warm waters from the North Pacific across the Indian Ocean, around the southern tip of Africa and north into the Atlantic Ocean. The flow continues northwards into the North Atlantic where the warm surface water is cooled and becomes increasingly salty due to evaporation. Off Labrador and north of Iceland, in the Greenland and Norwegian Seas, the cool, dense water sinks deep into the ocean and flows back southwards. The formation of sea ice in the Greenland Sea also makes the surface waters saltier and denser as salt is expelled from the ice as the water freezes.

This overturning circulation plays a crucial role in the global climate system and is of particular interest to us here in Scotland. In the North Atlantic region, the warm water is carried northeastwards by the Gulf Stream and the North Atlantic Drift, which are primarily near-surface, wind-driven currents. These release heat and, in part, help to maintain a milder climate in Scotland and northern Europe than would otherwise be the case for their latitudes. However, under certain conditions, increased flow of fresh waters into the North Atlantic may reduce the density of the surface ocean waters and weaken or shut down the overturning circulation, with potentially significant impacts on the climate of Europe. This happened at the end of the last glaciation, following huge inputs of meltwater into the North Atlantic from the sudden discharge of vast ice-dammed lakes that had formed along the southern margins of the North American ice sheet. One such event halted the circulation just after 13,000 years ago, chilling western Europe for nearly 1,500 years and allowing glaciers to reform in Scotland at the time of the Loch Lomond Readvance (see chapter 4).

Ice Age Landscapes

During the coldest parts of the Ice Age, glaciers grew in size in the world's mountains, and vast ice sheets spread south across the middle latitudes of Europe and North America. At its greatest extent, the North American, or Laurentide, ice sheet covered most of Canada and the central and north-eastern United States. The maximum thickness of the ice may have reached 4 kilometres. In Europe, glaciers expanded from centres in the Scandinavian mountains. They moved north over the Barents Sea, east into northern Siberia and south into central Europe, Poland and Germany. At its largest, the Scandinavian ice sheet merged in the North Sea with a smaller ice sheet in Britain, which covered all of Scotland and extended almost as far south as London. Mountain areas like the Alps, Andes, Himalaya and Rockies also had expanded icefields, and the ice sheets in Greenland and Antarctica were larger than today. Areas immediately beyond the ice sheets experienced intensely cold, periglacial climates, like parts of Alaska and Siberia today, and frozen ground (permafrost) was widespread. The last ice sheets were generally at their most extensive sometime between 30,000–20,000 years ago. Rapid climate warming after about 15,000 years ago saw them shrink significantly, although the North American and Scandinavian ice sheets did not finally disappear until some time after the further period of sustained warming around

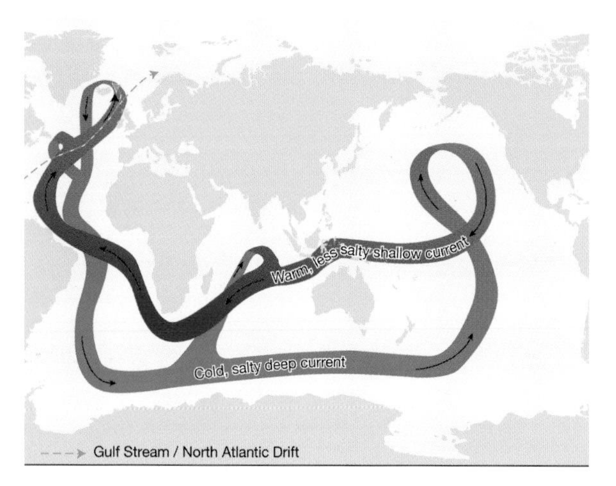

Left. The thermohaline circulation, or 'ocean conveyor', acts as a central heating system for the North Atlantic and North-west Europe, transporting heat from the tropics to higher latitudes. The blue dots indicate the two main areas of deep water formation in the North Atlantic. The wind-driven currents of the Gulf Stream and North Atlantic Drift play an important part in the heat transfer.

Right. The maximum extent of Northern hemisphere ice sheets and the extent of winter sea ice during the last glaciation. The glaciers reached their maximum positions at different times in different areas, between about 30,000–20,000 years ago.

Changes in the relative levels of the land and sea accompany the growth and decay of ice sheets on land.

Left. During glaciations, global sea level is lowered by the transfer of water from the oceans to expanded ice sheets on land. Water evaporated from the oceans falls as snow and is stored frozen in the ice sheet. At the same time, the weight of ice on the land depresses the continental crust. The depression of the crust due to the ice loading is approximately one-third that of the ice thickness.

Right. When the ice melts, sea level rises quickly as the meltwater is returned rapidly to the oceans. Rebound of the crust is slower, lasts longer and is greater where the ice was thicker. Relative sea level therefore varies over time and from place to place.

11,500 years ago that finally ushered in the present interglacial.

The great ice sheets changed considerably the landscapes of large areas of the northern and middle latitudes and left behind a legacy of distinctive landforms (see chapter 4). They deepened the pre-glacial river valleys and sometimes even created new valleys where none existed before. They carved out the corries and glens of Scotland with their famous lochs, such as Loch Ness and Loch Lomond. In some mountain areas, the effects of ice erosion were 'selective'. In the Cairngorm Mountains, the glaciers excavated deep valleys through the mountains but left the plateau surfaces in between relatively unchanged. In lowland areas where the ice sheets were sliding on rocky beds, they scoured the landscape, forming ice-scraped hills and lake-filled depressions. Such landscapes are typical of North-west Scotland. Elsewhere, the ice sheets covered large areas with deposits of till (an unsorted mixture of stones, sand, silt and clay) on which many of our agricultural soils have formed. These deposits were often shaped underneath the ice into a variety of distinctive landforms, such as drumlins (low oval hills). Sometimes the glaciers also left behind large 'erratic' boulders different from the local bedrock and transported many kilometres from their source areas. As the glaciers melted, they fed large rivers of meltwater which deposited extensive spreads of sand and gravel and sometimes mounds and ridges of the same material.

Dramatic changes in global sea level accompanied the waxing and waning of the great ice sheets: sea level fell as the growing masses of ice locked up the world's fresh waters, and rose again during interglacials when the glaciers melted and released their water back into the sea. This process is known as glacioeustasy. At the height of the last glaciation, global sea level was as much as 120 metres lower than at present, so that large areas of the continental shelf were dry land. For example, at the end of the last glaciation it would have been possible to walk from Britain to France, since the rising sea level did not flood the English Channel until around 8,000 years ago. The level of the land has also varied, the crust sinking and deforming under the weight of the growing ice sheets and rising up again when they melted. This process is known as glacioisostasy. Its effects are now evident in the presence of raised, or emerged, shorelines around the world's glaciated coasts.

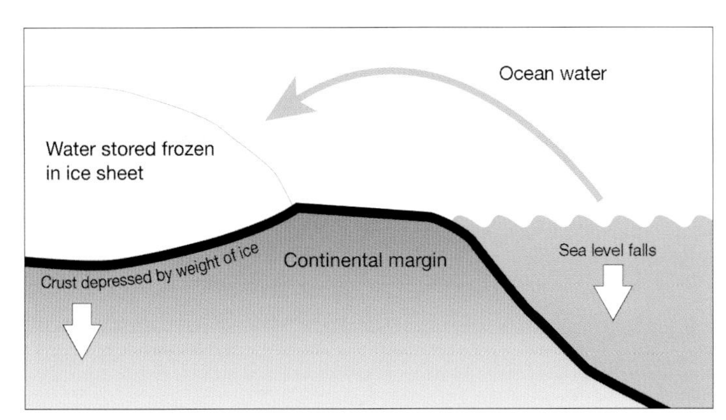

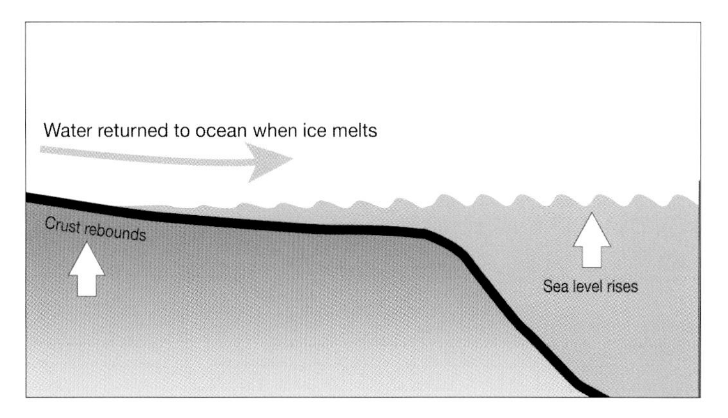

Modern glaciers in the Antarctic show how the landscape of Scotland might have appeared at different times during the Ice Age.

Left. Whole mountain ranges lie buried or nearly buried beneath the Antarctic Ice Sheet and fast-flowing glaciers fill the valleys. Similar ice sheets have covered Scotland a number of times during the Ice Age.

Right. During periods of more restricted glaciation, smaller icefields existed in the West Highlands. Glaciers from these icefields extended down the glens to the west coast, like this glacier today on the island of South Georgia.

Glaciers

Glaciers are moving bodies of ice. They come in various shapes and sizes. Mountain glaciers include small corrie glaciers, which occupy shaded basins at the heads of alpine valleys, and valley glaciers flowing out from corries or mountain icefields. On a much larger scale are the vast, dome-shaped ice sheets, like those that cover Greenland and Antarctica today, which bury complete mountain ranges. Where they are frozen to their beds, glaciers are said to be 'cold-based', and movement takes place by slow 'plastic' deformation in the basal layers of the ice. However, where water is present, glaciers are additionally able to slide over their beds; such glaciers are said to be 'warm-based' and flow much faster. Cold-based glaciers generally occur in polar and continental areas (for example, in Antarctica and the north of Greenland) where temperatures and snowfall are very low. Warm-based glaciers are typical of more oceanic areas where temperatures and snowfall are higher (for example in Norway, Iceland and south-east Alaska).

Ice sheets may contain zones of warm *and* cold-based ice. Where temperatures and snowfall are low or the ice is relatively thin, for example over hills and mountains, the ice may be cold-based and moving very slowly at only a few tens of metres per year. Where temperatures and snowfall are higher, or the ice is thicker, for example in valleys, then the ice may be warm-based and form 'ice streams', rather like solid rivers of ice, which can flow much faster at speeds of 2 kilometres a year or more. Ice caps are smaller versions of ice sheets and are usually drained by warm-based glaciers, which tend to follow the valleys that lie beneath the ice; examples are Vatnajökull in Iceland and Jostedalsbreen in Norway.

Generally, warm-based glaciers are able to alter the landscape much more than cold-based glaciers because they are sliding over their beds and are faster moving. Such variations have produced marked contrasts in landforms across Scotland (see chapter 4).

Scotland and the development of the glacial theory

Today we take the existence of the Ice Age for granted. However, the concept is less than 200 years old and represents one of the major scientific advances of the nineteenth century. The realisation that large areas of northern Europe and North America were formerly covered by extensive land-based ice sheets in the recent geological past has proved crucial to understanding the origins of the present landscape. Although the former existence of more extensive glaciers and their effects on the landscape had been recognised earlier in areas close to the Alps, it required a major conceptual advance to apply the glacial theory more widely, and particularly to those regions where glaciers were no longer present and that lay far from high alpine mountains.

The mid-nineteenth century saw a revolution in geological thinking about the processes that shaped the present landscape, and Scotland and Scottish geologists played a key part. Until then, most geologists explained surface landforms and deposits in terms of a great submergence, akin to the Biblical Flood. Sir James Hall, the eminent Edinburgh geologist and physicist, envisaged great diluvial waves washing over the Edinburgh area from the west, smoothing the rock and forming the asymmetric, crag-and-tail-shaped hills such as Corstorphine Hill, and Castle Rock and the Royal Mile, with their steeper western flanks and gentler, streamlined eastern slopes. Other influential figures, such as Sir Charles Lyell, assigned an important role to floating ice or icebergs in forming surface, or 'drift', deposits. By the end of the century, however, these ideas had been dismissed and it was generally accepted that glaciers had played a fundamental part in shaping the modern landscape during the course of repeated glaciations.

In the late eighteenth century, James Hutton and John Playfair already appreciated the significance of modern Alpine glaciers in transporting rock debris. Playfair later wrote in 1802 that: "for the moving of large masses of rock, the most powerful engines without doubt which nature employs are the glaciers . . . These great masses are in perpetual motion . . . impelled down the declivities on which they rest by their own enormous weight, together with that of the innumerable fragments of rock with which they are loaded." From the distribution of erratic boulders, Hutton and Playfair also recognised that the Alpine glaciers had formerly extended further northwards. Others working in the Alps and Norway soon proposed that glaciers had been even more extensive. This idea of more extensive glaciation was certainly familiar to Scottish geologists. Indeed in the 1820s, Robert Jameson, Professor of Natural History at Edinburgh University, expressed the view in his lectures that glaciers had once existed in Scotland. However, as with many new discoveries, it required a charismatic figure to mount a compelling challenge to the prevailing orthodoxy. In the case of the glacial theory, it was the Swiss geologist Louis Agassiz, already an eminent figure from his work on fossil fish, who most effectively developed and promoted the concept of continental-scale glaciation in the Northern Hemisphere during a great Ice Age.

Agassiz initially conducted his research in Switzerland but it was important for him to convince the geological community in Britain, then at the forefront of geological science. In 1840 he visited Scotland, an ideal test area for his ideas since there are no present-day glaciers and the mountains are relatively low in comparison

In the early nineteenth century, the idea of a catastrophic flood of biblical proportions dominated geological thinking on the shaping of present landforms. Sir James Hall, a leading proponent, advocated that great torrents of water were responsible for the formation of the surface deposits, smoothed rocks with scratched surfaces, asymmetric hills and scattered boulders in central Scotland. Other eminent geologists, including Sir Charles Lyell, favoured a less catastrophic interpretation, but still involving marine submergence and the transport of boulders and other rock debris by icebergs drifting down from the far north. This view was supported by contemporary observations of the presence of such debris on icebergs in the polar regions. Consequently, the word 'drift' was used as a general geological term for unconsolidated surface deposits.

with the Alps. Following a presentation of his ideas at a meeting of the British Association in Glasgow, Agassiz departed on a tour of the West Highlands, accompanied by the Rev. William Buckland, Professor of Geology and Mineralogy at Oxford University. He soon found abundant traces of former glaciers, in addition to those he had seen earlier around Glasgow. The clinching evidence was in Glen Roy and Glen Spean where Agassiz recognised the Parallel Roads to be the shorelines of former ice-dammed lakes, similar to modern features he had observed in Switzerland. From Fort Augustus, he wrote a letter about his findings to Robert Jameson in Edinburgh, intending that it be published in the *Edinburgh New Philosophical Journal*. Jameson immediately recognised the significance of Agassiz' discoveries. As the latest issue of the journal was already in press, Jameson passed the letter to Charles Maclaren, editor of the *Scotsman*, who was also a geologist. Thus on 7 October 1840, under the headline, 'Discovery of former glaciers in Scotland, especially in the Highlands, by Professor Agassiz', the Ice Age was first announced to the wider public.

Two weeks later, Agassiz arrived in Edinburgh where a party of local geologists took him to inspect scratches and grooves on the rocks at Blackford Hill. Agassiz confirmed these were glacial striations and he is said to have proclaimed: "That is the work of ice!" The site is now known as 'Agassiz Rock' (see chapter 6). Although it is not the first site where the existence of former glaciers was recognised in Scotland by Agassiz, it is nevertheless a place of historical significance. The location of the first site is obscured by the passage of time, but it is clear that Glen Roy and Glen Spean provided compelling evidence for Agassiz and deserve an appropriate accolade.

Agassiz recognised the value of his experience in Scotland. He later wrote in 1842: "It was in Scotland that I acquired precision in my ideas regarding ancient glaciers. The existence in that country of so considerable a network of these traces, enabled me to appreciate better the geological mechanism of glaciers and the importance of many facts of detail observed in the neighbour-

Above. Louis Agassiz studied Alpine glaciers and proposed that much of northern Europe, including the British Isles, had been covered by a vast ice sheet in the geologically recent past. The landforms he saw during a tour of Scotland in 1840 helped to confirm his ideas about the glacial theory.

Right. Part of the smoothed and striated rock surface at Agassiz Rock, at Blackford Hill in Edinburgh, which Louis Agassiz interpreted as the "work of ice" when he visited the site in 1840. The form of the rock surface is typical of glacially abraded surfaces elsewhere in Scotland.

hood of those which now exist." However, despite the initial persuasiveness of Agassiz' arguments, many eminent geologists continued to promote the iceberg hypothesis for several decades. Some accepted the former presence of glaciers in the mountains of Britain, but remained unconvinced that an ice sheet had covered the lowlands, moulding the rocks and forming the surface deposits. Eventually, however, the glacial theory prevailed, and a number of Scottish figures contributed greatly to its development and application. They included Charles Maclaren, Thomas Jamieson, Archibald and James Geikie, Andrew Ramsay, James Croll and James Forbes. Together, they provided compelling support for the glacier theory and the role of land glaciers in shaping the landscape; between them, they developed a basic sequence of events during the Ice Age (Jamieson and the Geikies), investigated the causes of climate change (Croll), identified relative sea-level changes (Maclaren and Jamieson), demonstrated glaciers to be a powerful agent of erosion (Ramsay) and advanced the understanding of glacier processes (Forbes). Thus, beginning with Hutton and Playfair, Scotland and Scottish geologists played a highly influential part in one of the major advances in geology in the nineteenth century.

This chapter has outlined some of the key processes underlying how the Earth works and which have a bearing on the shaping of Scotland's landscapes. It provides a foundation for reading the story of the rocks and landforms and revealing the momentous events and changes in Scotland's past. This is taken up in the following chapter with the remarkable story of Scotland's journey through time and across the surface of the globe.

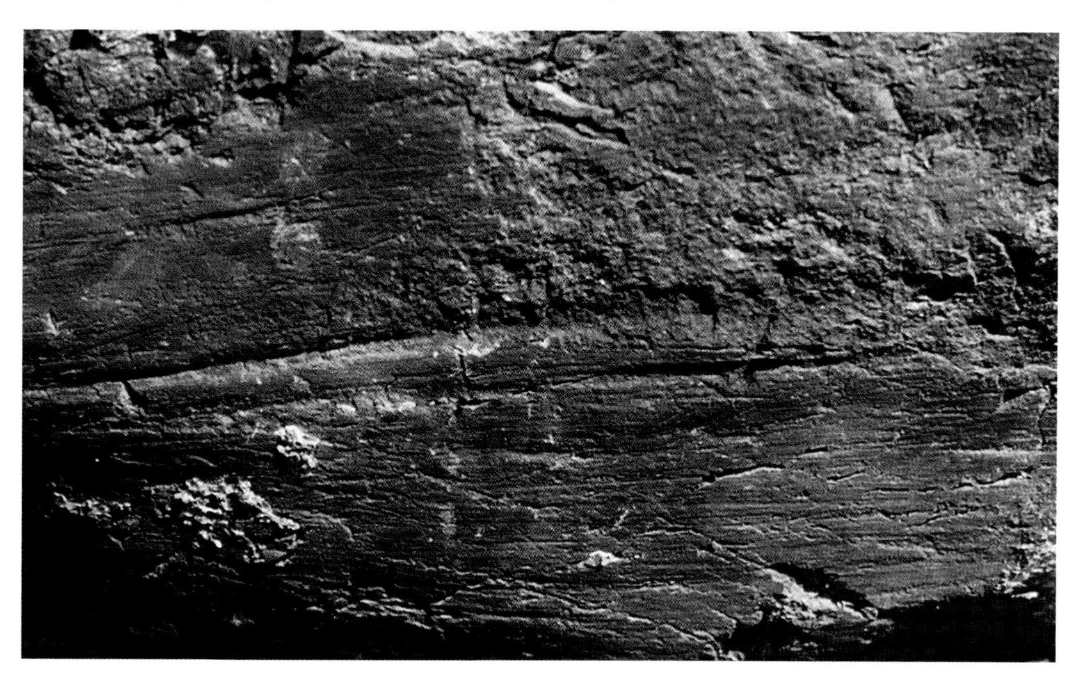

Scots pioneers in the study of the Ice Age

During the second half of the nineteenth century, Scottish geologists were in the vanguard of the development and application of the Ice Age theory.

Sir Archibald Geikie (1835–1924) was born in Edinburgh, the son of a businessman and composer. He began a career in banking, but abandoned it to go to university before joining the Geological Survey. He published a benchmark paper in 1863, 'On the glacial drift of Scotland', effectively dispelling the theory that surface landforms and deposits were the products of marine submergence and iceberg action. This reinvigorated the glacial theory at a time when influential figures, such as Sir Charles Lyell and Sir Roderick Murchison, strongly advocated the theory of submergence. Archibald Geikie was also the first to recognise multiple glaciations. He established that remains of plants and mammals found interbedded with the glacial deposits indicated multiple ice advances separated by significantly warmer intervals. Soon after in 1865, he published the first edition of his classic book, *The Scenery of Scotland Viewed in Connection with its Physical Geology*, based on his extensive field knowledge of Scotland. In 1867, he was appointed Director of the Geological Survey in Scotland and later became the first incumbent of the Murchison Chair of Geology and Mineralogy at the University of Edinburgh. Subsequently, in 1888, he became Director-General of the Geological Survey of the United Kingdom, succeeding Sir Andrew Crombie Ramsay. As well as being renowned for his work on the Ice Age and the scenery of Scotland, he was also an international authority on volcanic geology.

James Geikie (1839–1915) was born in Edinburgh and followed his brother through a career with the Geological Survey in Scotland. He, too, was a strong adherent of the role of glaciers in shaping the landscape, and set out his ideas in a number of influential textbooks. He made a major contribution in recognising the field evidence for multiple glaciations, which supported the concept of cyclic climate change during the Ice Age. He recognised five interglacial stages in Britain. His pre-eminent book was *The Great Ice Age and its Relation to the Antiquity of Man*, first published in 1874. In 1882, he succeeded his brother to the Murchison Chair of Geology and Mineralogy at the University of Edinburgh.

Sir Andrew Crombie Ramsay (1814–91) was born in Glasgow, the son of a manufacturing chemist. He became interested in geology during holidays on Arran and later published his first book on the geology of the island. In 1841, he was appointed to the Geological Survey, in which he served for forty years. He was also Professor of Geology at University College, London, and at the Royal School of Mines. He became Director-General of the Geological Survey in 1872. Ramsay published extensively on the geology of Wales, where he was initially posted by the Survey. He was interested in the origins of the landscape, and in particular the formation of lake basins and valleys in glaciated areas, his attention apparently being awakened to the effects of glacial erosion during his honeymoon in Switzerland. He attributed such features, including the many classic examples in Scotland, to powerful erosion by glaciers and he played a key part in developing the idea of large-scale

Sir Archibald Geikie played a key part in establishing the role of glaciers in shaping the landscape of Scotland.

James Geikie demonstrated that the Ice Age consisted of numerous cold and warm periods.

Sir Andrew Ramsay was a powerful advocate for the role of glaciers in shaping the landscape.

Charles Maclaren played an important part in promoting the glacial theory.

landscape modification by glacial erosion. Along with Thomas Jamieson and Archibald Geikie, Ramsay was one of the key figures who helped to establish the primacy of the glacial theory in the 1860s.

Charles Maclaren (1782–1866) was born at Ormiston, East Lothian, son of a farmer and cattle-dealer. He was almost entirely self-educated and became a clerk in Edinburgh. With others, he established the *Scotsman* newspaper in 1817 and became its first editor. He was also interested in geology and published several works, including *A Sketch of the Geology of Fife and the Lothians* (1838). He was actively involved in geological circles in Edinburgh and when Robert Jameson passed to him Agassiz' letter announcing the glacial theory, Maclaren achieved a famous journalistic scoop for the *Scotsman* noted earlier. He was also among the party of Edinburgh geologists who conducted Agassiz on his excursion around Edinburgh two weeks later. The following year, Maclaren published a series of his own articles in the newspaper, outlining further aspects of the glacial theory. In one of these articles, he first introduced the key concept, now known as glacioeustasy, that sea level would rise and fall as the ice sheets expanded and contracted and as water was locked up in the ice and then released.

Maclaren is also credited with the dubious honour of being involved in a duel in response to journalistic attacks from Dr James Browne, editor of the *Caledonian Mercury*. The two men met at Ravelston, exchanged shots and missed. They apparently refused to shake hands and parted without apology.

Thomas Francis Jamieson (1829–1913) was one of the leading early figures in the study of the Ice Age in Scotland. He was factor at Ellon Castle Estate in Aberdeenshire for many years, and lived at Mains of Waterton, in sight of the raised beaches of the Ythan valley. He was later appointed Fordyce Lecturer on agricultural research at the University of Aberdeen in 1862.

From his detailed field observations, Jamieson established the basic sequence of glaciation and changes in sea level during the Ice Age in Scotland. In particular, he was the first to state the fundamental concept of glacioisostasy in 1865, recognising that the land surface was depressed by the great weight of ice upon it and that it rose back up again following the disappearance of the ice. In the Forth valley and other east-coast estuaries, he identified the existence of a former land surface, comprising a layer of peat with the remains of trees, which was buried by estuarine carse deposits when the sea flooded inland during the postglacial (see chapter 4).

Jamieson also made other notable contributions. He was the first to unravel the details of the formation of the Parallel Roads of Glen Roy (see chapter 4 and chapter 6). Along with Archibald Geikie and Andrew Ramsay, he also played an important part in providing detailed field evidence in support of the glacial theory. In particular, in a series of papers he worked out the history of glaciation in North-east Scotland, identifying the role of multiple glaciations in shaping the landscape. In addition, he recognised the role of ice sheets in eroding the landscape and drew a powerful analogy with modern examples in Greenland and Antarctica.

Thomas Jamieson worked out the sequence of glaciation and sea-level changes in Scotland during the Ice Age.

James Forbes was a pioneer in the study of modern glaciers in the Alps.

James David Forbes (1809–68) made a major contribution to the emerging science of glaciology. He was born in Edinburgh, fourth son of Sir William Forbes of Pitsligo and grandson of the banker Sir William Forbes. He entered Edinburgh University where he attended lectures on geology by Robert Jameson, which included reference to glaciers and their former existence in Scotland. However, his early interests were in physics and meteorology. He quickly established an international reputation for his work on radiant heat and was appointed Professor of Natural Philosophy at Edinburgh University at the age of twenty-four. Forbes met Louis Agassiz during the latter's famous visit to Scotland in 1840, and was invited to join Agassiz at his research station on the Unteraar Glacier in Switzerland in 1841. This proved a turning point for Forbes and stimulated his interest in the structure and movement of glaciers. Unfortunately, the two men fell out after the visit, as Agassiz claimed precedence for recognising the significance of a type of glacier structure, which Forbes rightly considered was his discovery. Consequently, in subsequent years, Forbes established his own research programme on the glaciers of the Mont Blanc area, in particular the Mer de Glace. Here he undertook the first thoroughly detailed measurements of glacier movement and thus put the study of glaciers on a firm scientific footing. He established that glaciers flowed imperceptibly and continuously, but at variable speeds from day to day and season to season. They moved faster in the centre than at the sides and although apparently brittle, were viscous under steady pressure. The results were published in 1843 in his book *Travels through the Alps of Savoy*. In an unusual application of his measurements of glacier flow, Forbes later estimated almost exactly the year when the bodies of several climbers swept into a crevasse by an avalanche on Mont Blanc in 1820 would emerge from the Glacier des Bossons.

In Scotland, Forbes climbed in the Cuillin and in 1836 made the first recorded ascent of Sgùrr nan Gillean with Duncan McIntyre, a local forester, as guide. In 1846, he published one of the first detailed studies of glacier landforms in Scotland, in which he recognised ice-scoured bedrock and moraines in the Cuillin as the product of local mountain glaciers (see chapter 4). Although he suffered from ill health, Forbes travelled extensively in the Alps, and later in Norway. He made the first British ascent of the Jungfrau and was elected an honorary member of the Alpine Club in London. During the final decade of his life, Forbes was appointed Principal of the United College at St Andrews.

CHAPTER 3

SCOTLAND'S JOURNEY ACROSS THE GLOBE

READING THE RECORD OF THE ROCKS

The titanic forces that built the Himalayas and all the other mountains of the world proceed so slowly that they are normally invisible to our eyes. But occasionally, they burst into the most dramatic displays of force that the world can show. The earth begins to shake and the land explode.

David Attenborough, *The Living Planet*, 1984

Like a historian interpreting ancient artefacts or the pages of a rare manuscript, geologists study rocks and landforms to piece together the Earth's history. And like the historian, we find that the geological record is incomplete and fragmentary. Reconstructions are often best guesses: a rationalisation of the epic events of the past with only a meagre few of the scenes preserved for posterity. Much of the evidence has been lost during the 3,000-million-year geological history of Scotland. It is also necessary to understand that interpretations will change as new information comes to light and analytical techniques improve, so an accepted 'fact' may later be called into question.

For the two centuries following James Hutton's seminal work, Scotland occupied a pivotal position in geological study and enquiry. As a consequence of this upsurge of activity and creative thought, many geological phenomena first described in Scotland by the early pioneers have now been recognised the world over. Many of the scientists who trained in Scotland went to work abroad, some founding geological surveys in their adopted countries. This accords Scotland a special place in the annals of geological study that

Scotland on the map

The Geological Survey of Great Britain was founded in 1835, and just over thirty years later, a base was established in Scotland. Their first office was at No.1 India Buildings, Victoria Street, just off the Royal Mile in Edinburgh. Britain was the first country in the world to set about mapping the geology of its land surface in a systematic fashion. The Survey's express purpose was to produce geological maps of the country, based on the Ordnance Survey maps as they became available. Archibald Geikie was the first director in Scotland and he was later joined by his brother James. Together, their influence on the early development of the Survey in Scotland was considerable. Archibald was also appointed as the first Professor of Geology at Edinburgh University. Later recruits to the Survey included John Horne who formed a productive partnership with Cornishman Ben Peach. The appointment of Peach increased the staff complement of Survey geologists in Scotland to a total of four.

Peach and Horne mapped the mountains of Scotland during the summer months, but confined their fieldwork to the lower ground during the spring and autumn. They were the ideal team; Peach supplied the geological insights and intuition. He also left a great legacy of landscape drawings and paintings, sometimes sketched in his field notebooks. By contrast, Horne was a more methodical worker, but reputedly lacked the penetrating insights of his more illustrious colleague. It was John Horne who wrote up their joint findings in the standard Survey format – the geological memoir and accompanying coloured map. Many such memoirs were produced by Peach and Horne, covering varied aspects of the geology of Scotland, but their greatest contribution was made in unravelling the geological complexities of the North-west Highlands; an achievement that has been celebrated by the construction of a monument to their memory at Inchnadamph.

Monument to Peach and Horne at Inchnadamph

The British Geological Survey, as it is now known, continues to thrive. Approaching 200 years since its inception, the Survey has still to complete its initial goal of producing up-to-date geological maps covering the whole country, but its current range of outputs, including those in digital format, are diverse and extremely impressive. Their interests now extend offshore to the continental shelf and to other topics, such as economic minerals, geological hazard evaluation and monitoring earthquakes worldwide. They also run a geoscience information service that seeks to increase the availability of their data.

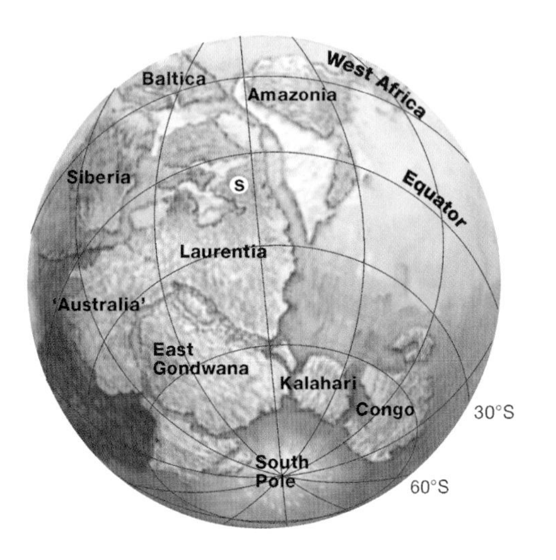

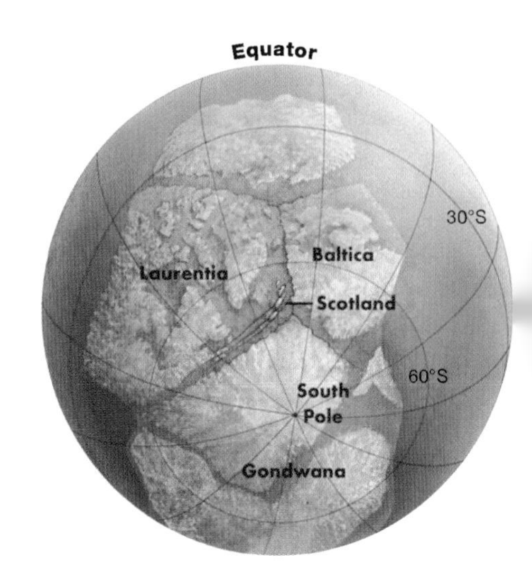

Scotland's Journey. Scotland has travelled widely across the globe for the last 1,000 million years and experienced every climatic variation that Planet Earth has to offer. From tropical climes, Scotland travelled towards the South Pole and back again to cross the Equator during Carboniferous times. The journey continues to this day as Europe moves further away from North America. (See chapter 5 for projections of how this journey will continue.)

750 million years ago. The land that was to become Scotland was just south of the Equator. The 'S' on a white dot indicates the position of Scotland. At this time, Scotland's near neighbours were South America (Amazonia) and North America (Laurentia).

600 million years ago. Over time, Scotland moved southwards towards the South Pole and formed part of a supercontinent that included most of the landmasses of the world at this time.

is unrivalled by any other country. A bold claim perhaps, but it is justified by the facts. These 'firsts' will be described as the story unfolds.

Scotland on the move

We can read the geological story of Scotland in the rocks and landforms. Over 3,000 million years of time have left a record of many treasures and rarities. They include the remains of long-disappeared oceans and ancient mountain ranges, the roots of a multitude of defunct volcanoes and the fossils of many long-extinct plants and animals. Evidence for an ice age that lasted 2 million years and the beginnings of life in Scotland are all present in the record of the rocks. But, most amazing of all is the evidence for Scotland's spectacular odyssey across the globe. From its earliest recorded position close to the Equator, Scotland travelled southwards towards the South Pole and then moved back northwards through equatorial latitudes to its current position 57°N.

Another remarkable aspect of Scotland's Earth history is that this landmass has spent more time in the Southern Hemisphere than in northern latitudes. Since its earliest beginnings, Scotland has been on an amazing journey that is described in some detail in the following pages. And revelation upon revelation, Scotland is not one land, but many separate continental fragments fashioned into one. At least four substantial fragments merged around 420 million years ago to form a proto-Scotland that has been much modified since that time. As if that were not enough, during its 3,000-million-year history, Scotland has been flooded to the point that nearly all the land has been submerged beneath the sea;

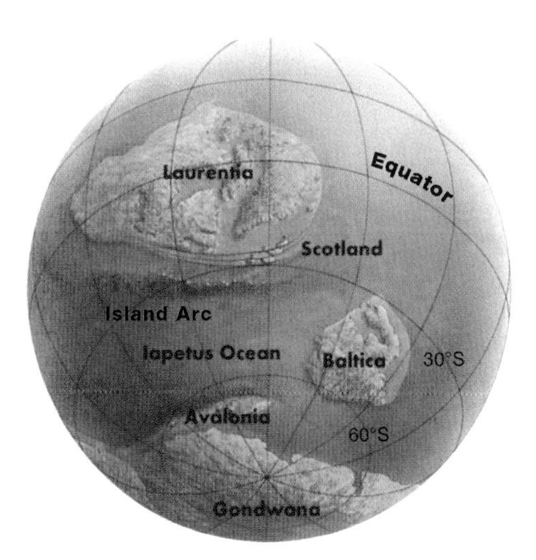

500 million years ago. The supercontinent split up into pieces and Scotland formed part of Laurentia. The Iapetus Ocean widened to its maximum extent and then started to close. Subduction zones and an associated island arc developed. The arc subsequently collided with Scotland, folding the Dalradian rocks in the process. England formed part of Avalonia at this point and was separated from Scotland by the width of an ocean.

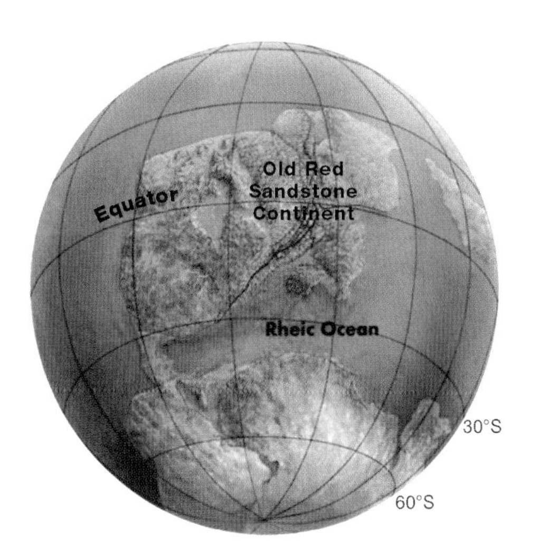

400 million years ago. The Iapetus Ocean closed as Baltica and Avalonia collided with Laurentia. The Old Red Sandstone Continent resulted from this collision.

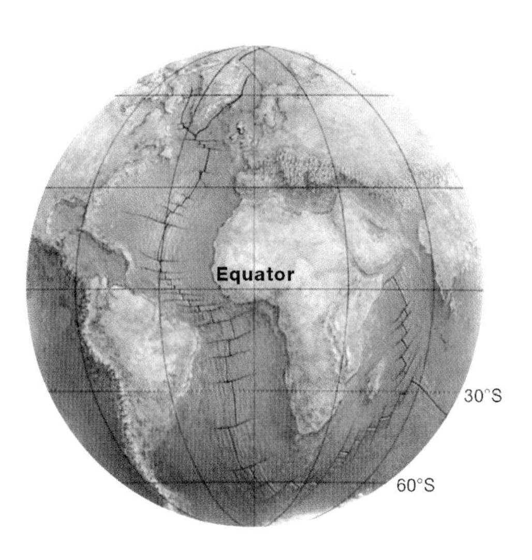

40 million years ago. The configuration of the continents took on a familiar look, as Scotland continued to drift northwards into the Northern Hemisphere. The opening of the North Atlantic Ocean marked the end of Scotland's long association with North America. Around 65 million years ago, Greenland split away from Europe.

racked, twisted and buckled by earthquakes; scorched in searing desert temperatures; and plunged into the icy grip of an ice age. This diverse and varied geological pedigree is a direct result of Scotland's itinerant history as it has wandered the globe, driven by the vagaries of the motion of the Earth's tectonic plates.

Reading the rocks, from the oldest Precambrian rocks through successively younger rocks, allows geologists to reconstruct Scotland's journey from earliest times from the Equator to the South Pole and back. The rocks formed at each stage of our geological development are a direct reflection of the various environments experienced at the time. Climates also varied in response to Scotland's changing geographic position. Whilst in equatorial latitudes, rainforests similar to those of today were widespread and the coal deposits of central Scotland now attest to those times spent astride the Equator. As Scotland moved northwards into latitudes similar to sub Saharan Africa today, shifting sand dunes were a key component of the landscape. Scotland has wandered the globe throughout its discernible geological history and that journey continues to this day. The fragments of crust that now comprise Scotland drifted like great arks, constructed of rock rather than wood, and driven not by tides, but by the movement of plates on the Earth's surface. On this journey, the crustal fragments carried a varied cargo of plants and animals. These early life forms had to adapt to the ever-changing environmental conditions, move to an area where life was likely to be more amenable or die out.

So these continental movements and periodic collisions with other landmasses have been the most fundamental process of all in shaping the planet we see today.

Scotland's geodiversity – a patchwork of rocks

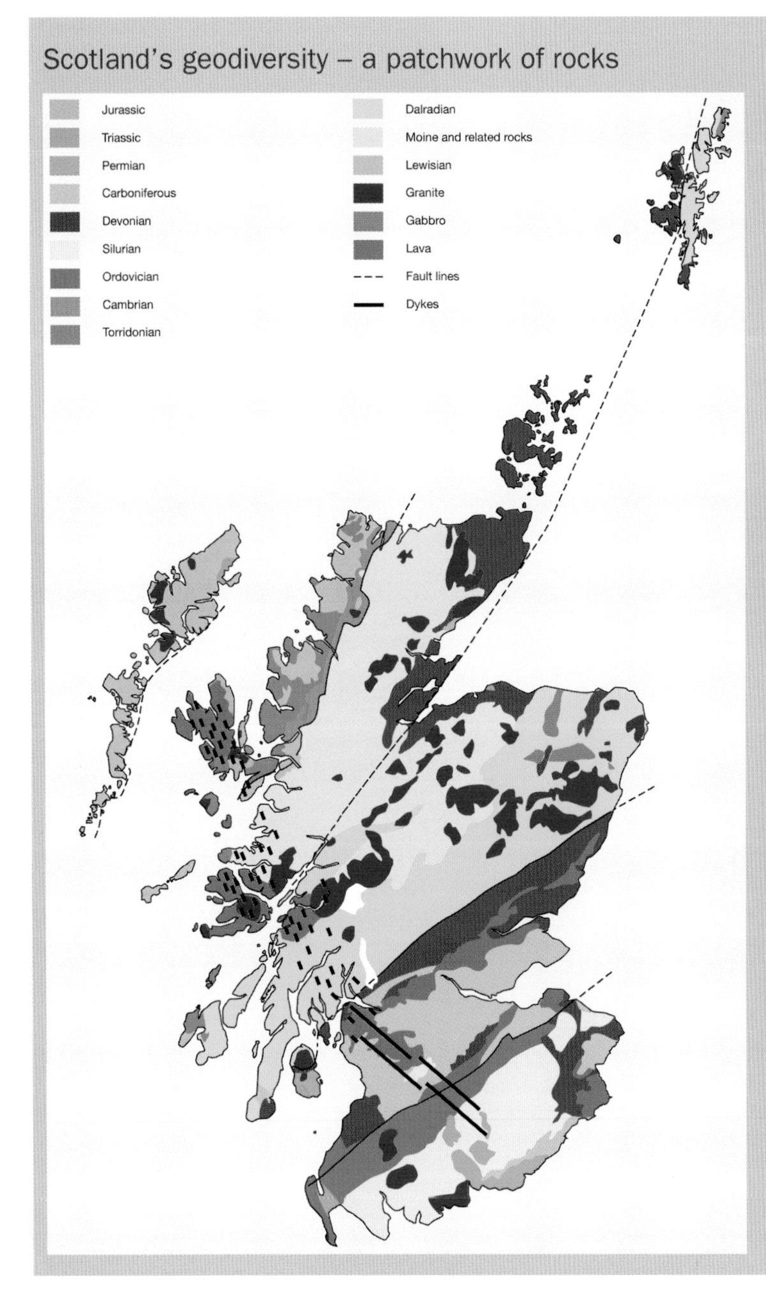

The global wanderings and associated collisions with other continents have given rise to Scotland's unrivalled geodiversity. Rocks of every kind and from every geological period are to be found in Scotland. The Western Isles and North-west Highlands boast the oldest rocks – the ancient crust that formed as Planet Earth cooled from its fiery beginnings.

The mountainous spine of the Highlands of Scotland is all that remains of a once towering mountain range. These mountains mark the site where three continents collided 420 million years ago, driven by movements of the Earth's crust. The mountain chain stretched to Norway in the north and southwards to the Appalachians in the eastern USA. It is only the creation of the Atlantic Ocean that has sliced this once continuous range of mountains apart.

The great mountain range that was northern Scotland was rapidly reduced in size by wind and water. Great rivers, comparable in size to the Mississippi of today, drained the high ground and transported boulders, sand and mud to the lower ground. These rivers and the lake environs of the Devonian age provided ideal environments for early life forms, such as primitive fish and some of the earliest plants to inhabit Planet Earth.

Rocks of Carboniferous age came next, as Scotland was positioned on the Equator. Great thicknesses of coal accumulated, sandwiched between layers of sandstone and limestone. The onset of desert conditions during Permian and Triassic times left scattered patches of sandstone across Scotland – from Dumfries to the Moray Firth. Next came Scotland's own Jurassic Park, complete with dinosaurs, crocodiles and plesiosaurs. Remnant deposits of sandstones, limestones and shales on Skye and the other Inner Hebridean islands are all that survive from this eventful period in Scotland's geological history. The Cretaceous Period was ushered in by a rise in sea levels of biblical proportions, with all but the highest ground of Scotland drowned under the advancing waves. Surprisingly, these dramatic events left little imprint on the landscape of Scotland.

About 65 million years ago, during the Palaeogene Period, there was yet another major rearrangement of the Earth's continental plates. Scotland had been 'a little piece of North America' for billions of years, but for no longer. We threw in our lot with Europe, as the North Atlantic widened. As the crust stretched under the strain of this continental break up, so a series of volcanoes erupted along the north-western seaboard. St Kilda, Skye, Rum, Ardnamurchan, Mull, Arran and Ailsa Craig were all active volcanoes at this time. Incredibly, during the succeeding Neogene Period, Scotland was at similiar latitude to today, but the climate was subtropical as the world was bathed in sunshine and high humidity. But that was not to last, as the climate cooled dramatically with the onset of the Ice Age around 2.6 million years ago.

The glaciers and ice sheets that covered Scotland carved the landscape into the familiar mountains and glens of today. As the ice melted, mounds of boulders, sand and clay that had been ripped up from the bedrock by the glaciers as they ground their way across the icy landscape, were sometimes dumped in regimented patterns. These deposits are widespread, but are not shown on the geological map opposite, which just records what geologists describe as 'bedrock'.

EARLY BEGINNINGS AND THE LEWISIAN GNEISS – ROCKS FORGED IN THE FIRES OF HELL

FROM 4,500 MILLION TO 1,600 MILLION YEARS AGO

The earliest time of the Earth's existence is called the Hadean – the fires of Hell. The planet's turbulent beginnings were characterised by violent volcanic outpourings of lava and sulphurous gases from an immature and unstable crust. These early environmental conditions were not conducive to supporting life in any form. But very gradually, the atmosphere became enriched in oxygen and the surface of the Earth cooled to a temperature where carbon-based life forms could thrive and diversify. The very earliest annals of Earth history go largely unrecorded in the rocks of Scotland. From planetary beginnings 4,500 million years ago to the time that the Lewisian gneisses (pronounced *nices*) of the North-west Highlands, Western Isles and Inner Hebrides made their appearance, we have no geological record. The Lewisian gneisses are intensively altered rocks, forged under great pressure and heat in the lower reaches of the early crust of the Earth and the oldest are dated at around 3,000 million years old.

The Lewisian gneiss of the Western Isles, seen here in its 'type area', has been extensively scoured by ice. These rocks are amongst the oldest to be found anywhere on the planet.

So approximately one third of Earth history, a period of around 1,500 million years, goes completely unnoticed and unrecorded in Scotland. After that, every other period of the Earth's history is faithfully, albeit incompletely and fragmentarily, recorded in the Scottish rock and fossil archive.

The wild landscapes of the Western Isles are like no other place in Britain. Similarly, the Lewisian rocks of the North-west Highlands have given rise to landscapes that have no equal anywhere else on mainland Scotland. In addition, Lewisian gneisses also occur on Coll, Tiree and parts of Skye, Rum and Iona. These rocks are described as 'basement'. It was on these rocks that later deposits were laid down.

Although the topography is subdued and none of the Lewisian hills are of any great height, this in no way detracts from the overall grandeur of the scenery. Ice has scraped the surface bare. Between the knolls of rock that resisted the passage of the ice are myriad lochans. These countless, shimmering, irregularly shaped pools of black, peaty water are the second key element in this dazzling tableau of rock and water.

Soils are generally thin and largely unproductive, so the land is marginal for agricultural purposes, supporting poor grazing and occasional arable crops. Geology has a direct consequence for the forms of economic activity that the area can support.

Bare rock with a characteristic banded appearance is exposed to the elements over broad acres, rising from the low-lying peat hags of Lewis to desolate heights of Clisham and Tirgamore on Harris. The predominant rock, known as gneiss, is characterised by alternating light and dark layers. Deep burial has caused the minerals to segregate into layers of light-coloured quartz and feldspar, alternating with layers of darker-

Walking on the Moon

The desolate summit plateau of Roineabhal on the Isle of Harris has more in common with the uplands of the Moon than could ever be imagined. In fact, both are built of exactly the same material – anorthosite. This type of rock is rare in Scotland, but not in other parts of the world. It is a variety of granite, comprised largely of one of the commonest rock-forming minerals, feldspar, or field rock from the German derivation of the word. The most common and perhaps familiar Moon rock is basalt, which covers much of the lunar surface in great dark areas called 'mare' or 'seas'. But rock samples recovered from the lighter-coloured highland areas indicate they are substantially made of anorthosite, similar in all respects to that found on Earth. So a greater geodiversity than was originally thought exists on the Moon. These anorthosite samples of Moon rock have been dated at over 4,000 million years old, which is similar to the rocks of Roineabhal, albeit older.

This mountain, which lies towards the south-eastern tip of Harris, was under threat for some time, as proposals existed to quarry away the mountainside to feed the voracious international appetite for crushed rock aggregate, but planning consent for this scheme has now been refused. So we will still be able to walk on the Moon, or the Hebridean version of it, for the foreseeable future!

The summit of Roineabhal

coloured minerals, mostly biotite, mica and hornblende. This grey gneiss, as it is known, is the signature rock type of the Western Isles and areas of the North-west Highlands that are underlain by the Lewisian.

Detailed study of the minerals that these rocks contain has revealed that they were intensively altered, almost certainly by deep burial in the Earth's crust. In places, the temperatures were so high that the rocks started to melt. Some estimates suggest that temperatures of almost 1,000°C and burial at a depth of up to 30 kilometres would be required to achieve the mineral assemblage that comprises these rocks. Over time, these rocks from the lower reaches of the Earth's crust were lifted closer to the surface by plate and other Earth movements. This process was assisted by the agents of erosion, such as ice and water, which stripped away the great thicknesses of overlying rock from the land surface. In this way, rocks previously deeply buried within the bowels of the Earth are now seen at the surface as prominent features of the contemporary landscape.

These Lewisian rocks took almost 2,000 million years to form. It is perhaps appropriate that these basement rocks, the foundation for all others, took longer than any later rocks to come into being. Recent interpretation suggests that these ancient rocks are the product of early continental collisions. At least four distinct areas of crust, which all started life as separate entities, have been identified in the belt of Lewisian rocks that runs from Cape Wrath to the Applecross area; each is separated from the next by a major fault or break in the Earth's crust. Each of these continental fragments, or geological terranes as they are known, had its own individual geological history before being 'welded' together as a single entity. This process of continental collision and welding together of fragments of the Earth's crust is a recurring theme in Scotland's lengthy geological evolution.

These ancient rocks are for the most part igneous in origin. They derive from a parent material that was formerly molten rock. A small part, however, is made up of strata that were originally limestones, sandstones and muds, now altered respectively to marble, quartzite and mica schist. Some of these sediments are thought to have been laid down in the ocean deeps, whilst others accumulated as river delta deposits.

Between 2,400 and 2,000 million years ago, the early Lewisian crust was being pulled apart and placed under great tension. Huge volumes of basalt were pumped into the crust from below over long periods of geological time as the surface of the Earth was stretched. These upwellings of magma solidified and in places now form prominent linear landscape features. They are known as Scourie Dykes, named after the place from where they were first described.

After the intrusion of the Scourie Dykes, the Lewisian crust was once again subjected to intense and all-pervasive alteration. Recent interpretations suggest that this was the result of collisions between terranes as the positions of the early Lewisian continental fragments rearranged themselves yet again. The Loch Maree sediments were cooked and squashed during this episode, transforming limestone to marble and sandstone into quartzite.

The penultimate event of significance in the formation of the Lewisian crust was the addition, or intrusion, of great volumes of granite magma. Around 1,700 million years ago, parts of the lower crust melted and the resultant melt material, being slightly less dense than the surrounding crust, rose like a bubble to a higher level. Great sheets of granite, dating back to this episode, occur around Loch Laxford and also

The mountains of Harris are largely built of granite and related rocks as the lower crust melted and generated huge quantities of molten rock.

form the core of the Harris mountains in the Western Isles.

The Lewisian gneisses have greater similarities and affinities with the rocks of the Canadian Shield and parts of Greenland than with any other part of the British Isles. They share a common history with these ancient rocks that are now to be found on the other side of the Atlantic Ocean. At the time of their formation, these landmasses were one coherent continent called Laurentia. This landmass was rent asunder as recently as 65 million years ago, when the North Atlantic Ocean started to form. Up until that point, and for a period of about 2,000 million years, this part of Scotland could have been accurately described as a 'little piece of North America'.

Patterns established in the early crust a few thousand million years ago have helped to shape the landscape of today. The major structures in these basement rocks have a predominant north-west to south-east orientation. They form lines of weakness that have been greatly accentuated by later erosion. These basement structures were picked out and scoured deeper during the recent Ice Age. The faults that marked out Loch Maree and most of the fjord-like lochs of the North-west Highlands illustrate this trend. It is incredible to think that events of over 2,000 million years ago have helped to create some of our most familiar and best-loved landscape features of today.

The final episode of these early times concerns the journey of the Lewisian gneisses

This view of Slioch looking northwards over Loch Maree reveals what this part of Scotland looked like over 1,000 million years ago. A gently undulating valley cut in Lewisian gneisses, forming the lower slopes of Slioch, was over time filled by Torridonian sandstones, which eventually buried any remnant of the Lewisian world. The thick cover of sandstones was subsequently stripped away to reveal a perfectly preserved surface, called an 'exhumed landscape'.

from the lower reaches of the Earth's crust to the surface. Some estimates suggest a figure of over 30 kilometres of rock having been planed off by erosion before the next chapter of Earth history began. We can make these estimates because the assemblages of minerals that make up the Lewisian gneisses that we see today could only have been formed at great depths in the crust. And, as the next oldest rock formation, the Torridonian sandstones, were deposited directly upon an eroded landscape of the Lewisian, these gneisses must have already been exposed at the surface.

THE TORRIDONIAN SANDSTONE – RIVERS OF SAND

FROM c.1,200 MILLION TO 850 MILLION YEARS AGO

By around 1,200 million years ago, the basement of Lewisian gneisses was exposed at the surface, battered and pock-marked by erosion. In the area of Loch Torridon and Slioch, just north of Loch Maree, we can actually see what that ancient landscape would have looked like. The surface of this early fragment of crust was

Great cliffs of Torridonian sandstone tower above the village of Torridon. Layers of sands and pebbles were dumped by rivers that flowed towards the continental edge of Laurentia.

undulating, uneven and carved into broad valleys; some over 300 metres deep. This ancient surface has been preserved for posterity by rapid burial under layer upon layer of Torridonian sandstone to a depth of almost 6 kilometres. It is remarkable that subsequent erosion of the Torridonian deposits has, in places, planed the land surface back down to reveal the contours of the original Lewisian landscape.

The turbulent times that gave rise to the Lewisian gneisses left a mountainous legacy. The high mountains created were built of rapidly eroding gneisses. We would probably recognise those mountains as part of Greenland and the Canadian Shield of today. This sliver of Lewisian gneiss is the only part of Scotland that existed in these far-off times and its affinities definitely lay with Laurentia – the major continental landmass that eventually gave rise to Greenland and North America. The land that was to become Scotland lay towards the edge of that supercontinent.

This landmass was drained by massive rivers, transporting huge volumes of mud, sand, pebbles and boulders of various sizes from the higher ground to the edge of the continent. As the rivers approached the continental edge, this burden was dumped. Great fans of sediment built up over time, reaching from the continental edge towards the sea. Four major fans have been identified and as these grew in size, they coalesced and amalgamated to form a continuous blanket of sand.

It is from this great wedge of sandstones that some of our finest and most dramatic scenery has been carved. The glacial corries of Beinn Bhan near Kishorn, the route of the winding road over to Applecross through Bealach na Bà (the 'Pass of the Cattle') and, of course, the majestic Glen Torridon are landscapes on a grand scale carved in Torridonian sandstone by the forces of nature.

Although the Torridonian sandstones are largely uniform layers of sandstone, they do hold a few surprises. Some horizons are thought to be desert sands, deposited in an environment akin to the arid zone of contemporary sub-Saharan Africa. A volcanic ash horizon has also been identified at a location near Enard Bay in Sutherland, indicating that there must have been some active volcanoes in the vicinity at this time. But no associated lavas or sites of ancient volcanoes have ever been found. The earliest fossils to be discovered anywhere in Scotland come from ancient muds of Torridonian age. These microfossils were first described in 1907 and were the first fossils to be found in rocks of Precambrian age in Britain. The rather unprepossessing appearance of these tiny structures, that are just a few microns in size, belies their importance and marks the beginning of recorded life in Scotland.

Torridonian age rocks also underlie the Sleat Penninsula on Skye, the western portion of Rum, a sliver of Iona, Colonsay and Oronsay and the northern area of Islay. The geological story of these strata is a similar one – river deposits laid down on top of an eroded Lewisian basement.

Rocks of Torridonian age contain evidence that suggests they were deposited around the Equator. We know that around 200 million years later, just before the beginning of the Cambrian Period, the Scottish landmass was near the South Pole, so within that extended period of time, there was rapid movement across the globe, from equatorial latitudes southwards towards the South Pole.

Summit of Stac Pollaidh

Suilven and Stac Pollaidh

Even amongst the scenic grandeur of the North-west Highlands of Scotland, Suilven and Stac Pollaidh stand out as aristocrats amongst Scottish mountains. They rise from a platform of eroded Lewisian gneiss and soar to a height of over 600 metres above the surrounding landscape to create iconic landmarks. Calendars, tourist brochures and colour supplements alike are decorated with dramatic views of these mountains. They are made of layer upon layer of Torridonian sandstone, transported from other parts of Laurentia (now Greenland and the Canadian Shield) by rivers laden with sand, pebbles and boulders. The great thicknesses of sandstone demonstrate the depth to which the basement rocks of Lewisian gneisses were buried during Torridonian times and also the extent of the erosion that has taken place since that time to exhume the ancient crust. Over a million millennia, the forces of erosion have ripped through the once continuous cover of Torridonian sandstones with the power of a gigantic chainsaw, leaving a few isolated patches, such as Suilven and Stac Pollaidh, as a reminder of the former blanket cover. Other peaks, such as Cul Mór, Cul Beag and the impressive Ben Mór Coigach, share a similar history. The serrated summit of Stac Pollaidh is a distinctive landmark for miles around. It is heavily eroded and regularly sheds lumps of rock that tumble down the mountainside. This ridge of crumbling rock, which some have compared to a row of rotten teeth, was created at the end of the last glaciation. As the glaciers thinned, Stac Pollaidh emerged above the surface of the ice and was exposed to the rigours of the Arctic climate. Successive episodes of freezing conditions followed by rapid thaws left the summit rocks severely frost shattered. And so they appear today. The lower slopes, moulded by the passage of the ice, are smooth and stable, whereas the higher ground is broken and fractured by exposure to the extreme temperatures that formerly gripped this part of the country.

Norman MacCaig captures the mood perfectly in his poem 'High up on Suilven':

Gulfs of blue air, two lochs like spectacles,
A frog (this height) and Harris in the sky –
There are more reasons for hills
Than being steep and reaching only high.

Meeting the cliff face, the American wind
Stands up on end: chute going the
  wrong way.
Nine ravens play with it and
Go up and down its lift half the long day.

Reasons for them? The hill's one . . .
  A web like this
Has a thread that goes beyond the
  possible;
The old spider outside space
Runs down it – and where's raven?
  Or where's hill?

THE MOINES – ROCKS OF THE MOOR

FROM *c.*1,000 MILLION TO *c.*800 MILLION YEARS AGO

As the rocks of the Torridonian were being dumped at the edge of the Laurentian supercontinent, thick sequences of sandstones and muds built up further offshore in a rift zone or deep trough cut into the sea floor. These sediments were to become the Moines. These were turbulent times and Earth movements buckled and wracked these layers of sediment shortly after they were laid down. Unlike the Torridonian sands of broadly equivalent age, these offshore layers were cooked and squashed by Earth movements and gave rise to the most important sequence of rocks in the northern Highlands – the Moines. These rocks were named after A'Mhoine in Sutherland which means 'the peat bog' in Gaelic.

We know about the power and pervasive nature of these Earth movements because the whole of the Moine sequence of rocks has been folded and metamorphosed to the point where in places the rocks have melted. Because of this extensive alteration or metamorphism, the early history of the Moines is very difficult to discern.

At this time, Scotland sat in equatorial latitudes. Adjacent to it lay Baltica, later to become Scandinavia, and, most improbably of all, Amazonia, which parted company to drift southwards to form South America.

The Moines have been described by some who have studied them in detail as 'monotonous', because they are largely made up of altered sandstones. But muds and occasional impure limestone bands were also part of the original pile of sediments that accumulated in the offshore rift

Left. Moine rocks are often associated with bleak, featureless moorlands as here at A'Mhoine in Sutherland.

Right. Around 1,000 million years ago, Scotland was part of a larger landmass. The location of the North-west Highlands is shown in relation to other continents. Perhaps most surprising is the proximity of Amazonia, which formed the core of the South American continent. Mountain chains that formed around 1000 million years ago are indicated by the tan-coloured areas.

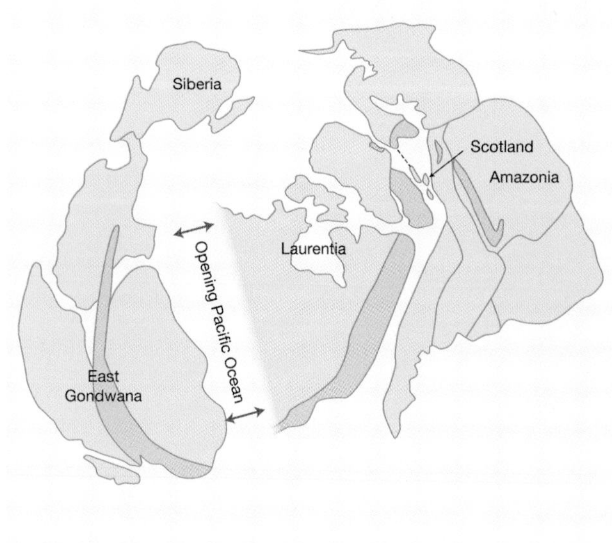

The Kildonan goldrush

The Helmsdale granite cut through the Moines of central Sutherland. Gold fever struck the Sutherland hamlet of Kildonan in 1840, when the Helmsdale granite was found to have shed nuggets of this precious metal into the streams that flowed across it. A nugget of gold, weighing more than half an ounce, was found in the Kildonan Burn. But it was only when local man R N Gilchrist returned from the Australian goldfields that exploitation started on a more commercial scale. The find was extensively reported in the local and national press. Some 600 prospectors were said to have descended on the area. This influx of hopefuls was perhaps on a smaller scale than the land grabs of the Wild West, but the discovery of gold sparked the same ambitions of instant wealth. The gold rush finished just as abruptly as it started when the Duke of Sutherland refused to issue any additional prospecting licences. Around 3,500 ounces of gold have now been recovered in total from the area. There is some 'fool's gold', otherwise known as iron pyrites, that may confuse the unwary. Any gold to be found today will be in flakes smaller than 1 millimetre in size, so do not expect to find any great nuggets.

zone over a period of many millions of years. Like the Torridonian sandstones, the Moines were, in part, deposited on a basement of Lewisian gneiss and slices of Lewisian are found throughout the Moines. But the metamorphic process has all but obliterated any trace of that original relationship.

These intensively altered Moine rocks are tough and are well able to resist the ravages of time and the forces of nature. It was in these far-off times, when the Moines were forged in the fiery depth of the Earth, that the roots of the Scottish countryside were created. Without this bulwark, the landscapes of Lochaber, north Sutherland, and indeed the whole area underlain by the Moine, would have been planed flat by ice, wind and water.

The distribution of the Moine rocks is defined by great cracks or faults in the Earth's crust. To the south and east, the Moines are bounded by the Great Glen Fault, which runs from Loch Linnhe to Inverness and beyond to slice through the Shetland Islands. To the north and west, the Moine Thrust defines the extent of the Moine outcrop. Movements of continental fragments along these fault lines are an intriguing part of Scotland's story, playing a vital role in the creation of land we know today.

THE ANCIENT ROCKS OF DALRIADA

FROM *c*.800 MILLION TO *c*.540 MILLION YEARS AGO

South and east of the Great Glen Fault lies Dalriada – the ancient Kingdom of Scotland – which gives its name to the rocks of the Monadhliath Mountains, Highland Perthshire, Argyllshire and the area north and east of the Cairngorm Mountains. The extent of these Dalradian rocks is defined by faults or cracks in the crust: to the north and west, the Great Glen Fault and to the south and east, the Highland Boundary Fault.

The Dalradian rocks underlie some of Scotland's most distinctive mountains. Celebrated by poets, artists and songwriters alike, this high ground is the product of many of the processes that shape the Earth's surface. The rocks that make up much of the highest ground in Scotland were formed during or shortly after a mountain building event called the Caledonian Orogeny. Although unimaginably ancient in terms of human experience, this event that started around 475 million years ago was the most significant to shape the bedrock of Scotland. Continental colli-

The Great Glen Fault runs from Shetland across Northern Scotland to Mull. On the mainland, the line of the fault is occupied principally by Lochs Ness and Linnhe. Here on Mull, the line of the fault is again picked out by water – Lochs Uisg and Spelve (centre of the picture). The fault line has been 'bent' by the later intrusion of the Mull volcano.

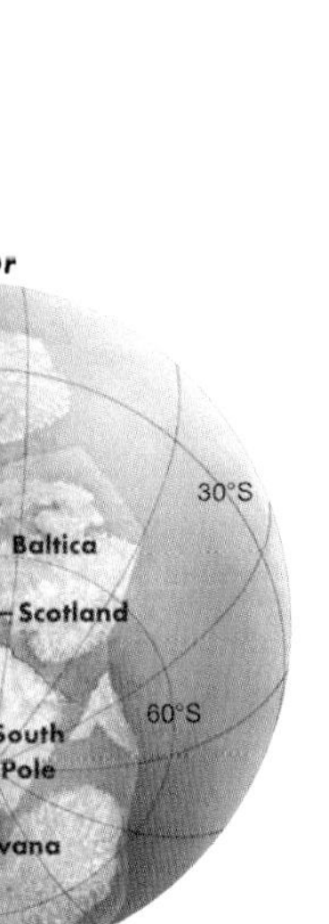

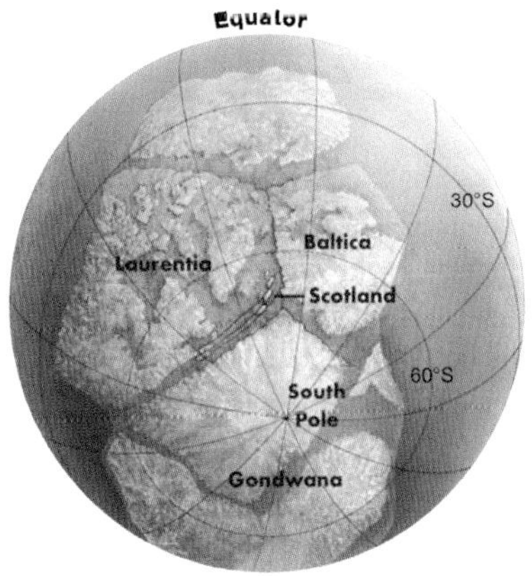

Above. Scotland arrived close to the South Pole around 600 million years ago as part of a larger landmass. But a major split appeared in the crust at this time and a new ocean was born – the Iapetus. Scotland then formed part of a smaller continent – Laurentia. This new continent drifted northwards as the new ocean continued to widen.

Right. Ben Lawers, in Highland Perthshire, is built of Dalradian rocks that were folded as continents collided.

sions and great upwellings of magma served to create a mountain chain that, when it was first formed, equalled the mighty Himalayan Range. These mountains, however, have long disappeared through erosion and the peaks we see today are merely their uplifted and eroded roots. Unravelling the complexity of the processes at work has taxed the best geological brains over many generations. Despite the intensity of study, the full story of how this mountain chain reared up to form peaks of towering proportions and was later planed back to meagre foothills has yet to be fully understood. To reconstruct events, we must rewind geological time and go back to the beginning of the chain of events that led inexorably to the creation of these mountains. This drama had one central 'character' – the Iapetus Ocean. This ocean takes its name from Greek mythology. Iapetus was the son of Uranus and the father of Prometheus and Atlas.

Birth of an ocean

Since the formation of the Moine and Torridonian rocks, the continents of the ancient world, which were largely joined to form one major landmass, had drifted southwards. Scotland then sat very close to the South Pole. In another major rearrangement of the continents, Laurentia split from the rest of the world's dry land and drifted away from the South Pole. A new ocean, known as Iapetus, began to open, as the plates shifted and new ocean floor was created. Scotland sat at the edge of the newly created continent of Laurentia. This new world order provided the setting for the accumulation, initially, of two major sequences of rock – the Dalradian and the limestones and related rocks of North-west Scotland. As the ocean widened, deep-sea sediments were laid down across the floor of the Iapetus Ocean. These strata, folded and buckled as the Iapetus Ocean eventually disappeared, were later to give rise to the Southern Uplands.

As the Iapetus Ocean opened, great thicknesses of sandstones, limestones and muds were laid down adjacent to the eastern flank of Laurentia. This sedimentary pile was over 25 kilometres in total thickness. Although these sediments have been affected by later metamorphism, it is still possible to determine their original nature and the conditions under which they were laid down. The sediments that gave rise to the Dalradian were deposited in a wide variety of different environments, including deep sea, near-shore delta, intertidal and even ice age conditions. As the layers of sediments built up in the deep parts of the ocean, so the crust was stretched to accommodate this thickening pile. But these deep-water basins eventually filled with sediment and shallow-water conditions resumed. Recognising these ancient environments helps considerably in reconstructing the distribution of land and sea that must have existed during these far-off times.

Recorded life from this part of the Iapetus Ocean is sparse, but the rocks of Islay provide a glimpse of an ancient ecosystem. Stromatolites, one of the first life forms to inhabit Planet Earth, have been described in chapter 2 as givers of life. These organisms were primitive bacteria that absorbed gases from the atmosphere, primarily

Schiehallion helps to 'weigh' the world

This beautiful mountain, located in the heart of Highland Perthshire, was the subject of ground-breaking work undertaken in 1774 by Dr Maskelyne, the Astronomer Royal to King George III. In this golden age of intellectual enquiry, thoughts of the scientific community turned to consider the weight or, more strictly speaking, the mass of the Earth. The pull that massive objects exert on bodies of a lesser size was beginning to be understood. We now recognise this force as gravity and it is this attraction that keeps us firmly rooted to the Earth's surface.

Maskelyne and his team of surveyors spent the summer of 1774 making measurements and astronomical observations from a number of locations on the slopes of Schiehallion. A plumb-bob was used to assist in these observations and it was noted that there was a significant gravitational displacement of the plumb-bob by the mountain.

Application of this principle led Maskelyne, and some twenty years later the eccentric scientist Henry Cavendish, to astonishingly accurate estimates of the Earth's mass. We now understand this figure to be around 5.9 billion trillion tonnes, which is only around 1 per cent different from the figure that these two eminent scientists derived from their pioneering experiments.

A mass of data was accumulated as the survey work proceeded. To rationalise the resultant muddle, the surveyors joined points of equal height around the slopes of Schiehallion. The contour line that figures on topographic maps of every country on the globe was thus invented.

carbon dioxide, and produced oxygen as a waste product. Over time, these unprepossessing organisms changed the composition of the atmosphere and paved the way for more complex life forms that ultimately gave rise to humankind. Remains of these mound-like structures that formed on the sea floor, are found near Bunnahabhain within the Dalradian rocks of Islay.

Volcanoes were also active during these times. Great piles of lava form an integral part of the Dalradian, belched out onto the sea floor through cracks in the crust, creating a mixed succession of lavas and other rocks laid down in the deep ocean. The great volumes of lava found at Tayvallich in Argyllshire indicate that the crust was very unstable at this time, probably as a result of it being stretched and pulled apart, creating deep-seated cracks that allowed lava to well up from the upper mantle. The Tayvallich volcanic rocks have a similar composition to the basalts that floor the oceans of today, which is not surprising as they were erupted from a similar source. As this volcanic event gathered momentum, so lavas were eventually erupted above the water level, suggested by the presence of volcanic ash deposits that were blasted directly into the air. The Tayvallich volcanics are regarded as significant in reconstructing the Iapetus Ocean story, as these lavas are thought to be a fragment of the rocks that floored this ancient ocean.

A Dalradian 'ice age'

Perhaps one of the most intriguing glimpses of past environments is to be found on Islay. A thick layer of sediments containing great boulders of varied composition in a matrix of finer material occurs at Port Askaig. The deposit is known as a tillite and has been interpreted as evidence for ice age conditions during Dalradian times. The boulders are thought to have been carried by icebergs which then melted and dropped them to the sea floor. This layer is an important time marker that occurs at a number of locations throughout the Highlands, including near Schiehallion in Perthshire. This ice age is thought to have been one of the longest running in the Earth's history, lasting for almost 200 million years. When these glacial deposits were formed, Scotland was close to the South Pole, so it is not surprising that glacial conditions would have prevailed.

Mineral collector extraordinaire

Matthew Forster Heddle (1828–97) left an extraordinary legacy to the nation. Trained as a medical doctor, he practised briefly in the back streets of Edinburgh, but abandoned his chosen profession for his other great passion, the minerals of Scotland. In 1856, he gained a position at St Andrews University and shortly thereafter was appointed as professor of chemistry. It was said of him that he "explored nearly every mountain and glen, and almost every part of the coast of Scotland in search of minerals". The Heddle Collection, which consists of over 7,000 separate specimens, was first displayed in the Scottish Geology and Mineralogy gallery of the National Museums of Scotland during 1895. An extract from the *Scotsman* newspaper dated 6 September 1895 captures the sense of anticipation: "The hall of the Scottish Mineral Collection on the upper floor of the west wing of the Museum of Science and Art has now been opened for visitors. For the greater part of nine months, Professor Heddle of St Andrews has been engaged in laying out and labelling the collection of Scottish minerals lately acquired from him, and which has been a great part of his lifework to gather together. This collection has now become national property and finds a permanent resting place in the Edinburgh Museum." It would have been the finest such collection of its day and still forms the basis of the National Museum's mineral exhibition. Heddle's treatise on *The Mineralogy of Scotland* and his many other published works are another invaluable legacy that this extraordinary man left to the world of science.

Minerals galore

Precious metals and minerals are found throughout the Dalradian and Moine rocks. In some locations, these rare metals are associated with granites, whilst in other places the rare metals were leached out of the host rock and concentrated by circulating fluids. Gold, silver, copper, lead, zinc and barytes have all been found in commercial quantities.

The earliest record of gold in Scotland comes from the time of King David I in the twelfth century, and since then mines have been dug to exploit a meagre resource in many parts of the country. A gold mine was developed at Cononish near Tyndrum in the 1990s, but the reserves proved to be insufficient to sustain a commercial operation.

A large deposit of the heavy mineral barytes (barium sulphate) was discovered near Aberfeldy and has been mined in recent years. It is used as a key component of drilling mud, primarily by the oil industry in the North Sea. It is also a key component of the barium meal used by doctors to help the diagnosis of some medical conditions. Deposits of barites have also been found in many other parts of Scotland, including Strontian in Argyllshire and Coalburn in Lanarkshire.

So in addition to being the basis for some of our most striking and treasured landscapes, the Moine and Dalradian rocks host considerable economic wealth.

A world first

The mineral strontianite was first described from a quarry just to the north of the village of Strontian in Sunart. This was a world first. An element that was new to the world of science was later extracted from these mineral deposits. The new element took its name from its type locality and was named strontium. There are some twenty-eight other minerals that were first described from Scotland, according to Dr Alec Livingston, our latter-day Professor Heddle, who was for many years curator of the mineral collection at the National Museums of Scotland. They include such rarities as lanarkite, brewsterite and leadhillite.

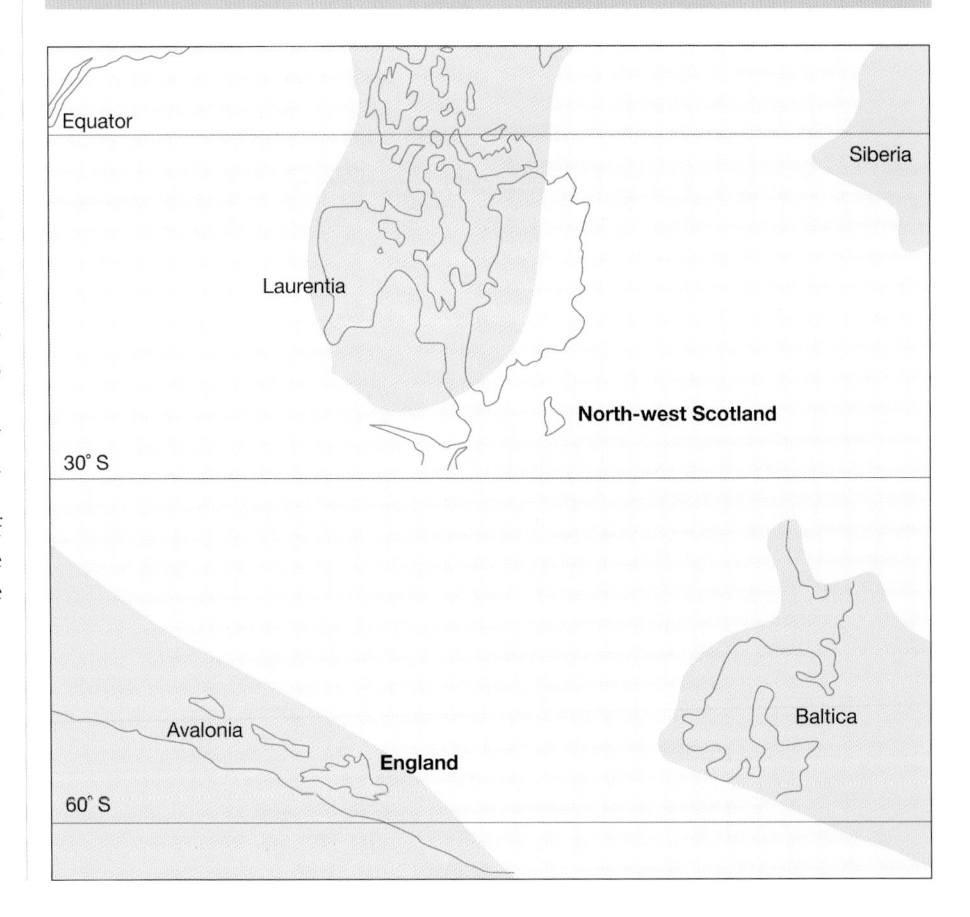

CAMBRIAN AND ORDOVICIAN ROCKS OF THE NORTH-WEST HIGHLANDS – EVIDENCE OF AN ANCIENT SHORELINE

FROM 542 MILLION TO 443 MILLION YEARS AGO

Opposite. As the Iapetus Ocean widened, so layers of sediment continued to build up in the ocean deeps and along the continental shelves adjacent to the Laurentian coastline. Thick layers of limestone accumulated during Cambrian and early Ordovician times on the relatively shallow continental shelf. We recognise these sediments today as the limestones and associated rocks of the North-west Highlands, otherwise known as the Durness limestone.

Right. Pipe Rock at Skiag Bridge, Sutherland. The vertical structures are interpreted as ancient worm burrows that were made in beach sands that fringed the Iapetus Ocean.

In Cambrian and Ordovician times, from about 540 million years ago and for a period of around 50 million years thereafter, sandstones and limestones were laid down along the coastline of the ancient continent of Laurentia. This occurred towards the end of the period when the most recent Dalradian sediments were accumulating further offshore in the Iapetus Ocean. These ancient sandstones and limestones of Cambro-Ordovician age were laid down in the inter-tidal and near-shore environments and share many of the characteristics of present-day beach sediments. The burrows of marine animals that lived in these ancient coastal habitats are well preserved at many places, as are the remains of some of the earliest and most primitive creatures to inhabit the Earth, including trilobites and the ancestors of snails, cockles and sea urchins. The beginning of the Cambrian was the time that life began to diversify rapidly across the primitive world and Scotland has its share of these early experimental life forms.

The beach sands were laid down on top of the eroded Torridonian sandstones and that juxtaposition can be seen at many locations, but perhaps most clearly on the hills to the south of Loch Assynt. In other places, these Cambrian rocks lie directly on the Lewisian gneisses, indicating the rapidity with which the once continuous cover of Torridonian rocks was eroded.

Over time, these pure sands were transformed into quartzites as the material cementing the individual quartz grains together recrystallised to form a very tough rock indeed. Deposited on top of these quartzites are relatively thin layers of silt, mud and grit, which were exploited during the

Second World War for their high potash content. The grit deposits have been interpreted as offshore sandbanks lying in shallow water.

A thick sequence of limestones makes up the final part of the succession of Cambro-Ordovician rocks. These are popularly known as the Durness limestones after the place from which they were first described. Their presence indicates that the environment had changed once again. From the beach conditions that gave rise to the underlying sands, these limestones indicate rather deeper-water conditions close to the edge of the continental shelf. So the sea had encroached further across the coastal plains of Laurentia and the basement rocks of Lewisian gneisses and Torridonian sandstones were then under a relatively deep sea. Some 1,000 metres of limestone accumulated in this shelf-sea environment.

Fossils are not particularly common in these rocks, but where they do occur, there is a marked similarity with those in limestones of the same age in Canada and Greenland. This affinity is further evidence for Scotland being part of the Laurentian supercontinent prior to the development of the North Atlantic Ocean. It is interesting to note that fossils of this age collected from Wales are very different in character from those in Scotland. It is therefore likely that the expanses of sea that supported these populations were distant from each other. As we will see later in this chapter, there is good evidence that both England and Wales at this time in geological history were located on the other side of the Iapetus Ocean. This ocean is thought to have been wider than the present North Atlantic Ocean. Detailed studies of fossil assemblages, such as those contained in the rocks of the North-west Highlands and Wales, have been used as vital evidence in establishing the distribution of ancient continents from Cambrian times onwards.

The limestones of the North-west Highlands

Below. Green pastures at Elphin are underlain by limestones, which give rise to a richer soil than on adjacent rocks.

Opposite. The quartzites of Beinn Eighe are a distinctive feature of the landscape.

are very visible landscape features, either as prominent rock outcrops or indirectly through the rich vegetation of the bright-green grassy pastures they support. They also form the host rock in which Scotland's main cave systems are developed in Assynt. The quartzites, too, make their presence known as ice-white caps on prominent hills such as Arkle and Beinn Eighe and also as dazzling aprons of scree around the summit of Foinaven.

Rocks were also being laid down elsewhere within the Iapetus Ocean during Ordovician times, notably those now exposed around Girvan and also those underlying the northern part of the Southern Uplands. Their place in the story will be told later.

The Iapetus Ocean begins to close

The Dalradian and Cambrian rocks were on the edge of the ancient continent of Laurentia. Beyond the continental margin lay open sea – the Iapetus Ocean.

The disparate lands on either side of the Iapetus Ocean were soon to be joined, possibly forever. After a period of relative quiescence and stability, the land that was to become Scotland was in for turbulent times. The Iapetus Ocean began to close slowly, making a continental collision inevitable. The spreading ocean that had reached the width of the modern North Atlantic slipped into reverse gear. The ocean floor started to shrink in size as a subduction zone developed off the coast of Laurentia (see p. 111).

The process of subduction drove a slab of ocean floor back into the upper mantle. As it descended, seawater from wet sediments and ocean crust was driven off and into the overlying mantle. This change in composition caused the mantle to melt, and the molten rock bubbled back to surface as magma. The island chain that formed as a result is known as an 'island arc'.

As the Iapetus Ocean continued to close, the island arc made landfall, colliding with the continental margin of Laurentia. It rode over the Dalradian sediments, burying them to considerable depths, which subjected the rocks to temperatures of over 650°C. This deep burial within the Earth's crust transformed the sands, shales, limestones and associated lavas into tough metamorphic

An underground world

The limestones of Assynt play host to a sub-terranean world that holds many surprises. The bone caves of the Creag nan Uamh are a series of chambers and linking passages that developed in the limestones of the Assynt area. The caves have developed over at least 250,000 years, so their history is long and complex. They are also an irreplaceable storehouse of information about the animals that inhabited this part of Scotland during and after the Ice Age. The weight of ice that scoured this landscape bulldozed everything in its path, removing any evidence of past life. But the layers of sediment that built up on the floors of these caves remained intact. By studying these sediments, layer by layer, a detailed picture has emerged of the animals that inhabited these frozen wastes as the ice sheets waxed and waned.

Peach and Horne were the first to appreciate the scientific value of the cave floor deposits and they excavated there in 1889. There was an organised dig in the 1920s and much of the bone material recovered then is now stored in the National Museums of Scotland in Edinburgh. Collectively, these finds have revealed a fragmentary picture of the Ice Age and the animals that inhabited the postglacial world. It is mainly the bones of large mammals, such as brown bear, wolf, lynx, arctic fox and reindeer that have survived, but the remains of a host of smaller animals such as wood mice, wildcat and water vole are also preserved.

Pavements of rock

Although of limited distribution, rock pavements built from limestone occur on some parts of the Durness limestone. They look like artificial constructs; great expanses of rock, swept bare by the ice and dissolved into regular shapes by the effects of weathering. The National Nature Reserve at Inchnadamph has perhaps the best and most extensive example of limestone pavements in Scotland. This habitat plays host to a number of botanical rarities that grow in the cracks between the limestone blocks.

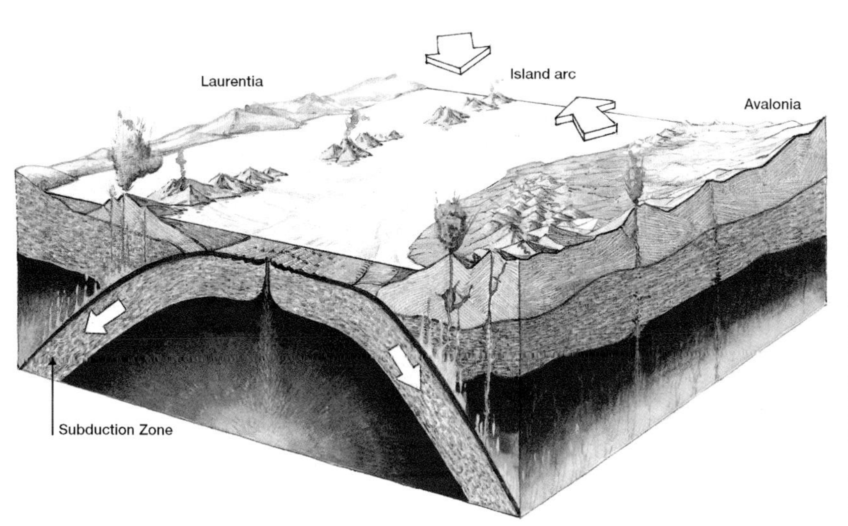

The Iapetus Ocean began to close in early Ordovician times, around 480 million years ago. A subduction zone developed offshore from the Laurentian coast and the ocean floor started to be consumed. After a period of relative quiescence and stability, the land that was to become Scotland was in for the most turbulent times in its history as the previously spreading Iapetus Ocean slipped into reverse gear. This brought the continental landmasses of Laurentia and Avalonia ever closer. An island arc developed above the subduction zone. It was this island arc that first collided with Laurentia, creating the rocks that now form the Highlands of Scotland.

rocks. It also caused the Dalradian rocks to become chaotically folded and buckled. This is one of the most significant events that shaped the bedrock of Scotland. It was a vital step in the creation of the land we know today. Great sweeping folds were created at this time, turning some rocks upside down as strata were bent double into great overfolds, called 'nappes'.

The final closure of the Iapetus Ocean occurred in a number of phases, but it was this collision with the island arc that primarily built the mountains of Scotland. The newly created Caledonian Mountains originally stretched from the Appalachians, through Scotland to Norway. This mountain chain was sliced through as the North Atlantic opened in later times.

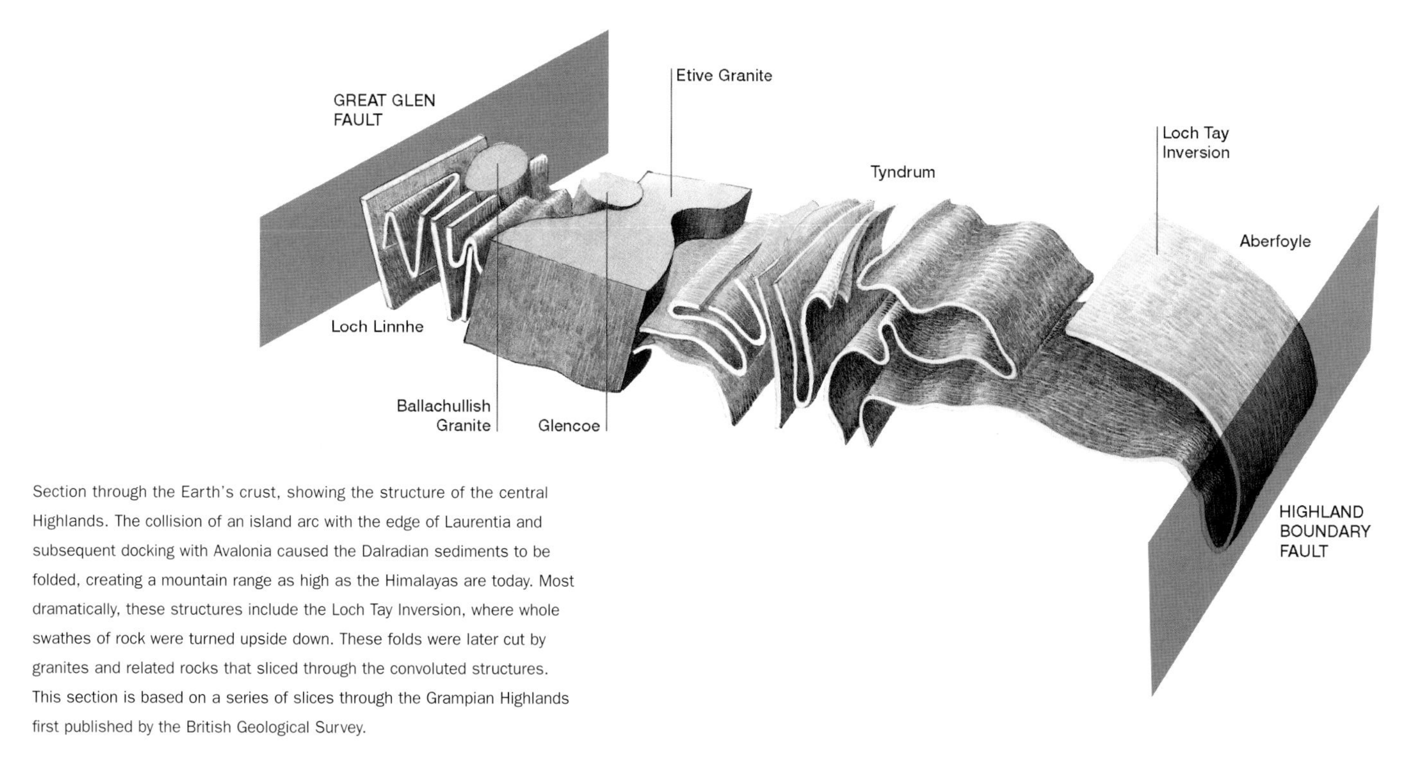

Section through the Earth's crust, showing the structure of the central Highlands. The collision of an island arc with the edge of Laurentia and subsequent docking with Avalonia caused the Dalradian sediments to be folded, creating a mountain range as high as the Himalayas are today. Most dramatically, these structures include the Loch Tay Inversion, where whole swathes of rock were turned upside down. These folds were later cut by granites and related rocks that sliced through the convoluted structures. This section is based on a series of slices through the Grampian Highlands first published by the British Geological Survey.

Ben Lawers – the upside-down mountain

Ben Lawers on the north shore of Loch Tay looks firmly rooted; the epitome of stability. But nothing could be further from the truth, as the rocks that build Ben Lawers are actually upside down. The inversion of these rock layers took place as the Iapetus Ocean closed and the Caledonian Mountains were created. The process of mountain building, described above, was not a simple one. The folds created, as the ocean closed and continents began to collide, were on a regional scale. The largest structures occupy many tens of square kilometres. Early folds were created and then these structures were refolded by later Earth movements. The rocks of Ben Lawers, initially laid down in the Iapetus Ocean as a series of flat-lying limestones and sandstones, were bent double as continents collided. The lower limb of the fold, known as the Loch Tay Inversion, was turned upside down or inverted as a result of the folding episode. Later erosion has completely removed the upper limb of this major structure, exposing the inverted limb below. More recently, it was from these upside-down rocks that the present-day Ben Lawers was carved by ice and water.

But this amazing complexity is not immediately apparent to the untutored eye; it can only be established by careful study of the regional geology.

Rocks from the ocean floor

The Iapetus Ocean existed for around 200 million years and the origins of many of the rocks throughout Scotland can be traced back to this long-disappeared expanse of water. As the ocean started to diminish in size, a subduction zone developed along the eastern coast of Laurentia. The expanse of ocean floor when the Iapetus was at its widest, must have been extensive; equivalent in area to that of the present-day North Atlantic Ocean. So where did all that ocean floor go as the continents collided? It was largely made up of basalts and related base-rich rocks, which were covered by thick layers of sediment that built up on the sea floor. This rock assemblage was forced back into the bowels of the Earth, as the plate descended under the emerging Scotland. Much was consumed by this process, disappearing into the lower reaches of the crust. But some fragments of this ancient ocean floor stayed at the surface, broken off or scraped from the descending plate. These assemblages of rock, which have been identified at five main locations throughout Scotland, are known as 'ophiolites'. They are all of a similar age and were formed early in the subduction process as the ocean started to close. All are between 500 and 470 million years old. The ophiolite locations are widely spread across Scotland, from Unst in the north to Girvan on the Ayrshire coast. The remaining three locations all lie on the Highland Boundary Fault, from Stonehaven to Conic Hill on the banks of Loch Lomond.

Unst, the most northerly of the Shetland Islands, is perhaps the best place to see rocks from the ocean floor that have made it onto dry land. The rocks of Unst are dark in colour, and rich in the heavy base-rich minerals that characteristically floor the oceans of the world, ancient

Conic Hill serpentine – a sliver of base-rich rock – follows the line of the Highland Boundary Fault which separates the Highlands from the Lowlands of Scotland.

and modern. As the Iapetus Ocean closed, fragments of ocean crust and related rocks were caught up in the mountain-building process. The Shetland ophiolite is thick, around 8 kilometres or so of base-rich material, and is one of the most complete sequences of its type in Britain. The rocks of the ophiolite complex also include economic deposits of chromite and talc. Relatively high levels of gold and platinum have also been recorded. The prized decorative rock serpentine is also an integral part of this complex. During the ocean closure process, when continents collided, this ophiolite material was greatly mangled and generally altered from its original form, but enough remains to determine the sequence of events that led to its formation.

As the new entity of Scotland was assembled, the fault lines along which the continental fragments, or terranes, slid into place were zones of great tension and upheaval. The rocks that underlie Conic Hill on the eastern shore of Loch Lomond were also derived from the ocean floor as the the Iapetus Ocean narrowed. The more base-rich rocks are known as serpentine. The presence of these rocks has had a significant effect on the local vegetation. Base-rich flushes on Conic Hill confirm that the bedrock differs substantially from the usual acid rocks of the area. Garron Point near Stonehaven is also a classic place to see the ocean floor rocks of the Iapetus Ocean. The basalt lavas show 'pillow' structures that are characteristic of molten lavas having been erupted directly into water.

The rocks of Ballantrae are another vital piece

of the ocean-closure jigsaw. This area has attracted interest from geologists for over 150 years and despite this intensive study, the rocks still retain some secrets. Many different rock types are juxtaposed – lavas, cherts and shales – but perhaps most intriguing of all is the occurrence of peridotite, now altered to serpentinite. This dark-green base-rich rock, which is greasy to the touch, is found in association with cherts and shales. As with the other occurrences throughout Scotland, this association of different rock types is interpreted as a fragment of ocean floor caught up in the continental collision zone.

Granites and gabbros

The many occurrences of granites, gabbros and related rocks found across Scotland are not all of the same age. Millions of years separate the earliest and youngest intrusions, but all are related to the closure of the Iapetus Ocean. As continents collided, so a huge variety of different environments were created where igneous rocks formed. Some are related to processes occurring deep within the crust, whereas others formed closer to the surface. These rocks also vary greatly in composition – from ultrabasic, where the rocks comprise base-rich minerals, such as olivine and pyroxene, to the other end of the scale where more silica-rich rocks, such as the widespread granites of North-east Scotland, have been noted.

Huge quantities of molten rock were generated as a result of the continental collisions. Many of our iconic mountain ranges, such as the Cairngorms and the Ben Nevis range, have their origins in this period of great upheaval. As the Dalradian rocks were being cooked and compressed, some were so intensively altered that they melted to form igneous rocks such as granites and gabbros. Granites predominate, although some rocks of more basic composition, such as gabbros, were also created. Many of these igneous rocks were generated at the height of the upheaval, and some developed later, cutting through the complicated series of folds that the alteration of the Dalradian had created. These granites and their blacker, more basic cousins, gabbros, developed at a fairly deep level in the Earth's crust.

These molten masses of granite and particularly gabbro had a significant effect on the geology of Scotland. Great volumes of molten rock sliced through the generally colder

Distribution map of granites and gabbros that were formed as the Caledonian Mountains were created.

Dalradian rocks. The temperature of base-rich gabbro magma was about 1,100°C when injected into the lower part of the crust. These phenomenally high temperatures caused the surrounding rocks to melt and this partial melting of the lower crust created even more magma, this time granitic in composition. The injection of such voluminous quantities of magma caused the crust to thicken and the Daldradian rocks that played host to these intrusions of new rock were further squashed. These high temperatures were matched by elevated pressures. The various stages of this complex process are recorded for posterity in the rocks of North-east Scotland.

The appearance of these rocks at the surface today indicates an aggressive regime of erosion that removed, in some instances, up to 25 kilometres of overlying rock. Some estimates put the

Glen Tilt provides proof that granites were molten

Arguments still raged in the late 1700s about the ultimate origins of granite and related igneous rocks. We now accept that granites were formed as once molten rocks that moved through the Earth's crust as a crystal mush. But during the time that James Hutton was formulating his *Theory of the Earth*, there were competing explanations for the formation of all rocks on the face of the Earth. The German mineralogist Abraham Gottlieb Werner proposed that the world was once blanketed by a 'universal ocean' and all rocks were supposedly deposited from this primeval waterworld. There were deliberate biblical references that gave this theory added credibility. Werner's ideas readily accommodated Noah's Flood and had strong echoes from the Book of Genesis. In these early days of science, explanations of the world at large had to be made within the context of a literal interpretation of the Bible. By contrast, Hutton's ideas were revolutionary. He proposed that granites were formed, not by precipitation from the sea, but by introduction in a molten state from below. This clash of ideas was characterised as the Neptunists (those who followed Werner's ideas) against the Plutonists, who favoured an igneous origin for granites and related rocks. We now know that the Plutonists' arguments were correct.

Glen Tilt, in Highland Perthshire, proved to be one of the key sites in this debate. Hutton searched the country for a place where the edge of a granite body was exposed. At Glen Tilt, he found the evidence he was looking for in the riverbed. Hutton and his field companion, John Clerk of Eldin, found the "granite breaking and displacing the strata in every conceivable manner". In other words, the granite was clearly veining the older rocks into which it had been intruded. This demonstrated a key principle – that the granite was in a molten state when it was introduced into the surrounding rock. A trip made to Galloway a year later confirmed his findings. Again he found granite injected into the surrounding rock. Hutton concluded "that without seeing the granite in a fluid state, we have every demonstration possible of this fact; that is to say, of granite having been forced to flow, in a state of fusion, amongst strata broken by subterraneous force, and distorted in very manner and degree".

This may seem a narrow, esoteric debating point of little note, but these elementary observations helped to torpedo the Neptunist geological theory that held sway throughout much of Europe in the late eighteenth century and had many adherents in the scientific community in this country. Hutton found another site that supported his ideas in Holyrood Park, within a few hundred metres of his home (see chapter 6).

The granites, gabbros and related rocks of Scotland have a long history of study. James Hutton's conclusions, that these rocks were formed from a molten state and not deposited from a primordial ocean, opened the way to their systematic study, and it was Archibald Geikie, who published the first exhaustive account in 1897, called *The Ancient Volcanoes of Great Britain*. James Nicol, a contemporary of Geike, also made a major contribution to the study of igneous rocks in Scotland. Later, in the early years of the twentieth century, C T Clough, H B Maud and Sir Edward B Bailey took the science forward with significant discoveries about the inner workings of the ancient supervolcanoes of Glencoe and Ben Nevis.

time taken for this erosion of the overlying rocks and exposure of the granites at the surface at about 5 million years. Geologists describe these granites that appear at the surface as having been 'unroofed' – a very apt description. For every granite that appears at the surface, there will be others that remain buried deep underground. Perhaps they will make an appearance at some distant time in the future if the present-day land surface is further excavated by water or ice.

The Caledonian igneous rocks of Scotland were emplaced in a series of bursts of activity. The earliest are of early Ordovician age, around 470 million years old, and are found around Ballantrae in Ayrshire and along the Highland Boundary Fault. In mid-Ordovician times, a series of base-rich gabbros and related ultrabasic rocks were emplaced in North-east Scotland – at Insch, Maud and Haddo House for example. In late Ordovician to mid-Silurian times, around 445 million years ago, a series of igneous rocks of unusual composition were emplaced in the Assynt area, including those at Loch Ailsh and at Canisp. Towards the end of the Caledonian mountain-building period, around 410 million years ago, wide-spread and voluminous granites were intruded into the metamorphic basement rocks of predominately Moine and Daldradian origin across Scotland, including Helmsdale, Ballachullish, and Peterhead in the north. The Criffel and Cheviot granites were intruded into the Southern Uplands in the south at around the same time.

Ben Nevis – Scotland's highest mountain

One of Scotland's best-known landmarks was formed shortly after the Iapetus Ocean closed. The spectacular peak and sheer, north-facing cliffs of Ben Nevis are made from granites and lavas that formed part of an active volcano around 425 million years ago. When the volcano was active, it collapsed in on itself. A central core around 2.5 kilometres in size subsided into the molten rock below as the volcano blew its top during particularly violent eruptions. A cap of lava floundered hundreds of metres downwards into a chamber of still-molten granite magma. Subsequent erosion of the ancient mountain has removed any trace of the volcanic depression, called a caldera, that would have developed at the surface during this major subsidence event. But the circular fault along which the lavas collapsed into the lower reaches of the volcano and the juxtaposition of lavas against granite are tell-tale signs of this type of collapse.

This area and the adjoining Glencoe were first investigated by geologists from the Geological Survey in the early 1900s. Later, Sir Edward Bailey made a case for this cauldron subsidence phenomenon that was new to science at that time. Since then, cauldron collapses have been recognised in the rock record from many countries around the world. Perhaps the best-known contemporary example of this type of hyper-explosive volcanic activity is Krakatoa in Indonesia. This volcano erupted in 1883 with catastrophic consequences. The shockwaves were felt around the world and global temperatures dropped on average by over 1°C because of the quantity of dust that the volcanic eruption pumped into the upper atmosphere. It seems likely that when the volcanoes of Ben Nevis and Glencoe were active, they would have had made their mark in similar fashion.

Right. Folded Lower Palaeozoic rocks along the Berwickshire coast – this disturbance was caused when the ancient Iapetus Ocean closed.

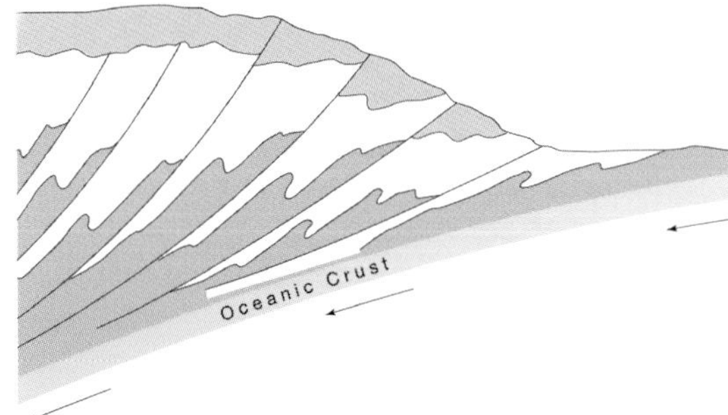

Formation of the Southern Uplands. Successive wedges of sediment were dragged down into the abyss of the subduction zone as the continents ground ever closer. Each wedge is separated from the previous one by a fault, presumably activated as it was thrust into place at the bottom of the pile of sediments. Such an arrangement of sediment wedges is called an 'accretionary prism'.

Death of an ocean

The penultimate chapter in the assembly of the British Isles was the docking of Scotland with England as the Iapetus Ocean closed. What had been one of the great oceans of the early Palaeozoic world had narrowed and finally disappeared in yet another major rearrangement of the continents. The huge thickness of mud of Ordovician and Silurian age that had accumulated in the ocean deeps of the Iapetus became dry land, as it was squeezed between the converging landmasses, like modelling clay caught in the jaws of a colossal vice. Subduction of the north-westwards-travelling plate under what is now the Central Lowlands brought the two continents of Laurentia and Avalonia into direct contact (see p. 123). It was a soft docking, as the continents gently nudged together, rather than a colossal collision. The ocean sediments were gently buckled and folded into broad whale-back ridges and swales.

The ocean floor sediments of the Southern Uplands were sliced into a series of packets, each separated from next by a north-east–south-west trending fault. These sediment slices, in general terms, get progressively younger from north to south, with each successive segment added to the pile as the downward movement of the descending plate dragged more ocean floor mud into the abyss. This process of subduction and accretion of more wedges of ocean floor sediments only stopped when the continents finally locked together. The results of this folding episode are clear throughout the Southern Uplands. These rocks are best exposed on the coasts – Berwickshire in the east and Kirkcudbrightshire in the west, where folding structures are beautifully displayed.

The structure of the Southern Uplands has been a matter of controversy since the area was first investigated by those pioneers of Scottish geology, Ben Peach and John Horne, at the turn of the nineteenth century, and before them by Charles Lapworth. Between 1872 and 1877, Lapworth lived in Birkhill Cottage, the equivalent these days of living above the shop, while he mapped the hills around Moffat. Amongst his considerable contributions to the unravelling of the mysteries of this geological puzzle was his use of fossils in dating the layers of rock. The beds of sandstone and shale that build the Southern Uplands are from different eras. They are distinguished by one main attribute – their fossil content. As they are deepwater sediments of some antiquity, traces of ancient life are likely to be few and far between. But, as luck would have it, the Iapetus Ocean teemed with one particularly peculiar creature – the graptolite. These creatures lived near the surface of the ocean and drifted with the currents. On death, they sank towards the ocean floor, caught up in the constant rain of fine-grained sediment settling in the ocean deeps. Over time, the remains of some of these animals were preserved as fossils. They resemble pencil smudges on freshly split rock surfaces.

Fossils aplenty. These creatures, known as graptolites, floated in the Iapetus Ocean and sank to the bottom of the sea on death. They evolved rapidly and proved to be particularly useful to Professor Lapworth as he tried to make sense of the sequence of sandstones and shales that build the Southern Uplands. He used the assemblages of graptolites to establish the stratigraphy or succession of strata from oldest to the youngest rocks.

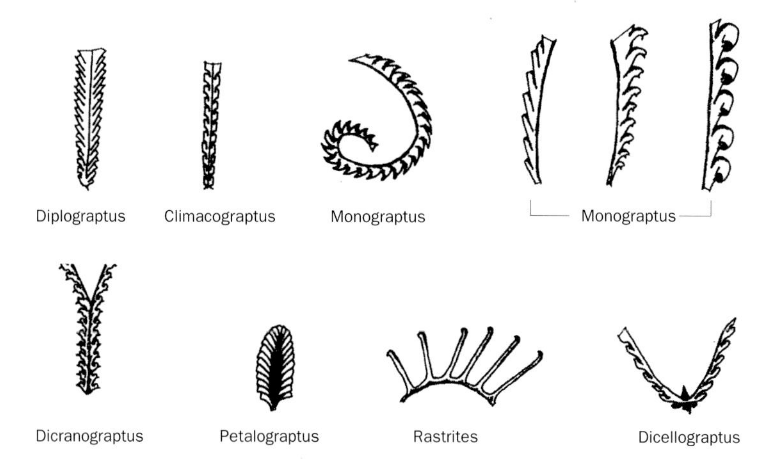

Fossils aplenty

Graptolites evolved quickly in geological terms into an array of different forms. *Monograptus*, *Dicranograptus* and *Diplograptus* and many other genera all shared the basic graptolite structure, but looked sufficiently different to allow the monotonous sequence of shale beds to be dated relative to each other.

Today the evolutionary sequence of these different species has been established, and it is relatively easy to ascribe a particular age to a rock outcrop on the basis of its fossil content. But in the nineteenth century, as no one had studied graptolites in any great detail before, none of these standards existed. Charles Lapworth threw himself into this painstaking work during the years he spent at Birkhill Cottage. He deserves enormous credit for his work in the Southern Uplands, which was genuinely groundbreaking in terms of the use of fossils in relative dating of rock layers.

The Golden Spike at Dob's Linn

Charles Lapworth's house was close to a place known as Dob's Linn which he helped establish as a locality of international geological importance. This rather unprepossessing gorge is now the internationally recognised reference section, or stratotype, for the transition between the top of the Ordovician and the base of the Silurian periods of geological time. Each period of geological time has such a reference section and the International Commission on Stratigraphy chose Dob's Linn as the Ordovician – Silurian time boundary. It has much to commend it for the specialist visitor. The rocks span around 25 million years of geological time from upper Ordovician into the lower part of the Silurian sequence. The strata, which largely comprise shales, are rich in graptolites, which have been used to date the succession with a high degree of precision. This is also the type area for a number of graptolite species; in other words, until they were described from here, they were unknown from anywhere else in the world. After Lapworth's seminal work here, celebrated Survey geologists Ben Peach and John Horne remapped the area and corrected a few details that Lapworth had misinterpreted. The site has also seen a steady stream of geologists in more recent times who have tried to refine further the ideas of the early pioneers.

There is no plaque to commemorate this international designation nor is there any signpost at the precise point where the transition between the Ordovician and Silurian takes place. Current estimates put the date of the base of the Silurian Period at 444 million years. It takes an expert to locate the exact position of the 'golden spike', as geologists call these internationally recognised transition points.

Granites and volcanoes of the Southern Uplands

The granites of Criffel and Glentrool in Dumfriesshire were formed as the ocean-floor sediments were dragged deep into the Earth's crust as the continents plotted their collision course. The fierce temperatures and high pressure of the lower crust melted the muddy sediments to form a melt of granite composition that then rose like a bubble towards the surface. Like the granites of Aberdeenshire, these masses have been unroofed as erosion has removed the layers of rock that lay between the top of the granite and the land surface.

The sedimentary layers of the Southern Uplands also contain many beds of ash that clearly indicate the proximity of active volcanoes. These fiery edifices blasted plumes of ash and steam high into the ancient atmosphere, only for the ash particles to land on the ocean surface and sink to the ocean floor. This carpet of ash was then buried by layers of mud and sand and thus preserved for posterity. This sequence of rock layers allows us to say with certainty that volcanoes fringed this ancient ocean, although their precise locations remain undetermined.

The great debate

Scotland's time on the rack was not finished. The final phase of these cataclysmic events involved yet another continental collision, this time with Baltica, a landmass that eventually gave rise to Scandinavia. The northern arm of the

Protecting a rare fossil site from thieves

The closing Iapetus Ocean also supported some other bizarre life forms. At Lesmahagow in Lanarkshire, complete specimens from an ancient jawless fish have been discovered. *Jamoytius kerwoodi*, as this creature is known, is thought to be an early precursor of the modern lamprey. The site was visited by Roderick Murchison in 1856 and many distinguished scientists have subsequently collected material from here. These fossil finds have made the site of international importance and specimens of this rarity have a considerable commercial as well as scientific value. As a result, it has been regularly targeted by fossil collectors, particularly from abroad, for the last twenty years or more. A successful prosecution was made of a group of German fossil hunters in the 1980s. However, later incidents that went unpunished clearly indicated that this site remained a target for these illegal activities. Detective work revealed that some *Jamoytius* specimens from Lesmahagow ended up in a museum in Germany. Strenuous efforts were made to recover these specimens, but to date the efforts of SNH to repatriate this material have not been successful. In a last-ditch effort to improve security at the site, what remains of the fossil deposit is now protected by a fence that is designed to deter unlicensed collectors. The Nature Conservation (Scotland) Act 2004 contains special provisions to safeguard our fossil heritage, making it an offence to collect material from special sites, such as this, without permission.

The fence around the *Jamoytius* site

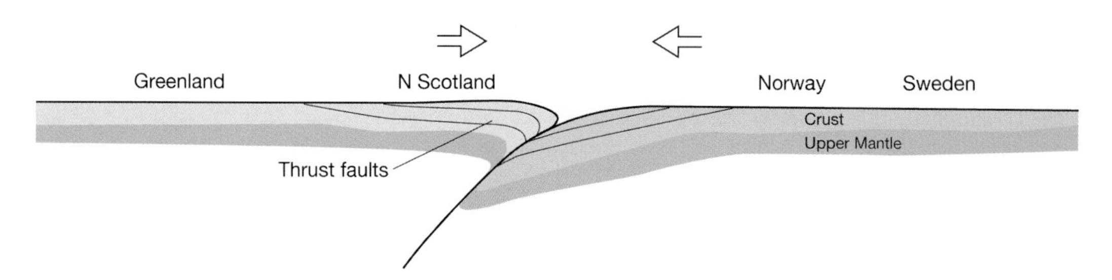

The final continent-to-continent crash occurred around 425 million years ago in mid-Silurian times, just 5 million years after the soft-docking of Laurentia and Avalonia had taken place to close the south-western arm of the Iapetus Ocean. The shock waves from this shuddering event were felt far and wide. A series of 'thrust' faults were formed as a result. These faults were low-angled and moved one block of rock over its neighbour, to create a series of segments of rock, each separated from the one below by a fault. In this way, rocks from great depth can be thrust over younger rocks. Geologists describe these low-angled faults as 'thrusts'.

Iapetus Ocean closed, bringing these areas of land crashing together. The effects on the North-west Highlands were dramatic. The force of the collision caused the crust to fracture in a way that led to the development of thrust faults. This type of fracture is low-angled, causing large masses of rock to override other sections of crust.

The net effect is for the crust to become shorter than before. The process of stacking slivers of rock one on top of another has a concertina effect, squeezing a large volume of rock into a smaller space. Many kilometres of crust were gobbled up in this way as Scotland and Baltica ground into each other. Great wedges of Moine rocks were driven westwards, thrust over the younger Cambrian quartzites and limestones. In the process, the natural order of things, where younger strata overlie older rocks, was disturbed.

The western edge of the area affected by this thrusting process is knife-sharp. It is called the

The location of a gently-dipping fault line, the Glencoul Thrust, is marked on the photograph. Below the fault line lie Lewisian gneisses overlain by Cambrian quartzites and limestones. Above the fault line are Lewisian gneisses, driven westwards by the massive continent-to-continent collision, which took place as the Iapetus Ocean finally closed. It is estimated that the overlying Lewisian was moved some 30 kilometres during this collision. The north shore of Loch Glencoul is one of the best places to appreciate the effects of these massive Earth movements on a landscape scale.

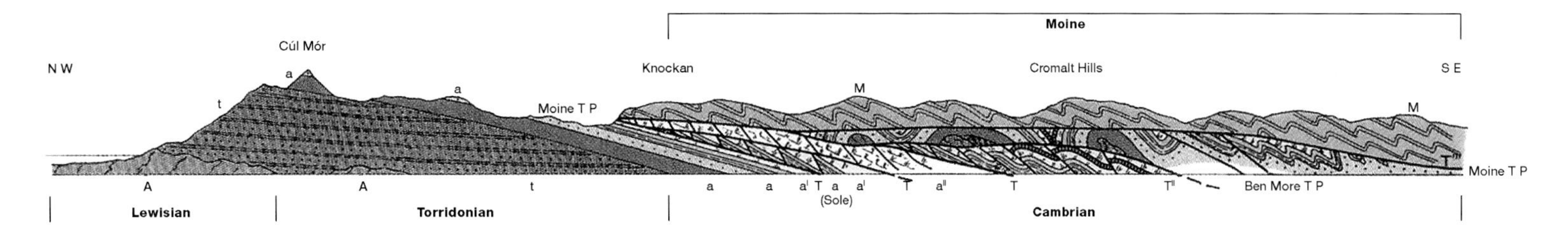

This cross-section of the Assynt area is part of a series of maps, sections and an accompanying memoir describing the geology of the North-west Highlands. This was Ben Peach and John Horne's finest work. This section from Cùl Mór to the Cromalt Hills clearly indicates the mechanisms by which older rocks were stacked on top of younger strata. The Moine Thrust is shown carrying the older Moine rocks over a crumpled and broken sequence of Cambrian quartzites and limestone. At that time, there was no known mechanism that could drive such prodigious displacements in the Earth's crust. Plate tectonics was not even a glint in the eye at this stage, so for Peach and Horne to see such a solution to the Knockan puzzle was visionary indeed.

Moine Thrust, where older Moine rocks have overridden younger Cambrian strata. At a place called Knockan Crag, just to the north of Ullapool, the Moine rocks comprise the upper part of a prominent cliff and are separated from the Cambrian rocks below by a major fault – the Moine Thrust. We now understand the circumstances that gave rise to this juxtaposition, but when the area was first studied, a great controversy raged. Sir Roderick Impey Murchison, Director-General of the Geological Survey, visited Knockan with his protégé Archibald Geikie, who subsequently pronounced that this was a normal sequence of rocks, with the oldest at the bottom of the cliff and the youngest at the top. James Nicol, Professor of Natural History at Aberdeen University, disputed this interpretation, saying that the sequence was not in the correct order, as older rocks lay above younger strata. Charles Lapworth, a schoolmaster who taught in Galashiels and had extensively mapped the rocks of the Southern Uplands, also studied the area in some detail and favoured the views of Nicol.

To resolve this great debate amongst these Victorian scientific pioneers, the Geological Survey sent in their crack troops, Benjamin Peach and John Horne. They were later joined by C T Clough, L W Hinxman, H Cadell and W Gunn – four other eminent geologists of the day. Together, they produced one of the most revered geological accounts ever written, *The Geological Structure of the Northwest Highlands of Scotland*, which was published by the Geological Survey in 1907. The field evidence favoured the

The Moine Thrust

Before the Moine Thrust controversy was finally resolved, Scottish geologist H Cadell modelled the behaviour of rocks under pressure. He constructed a 'pressure box' in which he laid down a series of coloured layers of wet sand and plaster of Paris. He designed the box so that one side could be forced inwards by a screw. This had the effect of compressing the layers of sand. As the jaws of the vice were tightened, so the layers of sand sheared to create fault planes, causing some layers to override others in a manner very similar to that found in nature. This experiment was also useful in demonstrating the role of water as a lubricant in this process of moving great blocks of rock around. In nature, thrusting takes place at depth and the high pore water pressures that exist at depth help support rock slices displaced by Earth movements. Cadell's work was completed in 1889 and he subsequently joined Peach and Horne in their pioneering efforts to solve the Knockan puzzle. His findings would have greatly helped the Survey team to understand the processes at work.

Peach and Horne at Inchnadamph Hotel in 1912 when the hotel was used as a base for the British Association field trip.

The Inchnadamph Hotel – steeped in geological history

This establishment, in the Assynt area of Scotland, is perhaps associated more than any other with the early development of the science of geology. In 1912, it played host to an excursion by the British Association for the Advancement of Science. The Edinburgh Geological Society's booklet on Assynt tells the story of the trip. A party of very distinguished geologists assembled in Dundee where the British Association conference was taking place. After the conference, they travelled by rail to Lairg and onwards to Inchnadamph by way of Strath Oykel. The party was led by Ben Peach and John Horne and included many of the leading geologists of the day from Britain, Europe and the USA. An old hotel visitor's book dating from this time, recently found in an outhouse, records the geological luminaries present. Amongst others were Sir Edward Battersby Bailey, Professor of Geology at Glasgow University, Dr Emile Tietze, Director of the Austrian Geological Survey, Dr Hans Henrik, Director of the Norwegian Geological Survey, and Dr Elisabeth Jeremine, University of St Petersburg. Peach and Horne led the trip and impressed the international audience with their new ideas on how mountains form. Dr Albert Heim, President of the Geological Commission of Switzerland, gave the vote of thanks. He thanked Peach and Horne for their leadership of the trip and invited them to visit the Alps to see a more recent mountain chain that had not been so intensively eroded as the Scottish Highlands. He described the folded rocks of the Alps as "much younger leaves and beautifully folded flowers". The Alpine fold mountain belt is much younger than the Scottish mountains and is perhaps what Scotland would have looked like before the forces of erosion planed our mountains down. The hotel now proudly displays these pages from the visitor's book in addition to photographs from the British Association trip.

Inchnadamph Hotel is still a mecca for geologists. The season starts around Easter as parties of students descend from all parts of Britain, using the hotel as a base to study what is one of the most intensively visited geological field localities in the country. The hotel also has another life – as a favourite haunt for fishermen trying their luck in the many streams and lochans around the Assynt area.

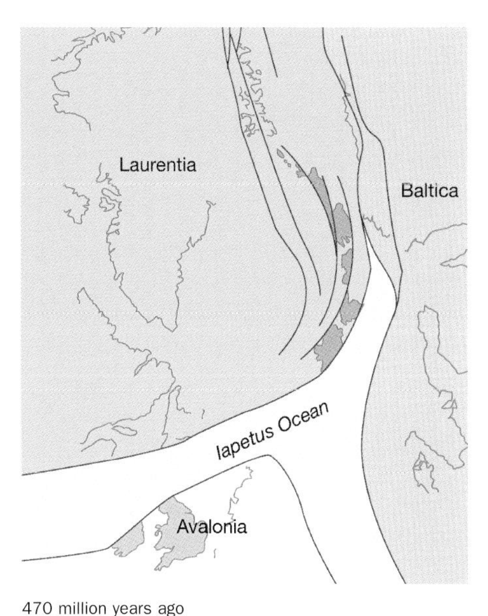

470 million years ago

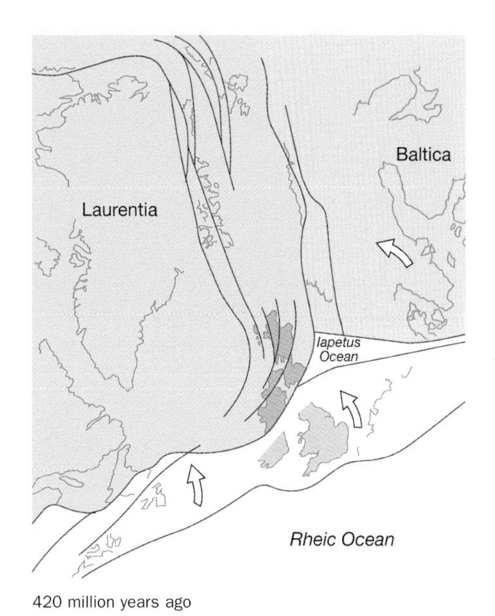

420 million years ago

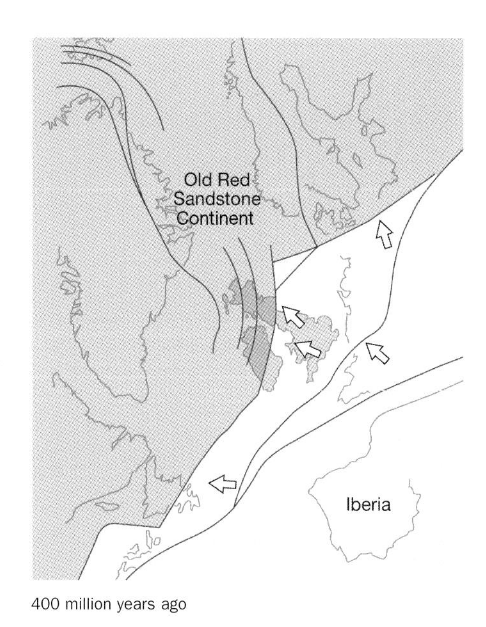

400 million years ago

The assembly of Scotland took place in many phases over a protracted period of time. This sequence of diagrams is a representation of how the coming together of the component parts could have taken place.

Nicol–Lapworth hypothesis for the existence of a thrust fault and that interpretation has stood the test of time.

Assembling the jigsaw

This chapter has described the development of the component parts of Scotland as it drifted northwards – the Western Isles and North-west Highlands; the Northern Highlands between the Moine Thrust and the Great Glen; the Central Highlands bounded by the Great Glen and the Highland Boundary Fault; the Central Lowlands and finally the Southern Uplands. Each is a terrane or a piece of continental crust that developed in its own right, separate and self-contained.

Perhaps the most remarkable aspect of Scotland's geological development is the fact that it was fashioned from disparate landmasses that collided by chance around the time that the Iapetus Ocean closed. The faults that bounded the terranes played a key role in this assembly process. The terranes were strung out along these parallel cracks that reached deep into the Earth's crust and only moved together to form a coherent whole as the Iapetus Ocean finally closed. Evidence of movement along these fault planes is abundant. Crushed and milled rock known as 'mylonite' occurs at many points along these large dislocations, indicating that each terrane ground against its neighbour as the pieces of this outsized jigsaw inched into place.

The Iapetus Ocean was no more. Many thousands of millennia before the Union of the Crowns and, a hundred years later, of the Parliaments, Scotland and England were first united as a physical entity.

THE DEVONIAN PERIOD – A LAND OF MOUNTAINS, RIVERS AND LAKES
FROM 416 MILLION TO 359 MILLION YEARS AGO

After the mayhem of continental collisions and the assembly of Scotland, what followed was played out at a rather more sedate pace. The new 'Scottish' landmass was now part of an expanded continent that consisted of Laurentia and Baltica. This landmass has been called the Old Red Sandstone Continent. Scotland was now largely landlocked and lay some 25°S of the equator. During the assembly process, the individual terranes that built Scotland travelled from the higher latitudes of the Southern Hemisphere to a subtropical location. The Devonian Period had dawned.

At this time around 400 million years ago, Scotland was an arid land ringed by mountain ranges. Erosion of the newly made mountains was spectacular in its rapidity. The forces of erosion tore the edifices down and planed the mountains to their roots within a few million years. That is quick work in geological terms. We know that this process happened at high speed because Old Red Sandstone is seen to sit directly upon the deeply eroded roots of the former Dalradian mountain range. So, many kilometres of rock must have been sliced away before the sandstones were deposited. As with any demolition process, huge quantities of rubble were created. Rivers and streams carried a bed-load of sand, pebbles and boulders to lower ground. Huge scree slopes formed on the rapidly eroding mountain slopes. The debris accumulated in chaotic mounds, fanning out across the lower slopes and flatter ground immediately adjacent to the mountain ranges. The debris was of mixed origin, reflecting the bedrock geology. Boulders of quartzite, granite, basalt and gabbro can all be found in the Old Red Sandstone deposits that consolidated from the products of this demolition process.

Crawton Bay and the cliffs around Dunnottar Castle near Stonehaven are two of the best places to see these conglomerates. Some of the boulders in the cliffs are up to a metre in size. It is envisaged that these deposits were laid down by an ancient river flowing from the rapidly eroding uplands. The bed-load of boulders and other debris would have been dumped on the lower ground when the currents slackened. The volume of water and thus the size of the river required to transport and knock the edges off lumps of rock that size would have been truly immense. The Colorado River that cut the Grand Canyon carries boulders of that size today. These boulders are rounded in a similar fashion to those at Crawton. So perhaps a river of this scale flowed across the landscapes of Devonian Scotland.

Detailed studies of the rocks in the conglomerate indicate that they are from a variety of

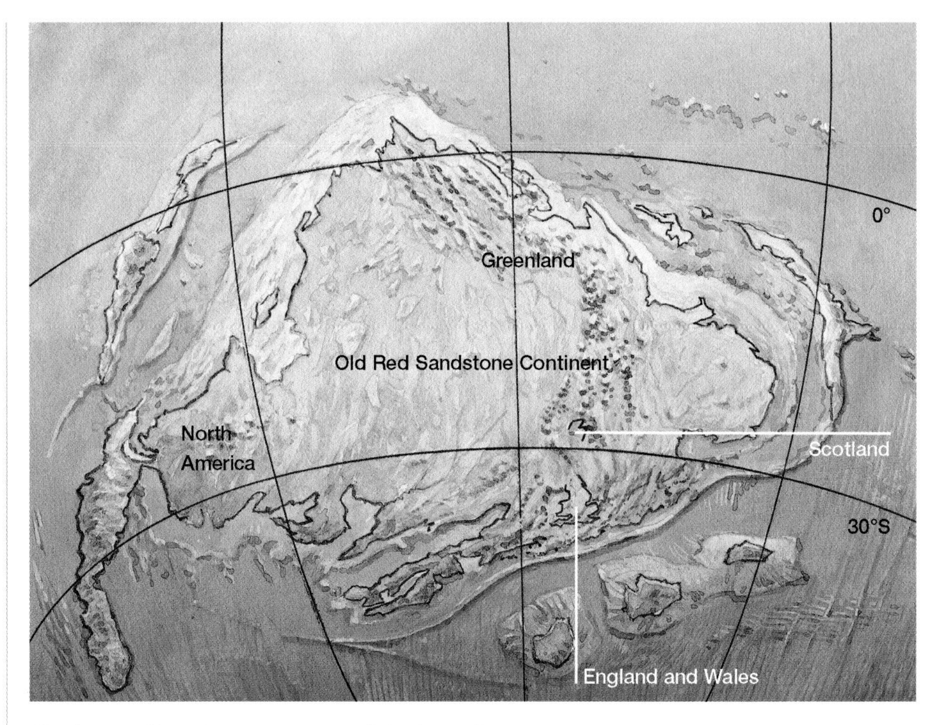

The world around 400 million years ago, shortly after the Iapetus Ocean closed. The Old Red Sandstone Continent lay to the south of the Equator and comprised North America, Greenland and Britain. Fold mountain belts traversed the continent from north to south, marking the site of ancient continental collisions.

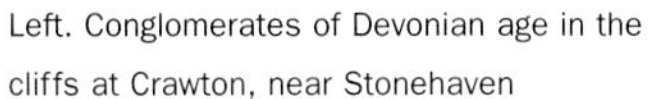

Left. Conglomerates of Devonian age in the cliffs at Crawton, near Stonehaven

Middle. The Colorado River. Huge, rounded boulders in the cliffs at Crawton are similar in scale to those from the bed of the Colorado River in the USA. From this, we can infer that the deposits at Crawton were laid down by a river of considerable size.

Right. Whiting Ness, near Arbroath. A break in deposition, or unconformity, is present in the cliffs (dashed line).

sources, some close at hand and others more distant. Thick lava flows are also found at Crawton and are thought to have temporarily diverted the rivers that flowed from the higher ground. The presence of lavas also suggests that the crust was far from stable during these early Devonian times. The road cutting adjacent to the Cairney Brae on the A9 between Auchterarder and Perth is another good place to look to see the scale of the forces at work. The structures in the sandstones there suggest that the river which created them must have been in excess of 15 metres deep, not dissimilar in scale to the modern-day Mississippi.

Thick deposits of Devonian rocks are found elsewhere, largely within the Central Lowlands of Scotland. A band of sandstones and related volcanic rocks runs from the Isle of Bute in the south-west to Stonehaven on the North Sea coast. Other deposits of this age are known in Ayrshire, around Biggar, and also in the south-east of Scotland from the Berwickshire coast southwards to the Merse. These strata were deposited in a series of three low-lying areas or basins that, over time, filled with sediments transported from the surrounding higher ground by a network of rivers. Erosion of the uplands was rapid, and through the Devonian Period, these basins gradually filled with sediments.

A time line

Deposition in these basins was not continuous. There were prolonged periods when no sediments were added to the existing pile. Where there are breaks in the record, features known as unconformities occur (see p. 126 for a fuller explanation). A well-known example is to be found near Arbroath at Whiting Ness. Here, the cobble-rich deposits of Upper Old Red Sandstone overlie the finer grained deposits of the Lower Old Red. Both deposits can be dated with relative accuracy – the overlying deposit at about 370 million years and the underlying rocks at about 410 million years. So the distinct line between the two represents around 40 million years of geological time. This provides an important demonstration of how rock sequences build up over time: deposition of sedimentary layers takes place over relatively short periods of geological time, almost instantaneously in some cases, but may be separated from the succeeding layer by a prolonged period of time. The breaks between successive layers often represent more prolonged periods of time than it took for the beds of rock to accumulate. At Whiting Ness, there has clearly been a substantial break in sedimentation between the deposition of these two distinct sequences of rock.

Hutton's unconformities

There are many sites around Scotland that are associated with James Hutton and his development of the *Theory of the Earth*. Hutton wrote in his seminal book, "We are led to conclude that all of the strata of the Earth have had their origins at the bottom of the sea, by the collection of sand and gravel, of shells, and of earths and clays. Nine tenths . . . have been formed by natural operations of the globe, in collecting loose materials, and depositing them at the bottom of the sea; consolidating those collections in various degrees, and either elevating those consolidated masses above the level on which they were formed, or lowering the level of that sea." These were groundbreaking insights at a time when it was generally accepted that the Earth was formed in seven days. After formulating his theory of the Earth, which was presented to the Royal Society of Edinburgh in 1785, Hutton then set about proving his ideas. The Old Red Sandstones first attracted his attention. He searched for the point where the sandstones were in contact with older rocks, as this would provide a clue as to how they were formed in the geological past. He found such a contact at two separate locations in South-east Scotland – at the famous Siccar Point and also at Allar's Mill, near Jedburgh. He also discovered a third 'unconformity' at the north end of Arran. The strata at Siccar Point, and an account of Hutton's visit there, are described in chapter 6.

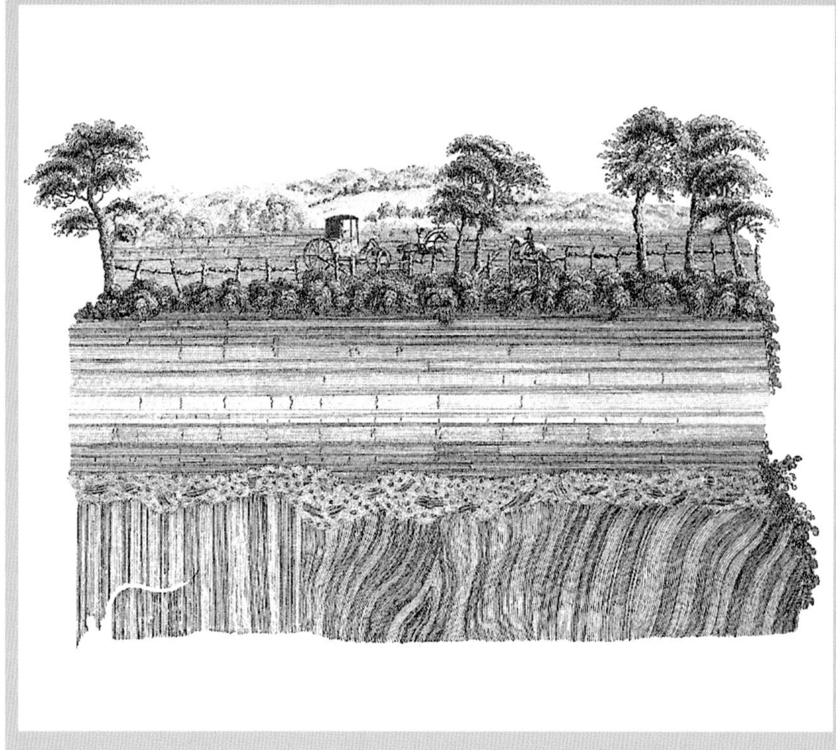

Allar's Mill, near Jedburgh. This drawing was made by John Clerk of Eldin, who accompanied James Hutton on many of his expeditions across Scotland. It shows the flat-lying Old Red Sandstones on top of the upturned strata of Lower Palaeozoic age.

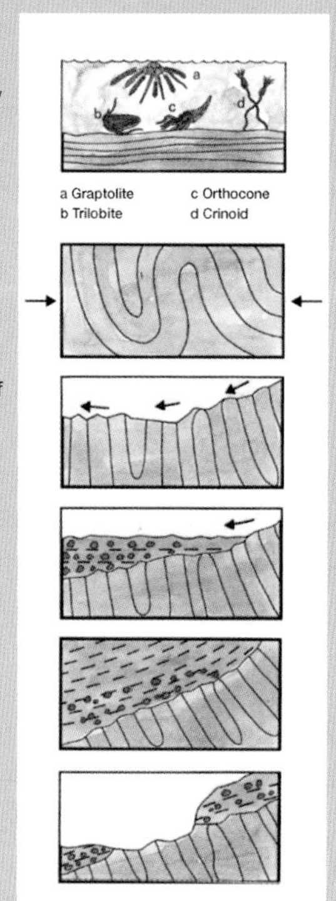

The development of an unconformity is a multi-stage process. The 'contact' between the rocks above and below the line of an unconformity usually represents a substantial period of geological time. The sequence of events runs from the top picture (oldest at the top) downwards.

Old Red environments

By studying the rocks carefully and observing the maxim 'that the present is the key to the past', we can piece together what the land that was to become Scotland would have looked like at the time these strata were being laid down. Scotland was continuing to move northwards. At the beginning of Devonian times, Scotland lay some 25°S of the Equator, but by the end of the period it lay 10°S, so the journey to present latitudes continued at some pace. Scotland's climate was hot during later Devonian times, which is consistent with its global position at that time. The landscape was a stark, red desert scene, cut by mighty rivers that meandered across wide floodplains. There was little vegetation to bind the thin soils. Although most of the sediments in the Central Lowlands were laid down by rivers, some were of aeolian origin – in other words they were deposited as wind-blown dunes. Sands were whipped up into impressive dunes by desert winds.

We also know that flash floods were common in Devonian times. Thick deposits of rounded conglomerate pebbles found in Fairy Glen, at Oldhamstocks, in East Lothian, attest to these violent deluges of water. Elsewhere, notably around Jedburgh, red sandstone cliffs are promi-

The Bunnet Stane, West Lomond, Fife. Weathering in recent times has exploited variations in the resistance of different layers of sandstone to produce strangely sculpted features.

nent features of the landscape. These deposits were laid down by fast-flowing rivers. The Bunnet Stane, in Fife, sculpted in more recent times into an unusual form, provides further evidence of the environment of the Old Red Sandstone world. The sandstone beds from which it has been eroded were deposited by winds blowing across an ancient desert landscape.

The Orcadian Lake

Great freshwater lakes lapped against the foothills of the eroding mountains. The largest of these occupied an area that included the present-day location of Orkney and Shetland and extended southwards to cover the Black Isle and much of the present Moray coast. It is referred to as the Orcadian Lake. At times there was a connection between this great expanse of freshwater that covered an estimated 50,000 square kilometres and the sea towards the south-east.

The Orcadian Lake existed for an estimated 10 million years. During that time, around 4 kilometres of sediments were deposited in one of the marginal areas of the lake, which now corresponds to Caithness. Further north in Shetland, this figure may have been even greater. The extent of the lake changed during that period, as the shoreline flooded the surrounding area and then retreated in response to changes in the climate. During times of retreat, rivers flowed across the marginal plains from the surrounding high ground. Wind-blown sands are also recorded in the rocks. The lake was never particularly deep, perhaps up to a maximum of 80 metres. The lake floor must have subsided at a fairly rapid rate to accommodate the great thicknesses of sediments that accumulated there.

Sediments built up on the floor of the lake quite slowly. Like the rings of a tree, each layer represents the sediment input to the lake in that particular year. The lake supported a diverse ecosystem, with armour-plated fish one of the many life forms preserved in the fossil record. The fish remains tend to be concentrated in particular horizons, known as fish beds. These beds are made up of fine-grained material, laid down in the deeper waters of the lake where conditions were ideal for the preservation of any animals that died and fell to the lake bottom. There were no predators or scavengers in the lower reaches of the lake, so the carcasses were little disturbed. These waters were also poor in oxygen. The resulting 'fish beds' occur at irregular intervals throughout the rock succession and

An area from the Moray Firth to as far north as Shetland was occupied by the Orcadian Lake. This body of water was flanked on its southern and western sides by the Caledonian Mountains. Large rivers drained into the lake, carrying sediments from the higher ground.

Rhynie – an ecosystem preserved in stone

The most intriguing rocks from the Devonian Period occur near the Aberdeenshire village of Rhynie. Buried beneath a grass field are layers of rock – the Rhynie chert – that are of international renown. Whole ecosystems, including primitive plants, insects, bacteria and fungi are beautifully preserved in deposits of chert. In particular, Rhynie has yielded some of the best-preserved land plants known from this age to be found anywhere in the world. The preservation is such that the details of the cell structures of these 400-million-year-old plants are still evident. The cherts were formed as hot springs erupted and sprayed silica-rich fluids into the air. This process created cones of silica-rich material, or sinter, around the vent or point of eruption. As the layers of sinter built up, plants and animals living in the vicinity became coated with this fluid and were incorporated into the deposit. Prolonged contact with the silica-rich fluid led to the organic structures of the plants and animals being replaced. Over time, they became rock. Burial of the sinter under succeeding layers led to its conversion to chert. The geysers in Yellowstone National Park provide a modern example of the type of environment that formed the rocks of Rhynie.

So around 400 million years later, the primitive plants and animals that inhabited this early world are preserved for posterity in exquisite detail. Similar to flies caught in amber, primitive plants that are quite unlike any that exist today, have been recovered from the chert deposits. An amazing collection of early terrestrial and freshwater arthropods, some making their appearance in the geological record for the very first time, have also been described. The presence of fossilised bacteria, which clearly thrived in the hot springs environment, have also been recorded.

The cherts at Rhynie were first described in 1912 by William Mackie, a doctor from nearby Elgin. He looked at the rocks in thin section under the microscope and found many examples of perfectly preserved plant stems. The Rhynie flora was studied in greater detail by Dr Robert Kidston and Professor William Lang and their findings were published in a series of five papers between 1917 and 1921. The fields overlying the Rhynie deposits were later purchased by Dr Lyon, an eminent palaeobotanist, who subsequently gifted the site to SNH. In recent years, the Rhynie chert has been intensively reinvestigated by a multi-disciplinary team from Aberdeen University, led by Professor Nigel Trewin. They used drilling rigs to recover cores from the deposit and undertook geophysical studies. Their work has added considerably to what was already known and the research team have made their findings available on the web in a very accessible format. The rocks of Rhynie are truly one of Scotland's geological gems.

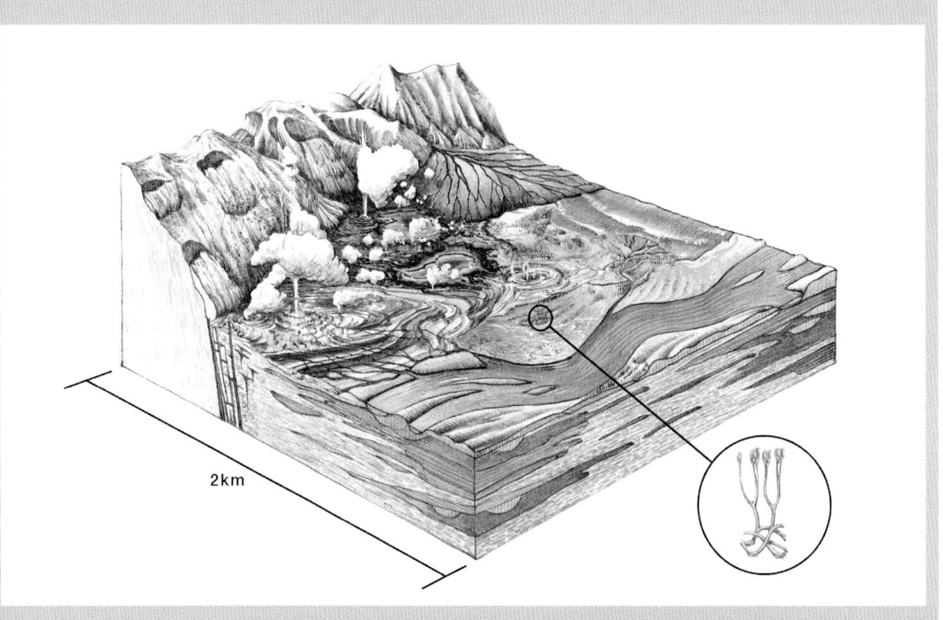

Top. The ecosystem around Rhynie in Aberdeenshire, as it would have looked around 400 million years ago. The early land plants for which Rhynie is internationally renowned are relatively small in stature, standing between 10 and 25 cm in height.

Above. A lump of Rhynie chert shows the fossilised remains of early plants that grew close to the hot springs.

Fish in the Orcadian Lake lived in shallow water. After death, their remains were carried by currents to the deeper reaches of the lake, below the warm oxygenated surface waters and into cooler unoxygenated water. Over time, their bodies were covered by a fine rain of sediment and the remains were fossilised as they gradually turned to stone.

a variety of different types of fish have been recovered from them.

The lake eventually filled with sediment and disappeared. Great arid plains replaced the freshwater lake and occasional deposits, believed to be of marine origin, are also recorded, suggesting a link to the open sea. Braided streams flowed near the margins of the ancient lake. Fossil wind-blown sand dunes are also present, indicating the build up of desert sands. So, after 10 million years, during which time it provided a watery habitat and a source for the bedrock of the Orkney Islands, Caithness, the Black Isle and many places besides, the Orcadian Lake was no more.

Devonian volcanoes

The quiet deposition of the Old Red Sandstones of the Devonian Period was accompanied by explosive volcanic activity. The Cheviot Hills, the Ochil Hills, Sidlaw Hills and part of the Pentland Hills, as well as the Lorne Plateau near Oban, and St Abbs Head are all formed from lavas of Devonian age.

The spectacular cliffed coastline of St Abbs Head is built of lava and volcanic ash from a vent of limited size, whilst the Cheviots bear testament to volcanic outpourings on quite a different scale. The lava fields associated with the Cheviot volcano occupy the ground near Kirk Yetholm, adjacent to the border with England. The underground volcanic plumbing cuts across the Iapetus Suture, the line along which Scotland and England collided, indicating that this volcanic episode postdates the collision. There may be some connection between the Earth movements that accompanied the assembly of the British Isles and these volcanic outpourings, although the evidence is not conclusive.

Scurdie Ness on the Angus coast provides further evidence of the great eruptions of lava during Devonian times. Around 70 metres of lava are now to be found in the coastal exposures there. These lavas are noteworthy as they are also full of holes. As the lava was erupted, the volatile components of the magma formed a series of gas bubbles, creating a honeycomb effect. After the lava cooled, these spaces were filled by minerals, such as calcite, agate and chalcedony. These filled

St Abbs Head, Berwickshire, is built from lavas and associated volcanic ash.

voids, known as amygdales, create a very pleasing effect and are a source of semi-precious stones.

The rocks of the Ochil and Sidlaw Hills were also formed around this time. Both ranges of hills are built of lavas of andesite and basalt composition. These volcanic rocks have a purplish hue in fist-sized specimens, with flecks of white. These lavas also have cavities filled with minerals, such as calcite and red chalcedony, formed in the manner described above. In a small quarry opened in the Ochil Hills at Pairney to quarry material for the adjacent A9 roadworks, more acid lavas known as rhyodacites have been found. These rocks are of interest as they carry fragments of lower-crustal material, known as xenoliths, which have been transported from depth as the molten rock ascended the crust to the volcanic vent.

Fossil hunter extraordinaire

Hugh Miller (1802–56) was born in Cromarty on the Black Isle in a thatched cottage that was built by his buccaneering great-grandfather. Miller is remembered for a rather more scholarly pursuit than that of his forebears – the study of fossils, particularly the fish remains of the Old Red Sandstone. From humble beginnings as a stonemason, Hugh Miller wrote voluminously on his beloved geology and became one of the first authors to popularise the subject beyond the narrow confines of the scientific community. When he discovered fossils in the sandstones of the Black Isle, he devoted his spare time to unravelling their secrets by collecting from the many places that yielded fish remains. Over time, he brought together a collection of specimens, many of which were new to science. He developed his knowledge of the anatomy of his specimens by dissecting modern fish that he caught locally.

Perhaps most remarkably of all, he worked in almost complete isolation from the mainstream scientists of the day who were researching similar topics. But the publication of Miller's first book, *Scenes and Legends*, brought him to the attention of Louis Agassiz, the famous Swiss geologist, and also Sir Roderick Impey Murchison, then Director of the Geological Survey in Scotland. They encouraged Miller to publish more of his work. Miller was a deeply religious man and one of his abiding moral dilemmas was that the literal interpretation of the Bible, and in particular its first book, was irreconcilable with the facts that had emerged from the work of James Hutton and subsequent authors. This troubled him greatly. He addressed this issue in his final work, *Testimony of the Rocks*, which has the subtitle *Geology in its Bearing on the Two Theologies, Natural and Revealed*. In this volume, which takes the form of a series of twelve lectures delivered to the Edinburgh Philosophical Institution and the Geological Section of the British Association, he explores these dilemmas in some depth. Shortly after this work was completed, Hugh Miller took his own life. The difficulty of reconciling his life's scientific work with his profound faith was finally thought to have overwhelmed him.

Dr Charles Waterston, former Keeper of Geology at the National Museums of Scotland and an eminent palaeontologist in his own right, summarised Miller's contribution thus: "Although the struggle ended in personal tragedy, his work had convinced thousands of ordinary people of the interest and value of geology, and thus did a great deal to establish a favourable climate of public opinion in which the maturing science could flourish unhindered by old prejudices." There could be no more fitting epitaph for one of Scotland's most prolific writers and scientific pioneers.

THE CARBONIFEROUS PERIOD – VOLCANOES, TROPICAL SWAMPS AND LIZARDS

FROM 359 MILLION TO 299 MILLION YEARS AGO

Scotland continued its northward journey during the Carboniferous Period, to sit astride the Equator. Tropical rainforests, with similarities to the contemporary extensive expanses of this habitat in Brazil and Africa, clothed the Central Lowlands of Scotland. Historically, the thick deposits of coal that developed from these ancient rainforests have been the most economically productive rocks in Scotland, so they have also been the most intensively investigated.

Scotland had a climate to match its position on the globe – tropical. The forms of life that inhabited this land became more diverse and exotic during the Carboniferous. Perhaps most famous is 'Lizzie the lizard' or *Westlothiana lizziae* to give it its full name. The fossil remains of this creature, which is thought to be one of the oldest ancestors of the reptiles, were found in a disused quarry near Edinburgh. Its discovery sparked international interest from the scientific community and also captured the public imagination. The money to purchase the fossil remains of Lizzie the lizard was partly raised by public subscription. The specimen is now on public display in the National Museum of Scotland in Chambers Street, Edinburgh.

The location at East Kirkton in the Bathgate Hills where Lizzie was found is one of the most intensively studied in Scotland. The find was made by Stan Wood, a commercial fossil collector and dealer, and sparked renewed interest in a place whose geological interest had been well known since the early nineteenth century. A wide-ranging research programme was developed to investigate all aspects of these Lower Carboniferous rocks. The results were presented at a symposium at the Royal Society of Edinburgh and later published as a volume in their prestigious *Transactions* series. The following account is drawn from the descriptions published in that volume.

The Bathgate Hills owe their origin to persistent volcanic eruptions that occurred during Early Carboniferous times. The rocks that gave rise to the strata now exposed at East Kirkton, were laid down in a shallow lake, set within a richly vegetated landscape formed of volcanic cones, some many hundreds of metres in height. There was little volcanic activity whilst the lake existed, although the water temperature may have risen sharply due to localised 'hot-spring' activity. The plants and animals that lived around this lake gave rise to some of the most exceptional fossils found anywhere in Scotland. Large eurypterids, that could live on land or in the water, have also been recovered from the rocks of East Kirkton. *Eldeceeon rolfei,* a new species of retilomorph, the ancestors of the reptiles, was also described from here. This primitive creature, which was new to science, was named after Dr Ian Rolfe, Keeper of Geology at the National Museums of Scotland for many years. The remains of fossil sharks were also discovered nearby, as were the remains of many primitive fish, known as acanthodian or 'spiny-finned' fish. These lake-dwellers had streamlined, shark-like bodies and sharp spines on their backs. Scorpions are not particularly common in the fossil record, particularly in Scotland, so the discovery of some complete specimens caused considerable excitement. *Pulmonoscorpius kirktonensis* is a species new to science and, of course, was named after its place of origin.

It is clear from the descriptions of these highlights from the Lower Carboniferous rocks that the landscape changed from the relatively barren uplands and plains of the Devonian world to a more verdant and biologically productive scene. As time passed, vegetation cover became more lush and extensive. Tropical rainforests were widespread throughout the Central Lowlands in these later Carboniferous times. It was the dead

A tropical limestone paradise

Top. Reconstruction, by the National Museums of Scotland, showing how these ancient reefs would have looked during Carboniferous times.

Above. Roscobie Quarry is a superb illustration of sea-floor reef structures. The reef limestones (below the dashed line) are overlain by thick layers of mud that built up as rivers dumped their sediment load in the near-shore waters. Thicker sandstones were laid down on top of the muds as the shallow seas were reclaimed and dry land was re-established.

During the early part of Carboniferous times, shallow tropical seas were widespread across the Central Lowlands. Reefs were commonly developed on the sea floors of this tropical paradise. These structures, built from microscopic algae and bacteria, stood around 15 metres high on the sea floor. Tropical shelled animals, bryozoans, corals and sea lilies, also known as crinoids, all flourished in this environment. But, from time to time, the clarity of the clear blue waters was clouded by the influx of sediments carried by rivers flowing into the sea. Thick muddy deposits were laid down as deltas built out over the sea floor. As more sediment was piled on top, so the sea became shallower and eventually areas of dry land were formed. These constantly changing environments of Carboniferous times are reflected in the cycles of sedi-mentary strata laid down and preserved in the record of the rocks. There are a number of sites across the Central Belt where these rock cycles can be studied.

Roscobie Quarry in Fife is perhaps one of the best places to see these reef structures. It is estimated that the Roscobie reefs must have stood at least 15 metres high, 46 metres wide and 230 metres in length. The site is very rich in crinoids: twenty-seven separate species have been described from here as well as many sponges and trilobites, and a wide variety of shelled creatures such as bivalves, gastropods and brachiopods. This former limestone quarry that is now substantially flooded, and so very dangerous to visit. The site is also in private ownership and arrangements must be made to gain permission for access.

Petershill Quarry in West Lothian shows broadly similar structures of Lower Carboniferous age. The site is managed as a local nature reserve and can be visited in relative safety. At the southern end of the quarry there is an impressive reef structure, which was built mainly from sponges, bryozoans and algae. For the most part, the rocks of Petershill would have been deposited in clear tropical seas. There is, however, also evidence for deeper, more muddy, waters. Many different corals have been found at this site and Petershill remains a classic place for the study of these creatures. When this ecosystem was flourishing, volcanic activity was close at hand. Several clay horizons within the limestones have been interpreted as altered volcanic ash derived from volcanoes that were erupting locally. In fact, the Petershill limestone is part of a sequence of rocks known as the 'Bathgate Hills Volcanic Formation', which predominately comprises volcanic rocks and associated ash deposits. So we have evidence for the existence of a tropical paradise in Central Scotland around 350 million years ago, but environmental upheaval, in the form of regular volcanic eruptions, was never far away during these times.

Scorpions are rare in Scotland's fossil record, but they were relatively abundant during Lower Carboniferous times.

and decaying remains of the giant club mosses and tree ferns that created the coal seams of today. The Central Lowlands was a valley during these times, bounded by the Highlands to the north and Southern Uplands to the south. The intervening land was a low-lying plain which the sea regularly swept across as sea levels periodically changed. During the Carboniferous, the climate oscillated between 'hot-house' and 'ice-house' conditions. During the colder periods, extensive polar ice caps formed. As water was locked up in the ice caps, so worldwide sea levels fell in response. During warmer phases, the sea levels rose once again. This process was repeated many times during the Carboniferous, so explaining the changing sea levels experienced throughout this period. The presence of giant insects, such as dragonflies, and the frequent occurrence of wildfires that tore through the tropical rainforests, has led some experts to speculate that the oxygen content of the atmosphere must have been particularly high at this time.

The rivers, draining from the higher ground, built deltas out across the coastal plains that were densely vegetated by luxuriant rainforest. However, the Carboniferous trees were very

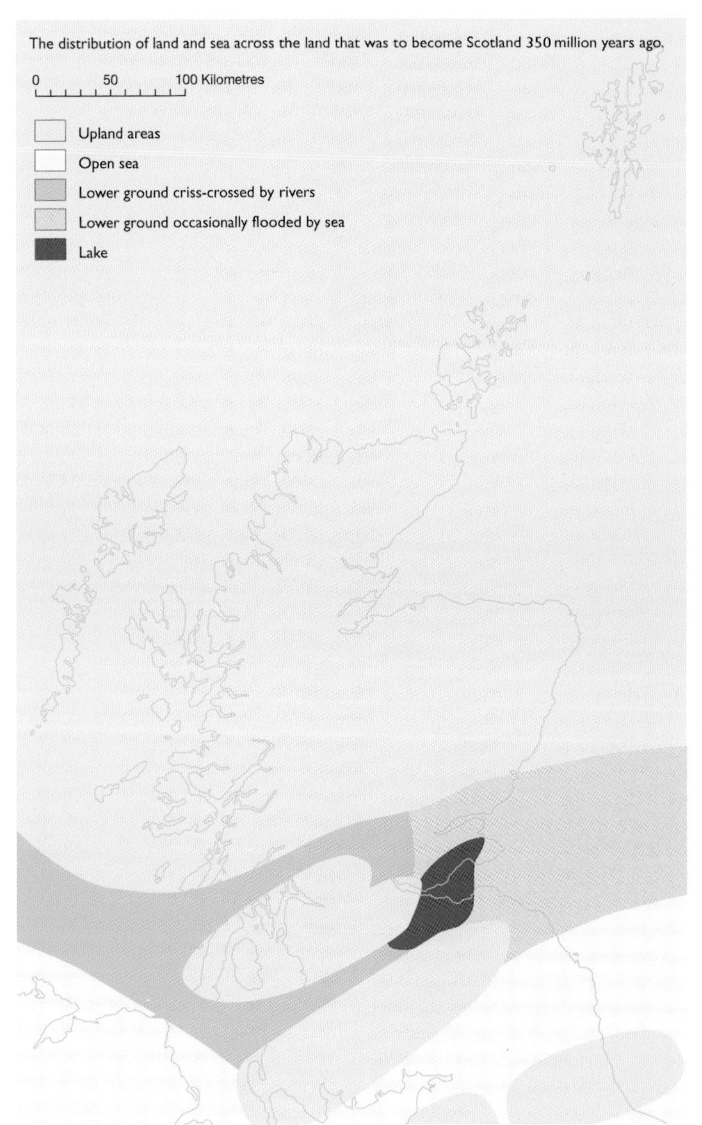

A reconstruction of the geography of Scotland 350 million years ago during early Carboniferous times. The Caledonian Mountains formed high ground to the north. The volcanic piles of the Clyde Plateau and Gargunnock Hills and folded sea-floor sediments of the Southern Uplands also formed higher ground. A flat alluvial plain separated these upland areas, which was periodically inundated by the sea to create a warm, shallow, tropical marine environment teeming with life.

Life in the Carboniferous seas

Life also diversified in the Carboniferous seas. Sharks and many other species of fish, some up to 5 metres in length, inhabited the tropical waters that fringed the land. A prize fossil find was made near Glasgow where the best-preserved fossil shark of its time has been excavated. From end to end, its fragile skeleton is almost intact, having been locked in the black shales of Bearsden for over 330 million years. It is so well preserved that the remains of muscles and blood vessels can still be seen. Researchers have even identified the remains of a last meal of fish in the shark's stomach. It has now been scientifically named as *Akmonistion zangerli* – a completely new creature to science.

Above. The Bearsden shark. This is how the Carboniferous seas would have looked. In addition to the Bearsden shark, primitive fish would have swum alongside a variety of invertebrates and corals.

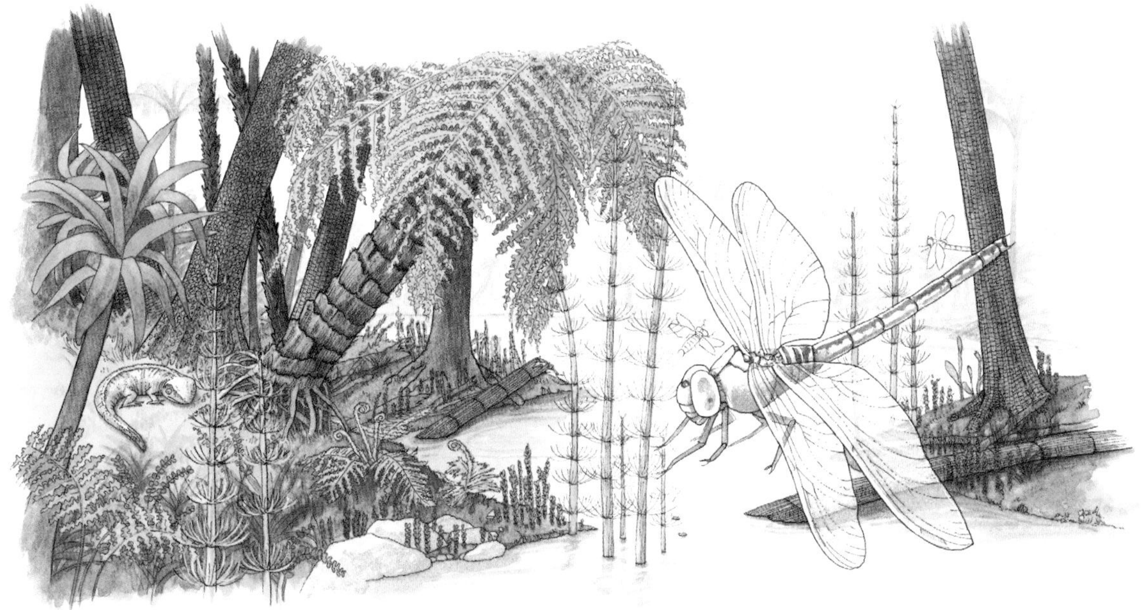

Right. A tropical rainforest of Carboniferous age as it would have looked around 310 million years ago.

different from their modern relatives and only a few had woody bark. They were simply constructed, small and rapidly growing. As the vegetation died and began to rot, it formed considerable layers of peat up to 20 metres in thickness. Meandering rivers flowed through the peat bogs, dumping sand and silt across their floodplains.

As the seas rose and fell in response to global climate change, so the forests were periodically inundated and layers of silt and sand were dumped on top of the peats. Where sea levels rose for a sustained period, limestones and muds were laid on top of the sands. Many of these layers contain fossils, such as bivalves, that we know lived in the sea, so the strong marine influence can be readily confirmed. This cycle of change was repeated many times during the Carboniferous, and each time an area of forest was inundated by the sea, the potential was created for another coal seam to form.

As layers of sand and limestone accumulated over the layers of peat, so the crushing weight of the overlying sediments had the effect of changing these organic rich layers into coal. The thickness of the peat layers was considerably reduced by this process and a 20-metre layer would be reduced to a fraction of its former thickness.

The distribution of these coal-bearing deposits influenced the location of many of the early settlements in central and southern Scotland. Mining towns and villages relied on this natural resource for their livelihood throughout the last century or more. Coalmining is a part of the industrial and cultural fabric of Scotland and has provided a way of life for many generations of Scots. Although all the deep mines are now closed, opencast coal working remains a viable economic prospect for the foreseeable future.

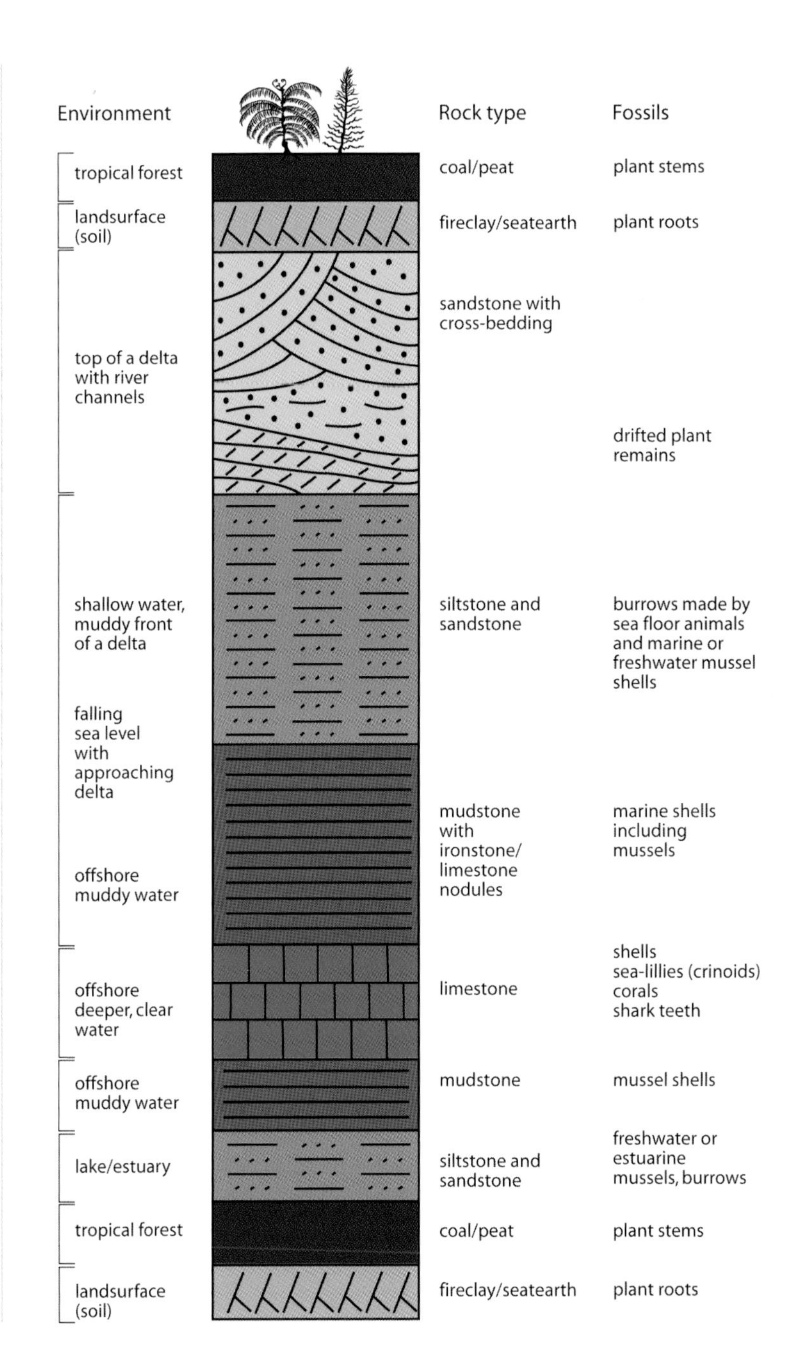

Opposite. Cycle of coal formation. As conditions fluctuated, the rocks laid down at each stage of the cycle varied in response. As forests flourished, so layers of peat formed that eventually gave rise to coal. As the sea level rose and marine conditions were re-established, layers of mud and limestone were laid down. River deltas then built out into the open sea to reclaim the area as dry land and soils formed on the newly stabilised land. This area was then colonised by tropical forests and the cycle began again. All of these changes in environmental conditions are preserved in the cycle of coal formation.

Early beginnings for the use of the microscope

Lennel Braes, near Coldstream in the Scottish Borders, has an interesting claim to fame. It was using fossil plant material of Carboniferous age from this site that scientific pioneer George Sanderson first developed a technique to make thin sections that could be studied under the microscope. This way of looking at fossil material opened the way for new and incisive studies of not only fossil material, but eventually all rocks. The technique was further refined by William Nichol who mounted his specimens on glass slides. His findings were published in the *Edinburgh New Philosophical Journal* in 1834 in a paper entitled 'Observations on the structure of recent and fossil conifers'. The Geological Conservation Review account of the site recounts that it was one of Nichol's friends, Henry Witham, who was the first person to use this technique extensively, and that he highlighted its value to the wider scientific community.

This technique of making thin sections of rocks or fossil material is now a standard approach that many scientists use. Thin slices of rock are mounted on glass slides and ground down to the point where they become transparent. By examining the rock slice under a powerful microscope, the smallest of structures, invisible to the naked eye, can be studied in detail. The technique has now been adapted to study rocks of every composition, and complex rock-forming minerals and fossil material can be fully investigated.

The development of the technology to look at rocks in thin section had its early beginnings at this site in the Borders and has been of immeasurable value to science as a whole.

Belching volcanoes and lava flows

As we have begun to see, volcanic activity was widespread, particularly during the early part of the Carboniferous. Over 1,000 volcanic vents that were active during this period have been identified throughout central Scotland. Great outpourings of basalt lava built the Kilpatrick and Fintry Hills and the Campsie Fells to the north of Glasgow and also the ground to the west and south of where the city now stands. In the Campsie Fells, great thicknesses of basalt and related rocks form an impressive escarpment that runs for over 7 kilometres. Single lava flows, some up to 10 metres in thickness, can be followed along this north-facing escarpment that is such a prominent landscape feature in West Central Scotland. A series of volcanic vents have been identified from Dumgoyne in the south-west to Dunmore in the north-east. Some now appear as prominent landscape features. These vents and related dykes are thought to be the feeders through which the lavas of the Campsies were erupted. In some instances, it is even possible to identify specific volcanic vents as sources for particular lava flows. This can be done by demonstrating that the chemical composition of the lavas matches that found in the neck of a specific volcanic vent.

Dumbarton Rock, on the north shore of the River Clyde, has been a fortified location from the fifth century. But its geological legacy can be traced back to around 300 million years ago when it was the site of an active volcano. What remains today is a shadow of the former structure that would have been built around a lava cone many hundreds of metres in height. Its precise nature, of course, will remain a matter for speculation because the superstructure has been removed by subsequent erosion. The Rock, as we see it today, formed deep in the throat of the feeder pipe that brought lava from the depths to the surface. Lava cooled in this feeder pipe, blocking further eruptions and creating a resistant feature that remained intact whilst the adjacent softer rocks were removed by the elements.

In the Borders, around fifty volcanic plugs have been identified, some of which are thought to be the source for the Kelso Traps, a thick sequence of mainly basalt lavas that floor part of the Tweed valley. Volcanoes were also very active in the area we now recognise as East Lothian and Edinburgh. Traprain Law, North Berwick Law and the Garleton Hills were all sites of active volcanoes during the Carboniferous. Perhaps the most iconic of all Scottish landmarks are Edinburgh Castle Rock and Arthur's Seat. Both were volcanoes that belched lavas and volcanic ash. Subsequent erosion has of course modified

Below. Dumbarton Rock was an active volcano some 300 million years ago. It was originally many hundreds of metres in height, but it has been cut down to its present size by subsequent erosion.

Opposite. North Berwick Law is the remains of another volcano that was active during early Carboniferous times.

the form of all of these edifices, exposing the lower reaches of the volcanic plumbing in most cases.

The magmas that fed these volcanoes tapped a deep source. Some contain rock fragments, called xenoliths or foreign rocks, that have been transported from the upper mantle of the Earth by molten magma – a remarkable journey of around 70 kilometres. Crystals of garnet, a ruby-red mineral, transported from the deep crust to the surface by rising streams of magma are now to be found on the East Fife coast in some of the eroded remains of ancient vents, known as 'volcanic necks'. A blood-red variety of the mineral garnet, called pyrope, forms part of these far-travelled pods of rock from the upper mantle. These minerals are known locally as the Elie Rubies, but have no economic value. They are, however, of great value to the scientist and should not under any circumstances be plundered.

The underlying geology of the Lomond Hills in Fife was also formed at this time, when great pulses of molten rock were pumped from depth into overlying layers of sedimentary rocks. The molten magma travelled sideways between the layers of sandstones and limestones, creating a great sheet of igneous rock around 200 metres thick in places under the Central Lowlands and known as the Midland Valley Sill. The Lomond Hills represent the northern edge of this once molten sheet, which lies between host layers of sandstones and limestones. Erosion has dug deep to excavate these layers that were once buried below the surface. The summit of West Lomond is interpreted as a feeder to a volcanic vent that was active during this time.

Perhaps the most visible expression of the Midland Valley Sill is to be found in the road and railway cuttings at the north end of the Forth Bridge. The road cuttings, which are lit at night

to dramatic effect, slice through a thick sequence of dolerites. At one place, on the east side of the carriageway about 300 metres north of the bridge, the base of the sill is exposed. Here we see the contact between the lowest level of the sill and the rock into which the molten magma pulsed. The layers of mud and sand have been cooked by this contact with the once-molten rock.

The lava flows also provide evidence of climatic conditions during Carboniferous times. There are thick deposits of bauxite clay at High Smithstone Quarry in Ayrshire. The formation of this mineral, an oxide of aluminium, is usually associated with the weathering of base-rich lavas under tropical conditions. It is thought that these deposits were formed as the weathered tops of the underlying basalt lavas were transported, perhaps by a river, and dumped some distance away.

The Rock and Spindle near St Andrews is an example of a small volcanic vent that has been carved into an interesting shape by the elements.

THE PERMIAN AND TRIASSIC PERIODS – A DESERT STORM
FROM 299 MILLION TO 199 MILLION YEARS AGO

Scotland continued to drift northwards as part of the Pangaean supercontinent. This continental landmass included modern-day North and South America, Greenland, Scandinavia, much of mainland Europe and Africa. Dry land extended almost from pole to pole. The name Pangaea means 'all-earth'. For the first time, and also the last, animals were free to roam across the face of the Earth without sea barriers to impede their progress. There was, however, a major glaciation in progress at this time and the southern continents of the world were covered by ice. These geological periods are also significant as the time that this early world lost most of its biodiversity. Two of the major mass extinctions affected flora and fauna as it existed during Permian and Triassic times. The end-Permian event was truly apocalyptic. Around 96 per cent of all marine species just simply disappeared and only 4 per cent of marine biodiversity survived into the Triassic. Trilobites, the 'woodlice' of the ocean floor, and many reptile groups simply vanished from the fossil record. As with all the Earth's great extinctions, the cause is not entirely certain, but climate change is identified as one of the key contributory factors. Since life began, this is the closest that Planet Earth has come to being completely bereft of life. This clear-out of animals that had formerly dominated the ecosystems of the world, opened the way for groups which had previously played only a minor role. The effects on land-based species were just as devastating. It took over 30 million years for ecosystems to diversify again and for many ecological niches to be reoccupied.

Global sea levels were generally low and Scotland was many miles from the sea. Desert conditions prevailed throughout much of the continent. Towering sand dunes, in excess of 100 metres in height and 3 kilometres from end to end, were driven across the arid plains of central Scotland by withering winds, creating blistering conditions, hostile to plant and animal life. There is also evidence for intermittent storms and localised flooding. Proof comes from the presence of occasional pebble beds within the sequence of wind-blown sandstones.

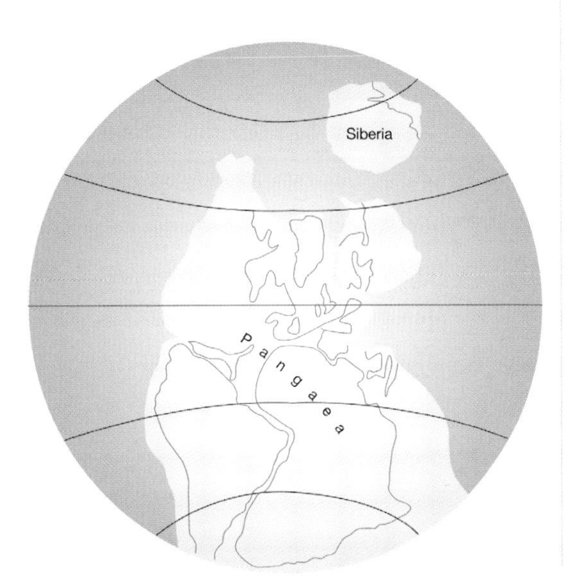

Pangaea straddles the Equator. The supercontinent of Pangaea included most of the world's dry land around 300 million years ago at the beginning of the Permian.

Life in the deserts of Scotland

Life, such as it existed at this time, was restricted to a few hardy species. Such arid desert conditions were far from ideal for preserving the remains of plants and animals that did survive in these unforgiving times. Despite the odds, however, some remarkable fossils have been found. This was the age of the reptiles. We know of their existence in the Scottish deserts of the Permian and Triassic because of the tracks they left in the soft sands, which were captured for

Early reptiles in Scotland

Some of the earliest reptiles to be found in Scotland's fossil record came from the desert sandstones of the Lossiemouth area. A number of species of archosaur, which are precursors to a diverse group of crocodiles and dinosaurs, have been found in a series of disused quarries worked for building stones. These quarries were active from as early as 1790 and employed around twenty masons. The account of these sites, published in the *Geological Conservation Review*, recounts the story of a workman finding a slab bearing many fossil remains. He passed it on to the Elgin town clerk and local geologist, Patrick Duff, who tried to get the remains identified by experts in Edinburgh. This proved fruitless, so he sent drawings to the eminent Swiss palaeontologist Louis Agassiz, who wrongly identified them as being of Devonian age. The truth was only later revealed when Mr Martin, an Elgin schoolmaster, "detected . . . a bone, possibly the scapula of a reptile". This find was made at a nearby location. Two celebrated natural scientists, Roderick Murchison and T H Huxley, later studied the area and confirmed Mr Martin's view that the remains were from an early reptile. Huxley suggested that the reptile remains were possibly those of an ancestor to the crocodile lineage.

Hyperodapedon, one of the reptile species found in these deposits, was a bulky lizard around 1.3 metres in length with four strong limbs. It had teeth with sharp edges designed to slice through tough plant material. It also had massive claws on its hind limbs and these were constructed in a way that suggested they were adapted for 'scratch digging'. So, from careful study of these remains, reliable conclusions can be drawn about what these early animals ate and how they lived their lives.

This scene is an imaginary one from the Elgin area in late Triassic times about 200 million years ago. Three *Hyperodapedons* feed on seed ferns in the foreground surrounded by other reptiles that existed during this period.

posterity through rapid burial by other layers of sand. Occasional whole skeletons and skull fragments have also been recovered from the enclosing sandstones – the last resting place for many of these Permian and Triassic reptiles.

The early reptiles that roamed the deserts of 'Scotland' during the Trias and Permian left footprints in the soft sediment. These rare finds occasionally turn up in the quarries that work these sandstone deposits, seen here at Clashach, near Elgin.

Permian and Triassic rocks across Scotland

Permian rocks are concentrated mainly in south-west Scotland, on Arran and on the Moray coast. That is not to say these were the only places where rocks of this age were deposited, but after 250 million years, this is all that is left. These deposits have been extensively quarried in all three locations for building stone. Locharbriggs Quarry near Dumfries still works the thick layers of well-sorted, fine- to medium-grained sandstone. The impressive cross-bedding structures that are such a feature of this quarry indicate the wind-blown origin of these sands. The vertical quarry faces are effectively slices through a series of ancient sand dunes, stacked on top of each other as the sand deposits accumulated over time. Locharbriggs Quarry is recognised as one of the best places in Britain to see the internal structures of these ancient sand dunes.

Ancient sand dunes are also preserved near Brodick on Arran. The sands are bright red in colour with strong internal structures confirming their wind-blown origins. The red colouration is also strongly indicative of a desert environment. These wind-blown layers are interspersed with river-laid deposits, so seasonal rivers clearly flowed across the arid desert plains. Further environmental indicators include marks on the sands that are interpreted as direct lightning strikes. These small pockmarks on the rock surface are just a few centimetres across and consist of fused rock, melted as the bolts of lightning struck.

In the Permian sands of Moray, tracks preserved in the dunes were made by creatures called dicynodonts. These odd-looking animals stood less than a metre in height. The name *dicynodont* means 'two dog teeth', referring to the two tusk-like teeth that protruded from the heavy skull.

Careful studies have been done at all of these Permian sites to work out the direction of the prevailing winds that blew across the desert plains. At this time, the shape of the Moray Firth existed in outline and the dunes were at the edge of an expanse of open sea. Establishing the direction in which the dunes were being driven by the wind can be done by noting the bedding directions and reconstructing the internal architecture of the dunes. At Clashach, near Elgin, the dominant wind direction was from the north-north-east, with secondary winds from the south-south-east. Studies of the rocks have revealed that only six major dunes existed in this area, but they were of colossal dimensions. Smaller crescent-shaped dunes migrated between larger 'star dunes'. Seasonal rivers flowed between the dunes after periods of heavy rainfall. It is thought that many of the preserved trackways were made in the areas between the dunes where the sediments were damp and more cohesive. The footprints made in the sand by passing animals would have been more likely to survive in these wetter environments.

Although limited in extent, Clashach sandstone has been used extensively for building stone for many of our historic and more modern buildings. The Hopeman sandstones from the Moray Firth coast are also excellent building stones. They clad the extension to the National Museum of Scotland in Edinburgh and also parts of the Scottish Parliament building at Holyrood. Stone used to build the Edwardian pleasure palace of Kinloch Castle, on the island of Rum, came from Arran.

The Stornoway Beds, in the Western Isles, were formerly thought to be of Torridonian or possibly Devonian age. With no fossil remains to confirm their age, it has been established from palaeomagnetic measurements (see chapter 2) that this 1,200-metre thick sequence of unfossiliferous sandstones and conglomerates was deposited at some stage during Permian or Triassic times.

Smaller patches of Triassic rocks are found throughout the Inner Hebrides and mainland Scotland, including at Aultbea, Gairloch, the Isle of Raasay, Rum, Ardnamurchan, Morven, Mull and Arran. With the exception of the occurrence on Arran, none of the other exposures is particularly extensive. After the end-Permian extinction, fossils in these deposits are rare, but they are known from Gribun Shore on Mull, where double-shelled marine organisms known as bivalves have been found. In addition, plant fossils are also present at this location. As with most of the strata of this age, they were laid down by rivers that flowed across the parched landscape in a braided fashion. The Triassic rocks of Arran demonstrate that some sediments were deposited close to the sea under what are interpreted as inter-tidal conditions.

Although exposures of rocks from the Permian and Triassic Periods are sparse on land, much of the North Sea and the sea bed to the north and west of Scotland are floored by rocks of this age. The sea floor just to the west of Orkney has been drilled as part of an oil exploration programme and around 1,500 metres of Triassic sandstones were logged in the borehole. Similarly, in the East Fair Isle Basin and in the Forth Approaches Basin that runs down the east coast of Scotland, great thicknesses of strata from the Permian and Triassic have been identified.

The 'age of dying'

Scotland's biodiversity was reduced to rock bottom at the end of the Triassic Period. Although the ecosystems of the world had partially recovered from the end-Permian event, this new 'age of dying' took biological diversity to a new low. As with all the great mass extinctions, the causes of this catastrophic event are unlikely to be established with certainty. The usual suspects, such as climate change, are definitely in the frame to explain these extraordinary events. Plants and animals that lived on land and also those whose natural habitats were at sea were both profoundly affected. Cephalopod populations, a group of animals including ammonites, squids and cuttlefish, were decimated. Bivalves also suffered and many species became extinct. Brachiopods and gastropods fared equally badly. Land animals, such as reptiles and amphibians, also suffered grievous losses. Many diverse families of tetropod failed to make it into the Jurassic. It is difficult to establish a full and accurate picture of the precise causes of this global event. Rapid sea-level fluctuations would have stressed some animal and plant populations. Changes in geography that would have followed either a rise or fall in sea level would inevitably alter the distribution of habitats – both marine and on land. But, in the midst of all this uncertainty, it is clear that this collapse in the world's populations across the board opened the door for the dinosaurs to dominate the world's ecosystems. The Jurassic Period, the age of the dinosaurs, was just about to dawn.

The Atlantic starts to open

From the Devonian onwards, there was a time of relative quiescence as most of the world's dry lands drifted northwards. But that peace was shattered as the Pangaean supercontinent began to break up. Great chasms developed in the Earth's crust as a fracture opened up that was much later to widen to become the Atlantic Ocean. This was one of the most significant events in Earth history, as it was at this time that the continental landmasses that we recognise today were blocked out in outline.

The Pangaean supercontinent broke up along a series of pre-existing lines of weakness into smaller continental fragments. The newly created continents were set adrift to chart their own course. North and South America, Africa, Antarctica, Australia, Asia and Europe were all roughly fashioned, although none was in its current position at this time. Thus the building blocks of the contemporary map of the world were formed from the ruins of Pangaea.

The words of James Hutton published over 200 years ago in *Theory of the Earth* have proved to be prophetic. He was correct in saying that "the strata which now compose our continents are all formed out of strata more ancient than themselves. The spoils or wreck of an older world are everywhere visible at the present."

The effects of these catastrophic global rearrangements on Scotland were substantial. Great cracks appeared either side of Scotland during this time, reaching down from the north of Norway. This process initiated the continental break up that would eventually lead to our current geographical position on the fringe of Europe and leave Scotland the width of an ocean away from North America. The proto-Atlantic sea was to widen considerably during the next 200 million years. The North Sea also widened at this time and Scotland parted company with Scandinavia.

A series of troughs were created on the sea floor between the two landmasses that were filled by huge volumes of sediments. Some of the layers of mud, particularly the Kimmeridge Clay of Jurassic age, were rich in organic material that subsequently decayed to liberate huge quantities of oil and gas, creating the basis of the North Sea oil industry.

The break up of Pangaea started in early Jurassic times and Scotland's long association with North America began to be severed. The forerunner of the Atlantic Ocean developed, as did a major fault-bounded valley or graben structure in the North Sea.

THE JURASSIC PERIOD – AGE OF DINOSAURS AND FLOOD

FROM 199 MILLION TO 145 MILLION YEARS AGO

The desert landscapes of the Permian and Triassic were drowned as sea levels rose dramatically at the end of the Triassic. What was arid and parched land disappeared under the advancing shelf seas that inundated all but the highest ground, which primarily lay in the north and east of Scotland. The Western Isles, which were part of a larger landmass that lay to west of the mainland, stayed dry. So too did Orkney and Shetland, which were part of the main landmass of Scotland at this point in geological history.

Although the distribution of Jurassic rocks is limited on land in Scotland, this period had a profound effect on the landscape and subsequent economic fortunes of the country. The seaway that was to become the North Atlantic continued to widen as Europe parted company with North America after an association of many hundreds of millions of years. The deep rift valleys on the floor of the North Sea continued to fill with oil-bearing sediments as the high ground of Scotland was drained by rivers that carried sand and muds to the continental shelf and ocean deeps.

The main Jurassic outcrops on land are found between Golspie and Helmsdale on the east coast and in a number of small patches on the Inner Hebridean islands, including Eigg, Raasay, around Elgol on Skye, and the largest occurrence on the Trotternish Peninsula of Skye. Although fairly limited in extent, they have been studied in great detail and reveal much about the environmental conditions that prevailed during this period. These sediments are also rich in the fossil remains of plants and animals of the time. By this stage, Scotland had drifted northwards to 30°N and the climate was wet and humid.

Much of the northern end of the Trotternish Penninsula of Skye is underlain by rocks of Jurassic age. As with many localities in Scotland, this area was first documented early in the nineteenth century. The youngest rocks in the Jurassic succession on Skye are to be found around Staffin Bay and northwards to Kildorais. Over 100 metres of richly fossiliferous shales and sandstones have been recorded in this vicinity. Staffin provides the best available section through rocks of the Upper Oxfordian, a subdivision of the youngest Jurassic strata. As such, it is an internationally recognised 'type locality' for rocks of this age. Ammonites are commonly found in rocks of Jurassic age and are used to subdivide

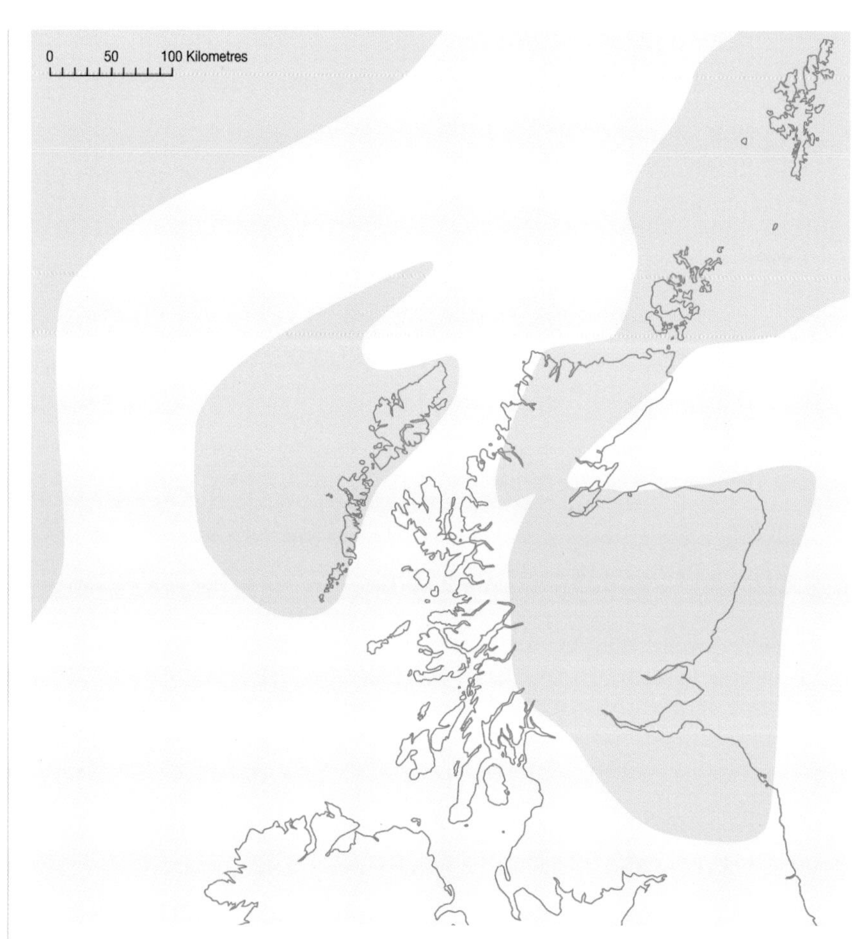

The geography of Scotland during the Jurassic Period. Much of the western part of Scotland lay under water during Early Jurassic times around 190 million years ago. The outcrop of Jurassic rocks on the mainland and islands of Scotland is limited, but extensive offshore deposits of this age have been identified during the extensive drilling programme for oil in the North Sea.

Dinosaur island

The Jurassic rocks of Skye made world headlines in 2002 when local resident Cathie Booth discovered a dinosaur footprint while walking her dog along the beach. Fourteen more tracks have now been found. Each footprint consisted of three toes, with the middle toe the longest. Guided by the size of the individual tracks, some of which are close to 50 centimetres in length, the dinosaur that made the tracks probably walked on two legs and was probably about 10 metres from head to tail. The toes are very narrow, suggesting that the animal was a carnivore. Plant-eating dinosaurs' footprints are much broader in comparison. Dinosaur remains of this age – Middle Jurassic, some 167 to 160 million years ago – are rare. Dr Neil Clark from the Hunterian Museum, who led the research team on this site, is only aware of "one or two places in the USA where remains of this age have been found". That leads him to the conclusion that the tracks may have been made by *Megalosaurus*. This beast was a large flesh-eater that stalked its prey on massive muscular hind legs. "It's impossible to be 100 per cent sure unless we follow the traffic and find a dead dinosaur at the end", Clark says, "but the *Megalosaurus* was the only large meat-eating animal known at that time."

A year later, more footprints were found. This time, it was of a plant-eating dinosaur with up to 10 smaller individuals. The footprints measured between 6 and 12 centimetres in length for the juveniles and 25 centimetres for the adult.

Far left. Jurassic dinosaur footprint from Skye, indicated by Dr Neil Clark, who led the research work on the dinosaurs found on Skye.

Left. Reconstruction of a plant-eating Skye dinosaur, *Cetiosaurus*

Jurassic rock succession – Elgol Sandstones of Jurassic age on Skye exhibit a distinctive honeycomb weathering effect.

the strata into successive time horizons. This allows one sequence of rocks to be compared with rocks of a similar age from other parts of the world. The preservation of the ammonites from Staffin is also particularly good. As with all other fossil-bearing sites, visitors are reminded to collect from this site responsibly, particularly bearing in mind its scientific importance. A huge variety of these sea-going creatures with characteristically coiled shells have been described from these rocks. The shell was a buoyancy chamber that allowed the squid-like creature that inhabited it to float above the sea floor. A tentacled head poked out of the final chamber of the outer whorl. This rather bizarre structure helped the animal forage for food. It is thought that ammonites were scavengers, rather than aggressive hunters.

Slightly older rocks of Middle Jurassic age are also found on Trotternish. These sandstones, shales, clays and limestones are rich in fossils – ammonites, oyster shells and bivalves. The indicators are that some of these deposits were laid down in lagoons close to the shoreline that existed in Jurassic times. Structures within the sandstones, known as cross-bedding, suggest that from time to time deltas built out across the coastal flats, covering the areas with sheets of sand. At Valtos, to the south of Staffin Bay, great cannonball structures have been noted in the sandstones. These unusual structures grew within the sandstone layers after the rocks were deposited and were revealed in more recent times as marine erosion cut into the cliffs.

Some amazing fossil discoveries were made by Hugh Miller from the Jurassic rocks of Eigg. He visited the island in 1844 whilst aboard the yacht *Betsey* that he used to travel around the Highlands and Islands of Scotland. Miller's reflections on the places he visited and the people he met are shared in his classic work, *The Cruise of the Betsey*, first published in 1858. This text has recently been republished by the National Museums of Scotland with notes from Professor T C Smout, Historiographer Royal in Scotland. Miller also suggests places to visit to see the key geological sites of Scotland. During his visit to Eigg, he found mainly reptile bones, and since then, the remains of other reptiles including marine turtles, crocodiles and plesiosaurs have been described from the island. An extraordinary site, now known as 'Hugh Miller's Reptile Bed', has been revisited many times since his initial expedition and the list of species whose fossilised remains have been discovered here continues to grow. The Reptile Bed is only a few centimetres thick. It has weathered red due to its high content of iron and is packed with fossil remains: whorled shells, known as gastropods; bivalves; fish scales, teeth and fin spines; and black reptile bones. Later studies have suggested that during Mid-Jurassic times, this area was a shallow lagoon that periodically dried out. The clue is that some of the mud horizons show desiccation cracks, suggesting that from time to time these mud flats were baked by the sun. Close by, a plesiosaur rib, a tooth that possibly came from a crocodile and a dinosaur vertebra have also been found.

On the east coast, the Jurassic rocks at Brora add considerably to our understanding of Scotland's Jurassic world. A rich fossil assemblage of corals, sea urchins, molluscs, brachiopods, ammonites, reptiles and fish has been recorded from the predominantly coastal rock outcrops. A single coal seam has been worked for generations, indicating the area was covered by thick forest during at least part of this period. The mine was active from 1590 and closed in 1972. This is the only commercially worked coal seam in Scotland that is not of Carboniferous age.

Further north at Helmsdale, a narrow strip of Jurassic strata holds further interest. Again, ammonites are to be found, along with bivalves and plant fragments, including ferns, cycads and conifer needles. A rather quirky feature of this site is the presence of a 'fallen stack'. It was originally interpreted as an ancient sea stack that had fallen on its side as a result of marine erosion during Jurassic times. Later work cast doubt on this interpretation. It is now accepted that this great column of rock was dislodged as the Helmsdale Fault moved and that it fell from the fault escarpment into deeper water. There is not another place in Britain where a similar occurrence has been observed.

Some of the oldest rocks from the Jurassic Period are to be found on the Ardmeanach Peninsula on Mull. Great thicknesses of flood basalt of Palaeogene age now cover the promontory. They were erupted onto a land surface built largely from Jurassic rocks. The strata of Lower Jurassic age consist of limestones and shales rich in fossils, such as ammonites and bivalves. This allows this small isolated patch of sediment to be correlated with rocks of a similar age elsewhere in the Hebrides and further afield. A thick sequence of Lower Jurassic rocks is also known from the Broadford area of Skye.

THE CRETACEOUS PERIOD – A FLOOD OF BIBLICAL PROPORTIONS

FROM 145 MILLION TO 65 MILLION YEARS AGO

Today, we have a genuine concern about sea-level rise, with global levels predicted to rise around 50 centimetres during the present century. Imagine then the effect of the sea-level rise that occurred during the later stages of the Cretaceous. Estimates are that sea levels stood for long periods at a staggering 300 metres or more higher than they are at present. Large parts of the British Isles, as they existed some 90 million years ago, were almost entirely submerged. Ireland, Wales and southern England lay almost entirely beneath the waves with only the higher ground of Scotland emerging from the warm, shallow tropical Cretaceous seas. The Highlands and possibly also the Southern Uplands emerged as tropical islands through the clear blue water. Intense volcanic activity in other parts of the world was responsible for these dramatic perturbations. Carbon dioxide and other greenhouse gases were pumped out as a by-product of this volcanic activity, leading to an overheated world. The huge rise in sea level may have resulted from the reduced volume of the ocean basins due to increased rates of volcanic activity along the mid-ocean ridges and the formation of vast lava plateaux in the Pacific and Indian Oceans. These dramatic events, however, have left precious little evidence in Scotland that they happened at all.

The Cretaceous seas left thick deposits of chalk in southern England and Northern Ireland, most famously building the White Cliffs of Dover, and on the North Sea floor, but their imprint on the Scottish mainland is almost invisible. Tantalising glimpses do occur, however, on Mull, Eigg and Skye. A lump of chalk the size of a semi-detached house and similar in appearance to the strata of southern England is also preserved on Arran. Should we interpret these occurrences as tiny remnants of a more extensive cover of Cretaceous rocks that have now been removed by erosion? It is possible.

The Cretaceous succession of southern England reaches some 500 metres in thickness, whereas the deposits in the Inner Hebrides of similar age have a maximum thickness of around 20 metres. The types of sediments deposited in the shelf seas that lapped around the upland areas of the western seaboard of Scotland are varied. White sands, thin limestones and chalk are found in association with layers of low-grade coal known as lignite. Some of the layers of sand are rich in oyster shells. At Gribun, on the south shore of Loch na Keal, Mull, rocks of Cretaceous age were first described by J W Judd in 1878. This find was significant, as it was the first time that rocks of this age had been described from the Western Highlands of Scotland.

Other patches of Cretaceous rocks are to be found at Beinn Iadain and Beinn na h-Uamha, in Morvern. Beneath lavas of later Palaeogene age, thin layers of fossil-bearing chalk have been discovered. Belemnites are most common. These are bullet-shaped fossils that accommodated a soft-bodied animal that resembled a squid. But only the hard parts of this animal have been fossilised. The chalk itself is made up of the remains of billions of tiny sea creatures that lived in the warm tropical seas. As these creatures died, their remains fell to the bottom of the sea like rain and built a white blanket that covered the sea floor. These oozes then hardened to form chalk.

The Lochaline Mines, also in Morvern, were opened to exploit sands of exceptional purity that were used in the manufacture of glass. The mines provide exposure of these rocks, which are out of character with the predominant geology of the surrounding area. There was some doubt as to the affinities of these deposits until a well-preserved starfish and other more poorly preserved fossils were recovered from them. The fossils indicate that these sands were deposited by the sea, probably in shallow-water conditions such as those found near to the shoreline. Such finds are vital in helping to assemble a picture of the distribution of land and sea during these far-off times.

The North Atlantic continued to widen during Cretaceous times as America and Europe drifted further apart. The nascent ocean inched ever wider, driven by the chain of sub-sea volcanic activity that added new material to the sea floor, pushing the continents at either side further apart. The northward drift also continued and Scotland now sat at around 40° north of the Equator. In the Rockall Trough, just west of the Scottish mainland, volcanoes were once again active during late Cretaceous times. This activity presaged more dramatic events to come. As the continents continued to drift further apart, so the Earth's crust was stretched to breaking point – and beyond.

The demise of the dinosaurs at the end of the Cretaceous Period has generated countless research studies and acres of scientific print. This is perhaps one of the aspects of the Earth's history that has truly grabbed the public imagination. What really happened to wipe out one of the most successful animal dynasties is currently a matter of conjecture, but what is certain is that it shifted the balance of power in favour of the emerging mammals and, in time, humankind.

THE PALAEOGENE PERIOD – ERUPTING VOLCANOES AND VERDANT FORESTS

FROM 65 MILLION TO 23 MILLION YEARS AGO

Just as quickly as it rose, so the sea level fell to reveal the broad outline of the British Isles for the first time. The Moray Firth and North Sea coastlines had already been roughly chiselled out, but the intricate fretwork of the west coast was still to be created. Things were far from quiet across the landscape that had just emerged from the bottom of the Cretaceous sea. Volcanic activity intensified dramatically during a 10-million-year period at the beginning of the Palaeogene. Great volcanic piles reared up and spat out fire, ash and lava. St Kilda, Skye, Rum, Mull, Arran and Ailsa Craig were all the sites of active volcanoes during this turbulent period.

This intensification was primarily related to events that were happening further afield. The North Atlantic continued to widen, stretching and thinning the crust in the process. Around 60 million years ago, a concentration of energy in

This view of the Skye volcano shows two of its key components – the lava in the foreground and the Red Cuillin, comprised of granite, in the background.

Aerial view of the ancient volcano at Ardnamurchan, showing the circular form of the intrusions of magma, called ring dykes.

the Earth's upper mantle, called a mantle plume, sat beneath southern Greenland. This ferocious font of energy brought a stream of superheated dense rock from the mantle to the base of the crust, influencing the rocks within a radius of 1,600 kilometres of the epicentre of the plume. The combination of this hot jet from below and the fact that North America was already pulling away from Europe accelerated the process and produced widespread volcanic activity.

A line of volcanoes developed down the western seaboard of Scotland. The Faeroe Islands, parts of Greenland and, most spectacularly, the beginnings of a new island, Iceland, also owe their origins to this most intense period of volcanic activity. Inevitably, the evidence for these dramatic events is far from complete. The volcanoes have all been substantially modified by subsequent erosion. As an illustration of the severity of that erosion, it is thought that over 2 kilometres of the Skye volcano have been sliced away to reveal the internal plumbing of the lower reaches of this structure. The Black and Red Cuillin of Skye formed part of a magma chamber that once lay beneath a volcanic superstructure that has long since disappeared. The lavas that build the Trotternish peninsula are thought to have welled out from great cracks or fissures in

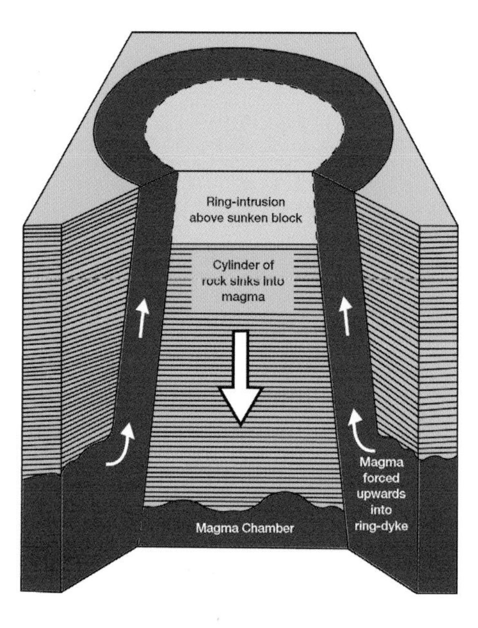

The formation of ring dykes. As the pressure rose due to upward movement of magma, so the overlying rocks were put under stress and circular cracks appeared at the surface. Over time, these grew wider and connected to the magma chamber below. This created a pathway for molten magma to rise to the surface as a series of concentric sheets. More substantial pulses of magma were subsequently forced upwards and these structures, known as ring dykes, are the circular landforms we see at the surface today. This is how the ring dykes of Ardnamurchan are thought to have formed (see p. 151).

the Earth's crust and not from a conventional volcano. The Skye volcano has an important place in the annals of geological sciences, as it was the first ancient extinct volcanic structure to be studied in any great detail and the understanding gained has been applied to unravelling the complexities of similar features elsewhere in the world.

The western tip of the Ardnamurchan peninsula is one of the most visually impressive remnants from this period. From the air, rings of once molten magma can be seen to define a series of almost perfect circles. Again, we are looking at the lower reaches of a volcano after the superstructure has been planed off by erosion. So what remains is the magma chamber, frozen in time. Pressure exerted from below by a build-up of magma and associated gases created the distinctive pattern of ring dykes. At Ardnamurchan, we also see how the focus of volcanic activity can shift over time. Three separate volcanic centres can be distinguished here, with the second cross-cutting the first, almost destroying the evidence for its existence, and the third centre cross-cutting both earlier structures. With no further substantial eruptions occurring in this area, the landforms faithfully follow the circular structures created when the magma pulsed into position from deep within the Earth's crust.

Just to the north of Ardnamuchan lies the Island of Eigg. The Sgurr of Eigg is one of the most distinctive landmarks in the country. The geology is also unique. This peak provides an intriguing glimpse into the final phases of volcanic activity in this part of Scotland. Great thicknesses of basalt lava had already been spewed out, creating a raw and inhospitable landscape. Thick layers of conglomerate containing water-rounded pebbles lie in a valley on top of these lavas, marking the location of an ancient watercourse. As volcanic activity came to an end in the area, a nearby volcano erupted sticky lava of unusual

Sgurr of Eigg dominates the skyline. It is formed from a distinctive type of lava, known as pitchstone.

composition that flowed slowly along this ancient riverbed. Over time, this lava, known as pitchstone, cooled and developed impressive organ-like columns many tens of metres high.

Other remarkable features formed at this time are the linear occurrences of igneous rock, or dykes, that are widespread across much of Scotland. They are often prominent landscape features that run across the country for many miles. Many of the ancient volcanoes of western Scotland are associated with a large number of these dykes. They can be seen to radiate from their parent volcanoes like spokes from a wheel. The dykes are of varying thickness, but 2–3 metres from side to side is an average size. But more remarkable is their length. A dyke associated with the Mull volcano appears as far away as the north of Yorkshire, a distance of some 370 kilometres. At its furthest point from Mull, the dyke has been quarried for road stone, so the linear excavation has revealed its internal structure and, in places, the contact with the surrounding rocks. A thin skin of rock glass is now intermittently exposed on the quarry faces, where the hot magma was chilled against the cooler country rock. It is thought that emplacement of these dykes would have taken no more than five days to achieve. Those associated with the main volcanic centres can also be seen across southern Scotland into Northumberland. And, in the other direction, dykes also associated with the Mull volcano extend as far as the Western Isles.

The Mull volcano

The upland areas of Mull were part of an active volcano around 60 million years ago. The magma built up at depth under a cap of lavas that had already been erupted. Pressure mounted as the magma chamber was fed from below and the volcano began to grow. Huge eruptions followed, as the overlying strata could no longer contain the pressure cooker below. As the eruption subsided, the upper part of the volcano collapsed back into the magma chamber below, which had been partially evacuated of liquid rock during these explosive events. This cataclysm created a huge depression, known as a 'caldera', which was around 10 kilometres in diameter. We recognise this feature today as a circular structure, centred on Glen More, which defines the limits of the collapse. A later and equally impressive caldera is centred on Loch Ba, south-east of Killiechronan, confirming that this pattern of explosive volcanic activity persisted over an extended period of geological time. Erosion has subsequently planed the upper layers away, but the evidence for these dramatic events is still evident in the record of the rocks.

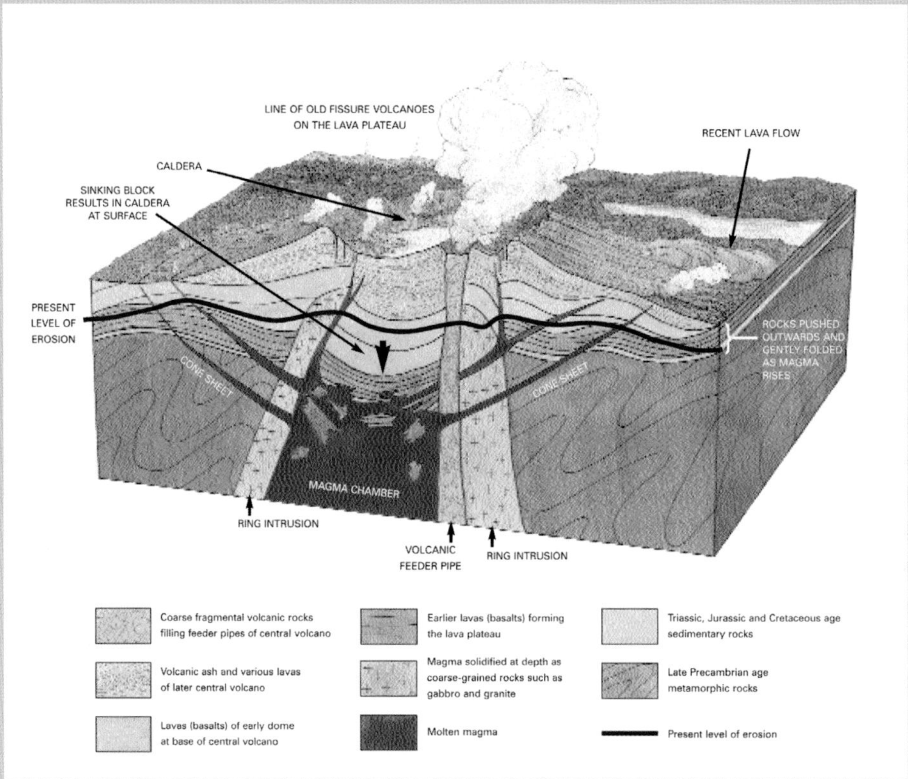

A reconstruction of the Mull volcano as it looked 60 million years ago.

LINE OF OLD FISSURE VOLCANOES ON THE LAVA PLATEAU
RECENT LAVA FLOW
CALDERA
SINKING BLOCK RESULTS IN CALDERA AT SURFACE
PRESENT LEVEL OF EROSION
CONE SHEET
CONE SHEET
ROCKS PUSHED OUTWARDS AND GENTLY FOLDED AS MAGMA RISES
MAGMA CHAMBER
RING INTRUSION
VOLCANIC FEEDER PIPE
RING INTRUSION

Coarse fragmental volcanic rocks filling feeder pipes of central volcano
Earlier lavas (basalts) forming the lava plateau
Triassic, Jurassic and Cretaceous age sedimentary rocks
Volcanic ash and various lavas of later central volcano
Magma solidified at depth as coarse-grained rocks such as gabbro and granite
Late Precambrian age metamorphic rocks
Lavas (basalts) of early dome at base of central volcano
Molten magma
Present level of erosion

The pressures required to drive molten rock such distances must have been truly colossal.

The Palaeogene was a world of subtropical to tropical conditions, with oppressive humidity. Traces of life that existed during this period in Scotland are few and far between, but occasional plant remains are preserved in thin layers between the lava flows. Many of these interleaving layers also contain thin soils that were formed as the lava surfaces were intensively weathered under the subtropical conditions. Visiting Skye on a dreich November day, it is perhaps difficult to believe that a tropical climate once prevailed in the area, but the evidence provided by the rocks is conclusive.

Fossil plants from Mull

Fossil plants were discovered at Ardtun on Mull in the mid-nineteenth century. The layers of soil and assorted sands and muds lying between

Left. Many of the lavas on Skye have red upper layers, which are interpreted as evidence of tropical weathering during Palaeogene times.

Right. Ardtun on Mull has yielded leaf fragments and pollen from a wide variety of plants including oak, hazel, plane and magnolia.

lavas from the Mull volcano have yielded a huge variety of remains of plants or their pollen, including oak, hazel, plane, magnolia and the ginkgo, also known as the maidenhair tree. This site is unique in Britain and is crucial for our understanding of plant evolution during Palaeogene times. The presence of deposits of rounded pebbles and cobbles between the lava flows indicates that fast-flowing rivers coursed across these lava moonscapes between eruptions.

This covering of greenery sprung up to clothe the lower slopes of the volcano between violent eruptions, only to be swept away as the next lava flow scarified the landscape. MacCulloch's tree stands as testament to this harsh environment. This 15-metre-high conifer was engulfed by lavas that flowed from the Mull volcano, but, remarkably, the trunk stayed upright. All that remains is the cast of this magnificent specimen tree, set into a sheer basalt cliff. The effects of Atlantic rollers and fossil collectors have seen to that. Columns of basalt have formed at right angles to the tree trunk (see p. 156).

Scotland remained a verdant tropical land for the remainder of the Palaeogene. And our journey across the globe to our current location 57° north of the Equator was almost at an end. Into the Neogene Period, the climate remained warm and humid and Scotland as a recognisable entity was soon to emerge. In the following chapter, we discover the processes that shaped the present landscape and the dramatic changes that occurred during the Ice Age.

John MacCulloch – maker of Scotland's first geological map

The man who first described the fossil tree from Mull was another geological pioneer of note. John MacCulloch was born in 1773 in Guernsey and later studied medicine at Edinburgh University. Like his illustrious predecessor, James Hutton, who graduated from the same medical school, MacCulloch had a greater interest in geology. MacCulloch is perhaps best known for his work on the first geological map of Scotland, which was published in 1836. This project was concluded over twenty years after William Smith had produced a geological map of England and southern Scotland. But MacCulloch had an immense advantage over Smith, as the map of Scotland was undertaken with support from the Treasury. He gathered the necessary information to construct the map during his extensive journeys across Scotland, often to some of its more remote corners. His *Description of the Western Islands of Scotland,* published in 1819, was well received and his lively text helped bring these further-flung parts of Scotland to a wider audience. Many of the specimens he collected are now housed in the Museum of Natural History at Oxford University.

The remains of MacCulloch's Tree on Mull. Little remains of the tree, which occupied the vertical groove in the cliff face just to the left of the figure. It is surrounded by columns of basalt formed as the rock cooled from a molten state.

CHAPTER 4

SHAPING THE LANDSCAPE

Glaciers, grinding West, gouged out
these valleys, rasping the brown
sandstone,
and left on the hard rock below – the
ruffled foreland –
this frieze of mountains, filed
on the blue air – Stac Polly,
Cul Beag, Cul Mor, Suilven,
Canisp – a frieze and
a litany.

Norman MacCaig, *A Man in Assynt*, 1969.

Rannoch Moor and the Black Mount hills. This area was a major centre of glacier growth and dispersal during the Ice Age.

The present landforms and landscapes of Scotland have largely been shaped during the time since the opening of the North Atlantic Ocean, 65 million years ago. This long and eventful period has been marked by uplift of the land surface and dramatic changes in climate that culminated in the Ice Age of the last 2.6 million years. Weathering and erosion over many millions of years have sculpted the broad outlines of the present landscape from the varied geological foundations; frost, glaciers, water, wind and the sea have all left their mark on the rocks and landforms. During the Ice Age, vast ice sheets frequently held all but the very tops of the highest mountains in their icy grip, and although some landforms can be traced back to the time before the Ice Age, much of the landscape we see today has been fashioned by the erosive power of the glaciers and their meltwaters and by the deposition of the rock debris carried by them. The glacial influence has also been fundamental in other ways. After the end of the last glaciation, 11,500 years ago, the shape of the coastline was greatly altered by changes in the relative levels of the land and sea that accompanied the growth and decay of the ice sheets. The glaciers also provided the source material for Scotland's beaches and many of our soils. In the last few millennia, a new force, human activity, has increasingly modified the natural landscape and continues to do so today. This chapter traces the story of these changes and the processes that have shaped the present landscape.

THE PALAEOGENE AND NEOGENE PERIODS – LANDFORMS AND LANDSCAPES BEFORE THE ICE AGE

FROM 65 MILLION TO 2.6 MILLION YEARS AGO

By around 415 million years ago, the geological foundations of Scotland were largely in place (see chapter 3). Over millions of years, erosion on a grand scale reduced the massive Caledonian mountains to their very roots and laid bare the once deeply buried granite intrusions, such as the Cairngorms. The land was periodically lifted up and tilted, and eventually worn down to a surface of low relief close to sea level in Mesozoic times (about 250 million to 65 million years ago) through prolonged weathering and erosion of the rocks. Later, upwelling of magma in the crust associated with the initiation of the Iceland mantle plume and the opening of the North Atlantic Ocean gave rise to uplift of the Scottish landmass during the Palaeogene and Neogene (about 65 million to 2.6 million years ago). This uplift formed the present Highlands. However, it was far from uniform and was generally greater in the west, with some tilting of the land surface towards the Moray Firth and the North Sea. The uplift was accompanied by weathering and erosion under warmer, and perhaps even wetter, climates than today. This helped to shape the gently inclined plateau surfaces and rolling hills of the eastern Highlands and Buchan, which were not greatly altered by later glacial erosion, and produced a number of distinctive features, including decomposed rock, tors and topographic basins. We can also trace the origins of the main rivers back to these pre-glacial times.

Decomposed rock

Around 55 million years ago during the Palaeogene, the climate was similar to the subtropical conditions in parts of western and southern Africa today. Under these hot, humid conditions and later under more temperate, humid conditions, the surface layers of the bedrock were deeply decomposed by chemical weathering. Lava flows on Skye were weathered, and elsewhere, rocks as hard as granite disintegrated from the effects of the warm, somewhat acidic rain and groundwater that penetrated cracks and joints. In places, the depth of disintegration was more than 50 metres. Remnants of this disintegrated rock can still be seen, especially in Buchan. Usually the rock has been transformed into a gravel-like material that can be scooped out by hand, even though it looks like solid rock; in some cases it has decomposed to clay. Sometimes, areas of more resistant rock have survived, either buried beneath the surface or standing proud as rock knobs or tors where the surrounding weathered rock has been removed. These upstanding masses of rock are typical of the granites of the eastern Highlands. It is remarkable that pockets of the soft, weathered bedrock and at least some of the tors have survived a number of glaciations, indicating that glaciers do not always erode effectively.

Differences in rock resistance to weathering also produced broad topographic basins – areas of lower ground surrounded by hills – that are a distinctive feature of the Scottish landscape. Most of these topographic basins owe their origins to the same deep chemical weathering of weaker rocks, followed by stripping of the weathered material by rivers and later by the Ice Age glaciers. Rannoch Moor is a good example of a large upland basin. It is formed in a type of granite that is less resistant to chemical weathering than the metamorphic rocks of the adjacent mountains. Many other examples, such as Howe of Cromar, Howe of Alford and Howe of Insch, occur in North-east Scotland, on basic igneous intrusions and granites that are susceptible to chemical weathering.

Weathered granite at Hill of Longhaven, near Peterhead. The disintegrated rock is crumbling into coarse sand and can be dug out by hand. The weathering has clearly etched out the joints in the granite.

Titanic masonry

Tors are the sentinels of the Cairngorm Mountains. Sub-horizontal sheet joints in the summit tors on Beinn Mheadhoin reflect the form of the pre-glacial land surface.

The rough crystalline granite has in many cases weathered into horizontal slabs, so well defined and so regular as to give the impression of titanic masonry.
Sir Henry Alexander, *The Cairngorms*, 1928

Tors are upstanding masses of rock rising abruptly above the adjacent surface of a plateau or ridge. They are lumps of locally more resistant rock that have survived weathering and erosion, usually because they have fewer joints and cracks than the adjacent rock. Tors are often thought to be pre-glacial landforms that began to form under the warm, humid climates before the Ice Age. Areas of densely jointed rock were decomposed by prolonged, deep, sub-surface chemical weathering, leaving intervening areas of less densely jointed rock more intact. The weaker, rotted rock was later eroded away during the last Ice Age, leaving the areas of sound rock upstanding as tors. However, recent studies suggest that the present tors are unlikely to be truly pre-glacial remnants, but some are nevertheless of considerable antiquity and pre-date several glacial episodes. They are thought to have formed during the last million years as a result of repeated weathering of the bedrock through shallow chemical decomposition and frost action, followed by removal of the weathered material. Weaker, heavily jointed areas of rock below the ground surface weathered faster, and as the weathered material was stripped by erosion, the more resistant rock emerged as tors. Some tors have also been modified by glaciers – blocks of rock have been toppled and dragged away by the ice – but others, particularly on Ben Avon, have undergone little modification.

Tors generally occur in areas where the landscape has evolved over long periods of time and where glacial erosion has been relatively minor. They are present on the granites of the eastern Highlands such as the Cairngorms, the Mounth hills of Lochnagar, Broad Cairn and Mount Keen, and further north-east on Bennachie and Ben Rinnes and also on Arran. In the Cairngorms, the tors on Ben Mheadhoin, Bynack More and Ben Avon are particularly spectacular. The highest rise up to 25 metres above the adjacent ground at the Barns of Bynack and Clach Bun Rudhtair. Some on Ben Avon have basins a metre or more across weathered into their surfaces, which are several hundred thousand years old. According to local custom, pregnant women used to sit in one of the potholes on Clach Ban, believing it would ease their labour.

As noted by Sir Henry Alexander, tors frequently comprise stacked layers of granite slabs. This layering, sometimes called sheet jointing, is usually parallel to the ground surface of the plateaux on which the tors occur and is believed to be due to the unloading and expansion of the granite in pre-glacial times following the removal of the overlying mass of rock by erosion. The edges of the slabs are typically rounded by granular disintegration of the granite due to frost weathering. The same layering can also be seen in the headwalls of some of the Cairngorm corries where it has been truncated by glacial erosion.

There is also an entirely different and quite unscientific explanation for tors. The Cairngorms and the eastern Grampian Mountains are well known in myth and legend for the giants who lived there. One such, a giant spectre called Am Fear Liath Mór – the Big Grey Man of Ben Macdui – is reputed to roam the plateau and follow the steps of hillwalkers. Maybe these giants used their immense strength to grab hold of lumps of the mountainside and heave them onto the mountain tops. This is certainly the legend behind the tor on Clachnaben, a very prominent feature that can be seen from the Cairn o' Mount road over the eastern Grampians. Here, the valley immediately below the tor is known locally as the 'deil's bite'. Anyone who has been in these hills in swirling mist can well imagine the shadows of mythical giants, brought readily to mind by the elongated shadow and halo effects of the particular light conditions known as a brocken spectre.

The Howe of Cromar, near Tarland, Aberdeenshire, is one of a number of basins that are characteristic of the landscape of North-east Scotland. It formed through chemical decomposition of the granitic bedrock and erosion of the weathered material. Remains of the rotted rock can still be seen in roadside sections and quarries.

Ancient land surfaces

Ancient land surfaces are distinctive features of the Scottish Highlands. The oldest example is the Lewisian landscape of North-west Scotland, which has been exhumed from beneath a 1,000-million-year-old cover of Torridonian sandstones. In eastern areas, hills, basins and major valleys, like Glen Rinnes, have emerged from beneath a cover of 400-million-year-old Devonian rocks. Hillwalkers will also have observed that there are extensive plateaux and generally uniform summit-levels in many parts of the Highlands. These 'tablelands' can be seen particularly well in the Glen Clova–Glen Esk area, the Cairngorms, the Drumochter Hills and the Monadhliath Mountains. There is no single summit-level but instead a 'staircase' of broadly level or gently sloping plateaux at different

View north-east across the Gaick plateau towards the Cairngorms (top right). The Gaick plateau is one of a number of land surfaces in the eastern Grampians and North-east Scotland that originated during pre-glacial times. These surfaces rise in a series of steps, culminating in the high plateau of the Cairngorms. The presence of weathered bedrock on its surface indicates that the Gaick plateau was not lowered significantly by glacial erosion. Locally, however, the glaciers carved through the pre-glacial watershed, forming the prominent glacial breach now occupied by Loch an Dùin in the centre of the photograph.

An upside-down landscape

Quartzite gravels at Windy Hills, near Turriff. The pebbles are very well rounded and were probably deposited in the bed of a pre-glacial river. Those in the upper part of the photograph are aligned vertically as a result of later frost action in the ground during the Ice Age.

There are several intriguing and unique gravel deposits in Buchan, which shed some light on the relative roles of rivers, and perhaps the sea, in the development of the pre-glacial landscape. At Windy Hills, south-east of Turriff, beautifully rounded and smoothed pebbles of quartzite were probably deposited by the River Ythan when its bed was some 90 metres above its present level. Further east beneath the Moss of Cruden, south-west of Peterhead, there are similar deposits of flint gravel. The origin of these latter gravels is still disputed: they have been variously interpreted as river, beach and glacial deposits. What is remarkable, however, is that at both locations the gravels are now stranded on the tops of hills because the surrounding land has been lowered by as much as 100 metres by later erosion. The flints are the last traces of a former cover of Cretaceous chalk in this area.

Along the flanks of a meltwater channel at Den of Boddam, south of Peterhead, the flint gravels were worked by Late Neolithic people around 5,000 years ago. This is one of the most extensive mining complexes of its type in Britain, with perhaps as many as 1,000 extraction pits. The material, however, was largely for local use and the extent of the working probably reflects the generally poor quality of the flints.

altitudes. They represent old land surfaces and are best preserved today in the eastern Highlands and North-east Scotland, extending from the coast inland to the Cairngorms. The surfaces were formed by uplift, tilting, weathering and limited erosion of a low-relief landscape during the Palaeogene and Neogene. The antiquity of these surfaces is indicated by the local survival of thick mantles of weathered rock and other deposits, particularly in Buchan, which is one of the best examples of a pre-glacial landscape preserved today within the glaciated regions of north-western Europe. The remarkable degree of preservation reflects the relative stability of the area and the limited erosion compared with western Scotland. Uplift was greater and rainfall higher in western Scotland, so that the landscape there was much more deeply dissected by the pre-glacial rivers and later by the Ice Age glaciers.

The origins of Scotland's rivers

Several factors have played a major part in the evolution of the drainage pattern of Scotland. Some valleys reflect the overall south-west to north-east geological grain of the Highland landscape inherited from the Caledonian mountain-building period, for example the Nairn, Findhorn, Spey and Deveron valleys directed north-east towards the Moray Firth; some, like the Glass and the lower reaches of the Dee and Don, follow the lines of ancient rock weakness; others, like the Tummel, however, are superimposed across the geological grain and may reflect later patterns of uplift. In very simple terms, greater uplift in the west and tilting of the landmass to the east during the Palaeogene and Neogene meant that the location of the main pre-glacial watershed in the Highlands, between west- and east-flowing rivers, was located near the west coast and running in a north–south direction in the Highlands. In addition, the Moray Firth has been an area of crustal sinking since the Devonian. As a result, a generally easterly drainage system became established, with rivers such as the Shin, Oykell, Nairn and Spey flowing towards the Moray Firth from both north and south of the Great Glen; rivers such as the Dee, Don, Ythan and Ugie flowing eastwards to the North Sea; and the Tay, North Esk and South Esk flowing south-east towards the Highland edge. The rivers draining to the west were shorter and had steeper courses than those flowing to the east. This contrast still exists today, with east-flowing rivers being three to four times longer than west-flowing ones. Rivers such as the Tay, Dee, Don and Spey all had their origins during the Palaeogene and Neogene and we can reconstruct approximately what the drainage pattern might have looked like. Extensive glacial breaching of the main pre-glacial watershed in the west has made it more difficult to work out the patterns of the early rivers there.

Thus, at the onset of glaciation the broad outlines of the Scottish landscape were already in place: extensive erosion surfaces in the eastern Highlands; more heavily and deeply dissected mountains in the west; a generally eastwardly directed drainage pattern in a well-established network of glens and straths often, but not always, following the underlying geological grain of the landscape; and a variable cover of deeply weathered bedrock. The form of the pre-glacial landscape played a significant part in influencing glacial activity, and the survival of many pre-glacial elements in the landscape today demonstrates that the erosive power of the Ice Age glaciers varied from place to place.

This reconstruction of the pre-glacial drainage pattern of Scotland shows the westerly position of the main watershed in the Highlands, which was caused by greater uplift of the land surface in the west and tilting to the east.

THE QUATERNARY PERIOD – THE ICE AGE IN SCOTLAND

FROM 2.6 MILLION YEARS AGO TO THE PRESENT DAY

During the many cold phases of the Ice Age (see chapter 2), mountain glaciers and ice-fields formed repeatedly in the Highlands. When the climate was at its very coldest, much larger ice sheets covered the whole of Scotland. This probably happened five or six times during the last 750,000 years. These ice sheets extended eastwards out across the floor of the North Sea, at times merging with expanded Scandinavian ice sheets, and westwards to terminate near the edge of the continental shelf west of the Western Isles. The scenery of Scotland has therefore been sculpted during the course of many glaciations. The last of these, between about 116,000 and 11,500 years ago, added the finishing touches (see p. 176).

The form of the bedrock on the floor of Coire Lagan in the Cuillin Hills gives the impression of a whale's back. As the glaciers moved over the bedrock, they removed all of the loose debris and moulded and polished the surface so that it is smooth and shiny. Scratches, known as striations, occur on the surface of the rock, formed by stones frozen into the base of the moving glaciers.

Footprints of the glaciers

The repeated formation of glaciers and their movement over the landscape has produced several distinctive types of landform. In some areas, the glaciers eroded their beds and transported the loosened rock away, like a giant conveyor. In others, they deposited this material, usually as a blanket of debris on the ground surface, but sometimes heaping it up into ridges and mounds known as moraines.

Glaciers erode the ground they flow over by processes of abrasion and quarrying. Rock debris in the basal layers of the glacier is dragged across the surface of the underlying rock, abrading it like a giant sander. This results in scratched or striated rock surfaces which have a smoothed, polished appearance; sometimes these are called whaleback forms. Excellent examples occur in the Cuillin Hills on Skye. The orientation of the striations on the rock indicates the former direction of flow of the ice.

As a glacier flows over a bedrock bump, it may weaken existing joints in the rock or create fresh fractures, particularly on the down-ice side (in the lee of the ice flow). The fragmented rock is then removed, or quarried, by the glacier. As a result, the bedrock bump becomes smoothed and abraded on its up-ice side but steep and cliffed on its down-ice side. Such asymmetric forms are called roches moutonnées because they were thought to resemble eighteenth-century wigs that were slicked down with mutton fat. They can vary in size from features a metre or so high to hills a few hundreds of metres high. There are good examples of the former in Glen Nevis and near Dulnain Bridge, where there is an interpretation site; and of the latter, between Braemar and Ballater and south of Aviemore near Loch an Eilein and Kincraig.

Coir' a' Ghrunnda and Sgùrr Alasdair, with the Inaccessible Pinnacle in the background.

The Cuillin Hills: a classic landscape of mountain glaciation

The whole opening of [Coir' a' Ghrunnda] which must have formed the channel of the glacier when it was present, is ground and shaven over in such a way, as to leave not a single protuberance in the direction in which the ice-flood must have passed over it.
James Forbes, *Edinburgh New Philosophical Journal*, 1846

The Black Cuillin of Skye is a classic landscape of mountain glaciation in Britain. The main peaks are conical, riven by the effects of ice and frost and are more akin in form to the Alpine peaks of Switzerland and Austria than the mountains in the Scottish Highlands. The ridges joining the peaks are textbook arêtes, narrow and with precipitous sides formed by the action of glacial erosion and frost-shattering in the corries on either side of them. It is likely that these peaks, and the ridges which join them, were nunataks (mountains standing above the surface of the ice) during the last ice sheet glaciation and later during the Loch Lomond Readvance when the mountains supported a local icefield. They were therefore subject to intense frost-shattering, to which the gabbro bedrock is very susceptible. In addition, whole mountainsides have been sliced off to form truncated spurs, and gaps have been etched out through the ridges by frost-shattering along the lines of weaker basalt dykes.

Many corries, with their classic textbook form, have been gouged out of the mountains. Some have been left hanging by more intense glacier erosion of the slopes below them, for example Coire Lagan and Coir' a' Ghrunnda along Glen Brittle. In addition, the adjacent glens have been enlarged and deepened to form troughs radiating out from the centre of the massif (e.g. those of Lochs Slapin, Ainort, Eynort, Sligachan, Harport and Brittle). Loch Coruisk occupies a classic glaciated valley, with its floor 30 metres below sea level and an ice-polished rock bar separating it from the current coastline. The southern extension of Glen Sligachan, between the main part of the Black Cuillin and Bla' Bheinn, is an example of both valley enlargement by ice streams and breaching of a watershed between two pre-glacial glens.

The Lairig Gartain is a spectacular glacial breach linking Glencoe and Glen Etive. It shows particularly clearly the parabolic cross-section typical of Scottish glens.

The junction between the area covered by ice and that exposed to the frost can be easily seen in many places and forms a classic periglacial trimline (the line marking the boundary between frost-shattered rock on the upper slopes and ice-moulded rock below). Below the trimline, there is no better sign of the moulding effects of the ice than the roches moutonnées and polished whaleback rocks complete with striations on the floors of the corries, most notably Coire Lagan and Coir' a' Ghrunnda, and at Loch Coruisk. As well as the presence of superb landforms of glacial erosion, there are excellent examples of moraines formed by the Loch Lomond Readvance glaciers in the Cuillin, notably the end moraines extending out from the corries along Glen Brittle and the distinctive expanse of hummocky moraine around Sligachan.

The Cuillin are also important historically in the study of mountain glaciation. In 1846, soon after the introduction of the glacial theory in Scotland, the pioneering Scottish glaciologist, James Forbes, published an account of the Cuillin, one of the first detailed regional studies of glacial landforms in Britain. This included the first map of the area, identifying the ice-moulded bedrock at Coire Lagan, Coir' a' Ghrunnda and Loch Coruisk, and the superb end moraine ridge below Coir' a' Ghrunnda.

Corries are probably the best known and most easily recognisable glacial landform in Scotland. These mountain hollows have a steep headwall and a more gently sloping floor. Although they have been buried and perhaps modified by ice sheets, their distinctive forms were largely produced during periods of less extensive glaciation when they were occupied by small mountain glaciers. Their steep headwalls were shaped by frost shattering of the exposed bedrock above the glaciers and by rockfalls; the lower headwalls and floors were eroded beneath the glaciers. Some corries contain small lochs that have formed in rock basins deepened by the ice or where the outlets have been dammed by moraines. The distribution of corries extends from Hoy in Orkney to the Southern Uplands, but their density is greatest in the western and northern Highlands, reflecting higher snowfall there during periods of mountain glaciation. Many examples occur in the Black Cuillin of Skye, in the mountains of Torridon and Applecross, and in the Cairngorms. Where corries occur close together, they are separated by narrow, rocky ridges, known as arêtes, formed by a combination of frost shattering and glacier erosion.

Where bigger, faster-flowing glaciers flowed along the glens formed by the pre-glacial rivers, they carved out deep glacial troughs, much deeper than normal river valleys and with wider floors. These are often described in textbooks as being 'U-shaped' valleys, although most Scottish glens have a more open, parabolic shape. Good examples of glacial troughs include Glencoe, Glen Docherty in Wester Ross, Glen Clova in Angus, Glen Muick and Glen Callater in the Mounth and Glen Avon and Gleann Einich in the Cairngorms. Spurs along the sides of the glens have frequently been sliced off, so that glacial

Moffatdale, in the Southern Uplands, is a glacial trough excavated along the line of a fault. Glacial erosion has deepened the valley and formed a series of truncated spurs along both flanks. The tributary valley of the Tail Burn, draining Loch Skeen, forms a hanging valley from which the Grey Mare's Tail waterfall spills down into Moffatdale some 200 metres below.

troughs are usually relatively straighter than the pre-glacial valleys would have been. The Devil's Point above the Lairig Ghru in the Cairngorms is a classic example of such a truncated spur, but there are good examples in many other glens.

Glacial troughs often contain deep rock basins excavated by the ice, which have since become occupied by lochs; well-known examples include Loch Ness and Loch Morar, which reach depths of 160 metres and 301 metres below sea level, respectively, demonstrating the power of the glaciers. Elsewhere, the floors of the troughs have been filled by postglacial sediments, so they appear to be flat; a good example can be seen from the viewpoint on the A832 near Corrieshalloch Gorge, looking down Strath More towards Loch Broom. All along the western seaboard of the Highlands, the sea has drowned the lower reaches of glacial troughs, forming spectacular fjords between Loch Eriboll and Loch Linnhe and between Loch Fyne and the Clyde. Loch Hourn, Loch Duich, Loch Broom and Loch Laxford are particularly good examples.

Hanging valleys are another distinctive feature of glacial erosion. They occur where a main trunk glen has been deepened more than its tributaries. The Lost Valley in Glencoe is a good example. The lips of hanging valleys are often the sites of spectacular waterfalls such as Steall in Glen Nevis and the Gray Mare's Tail in Moffatdale.

Glacial troughs frequently follow the lines of major faults or other lines of geological weakness. This is the case in the North-west Highlands

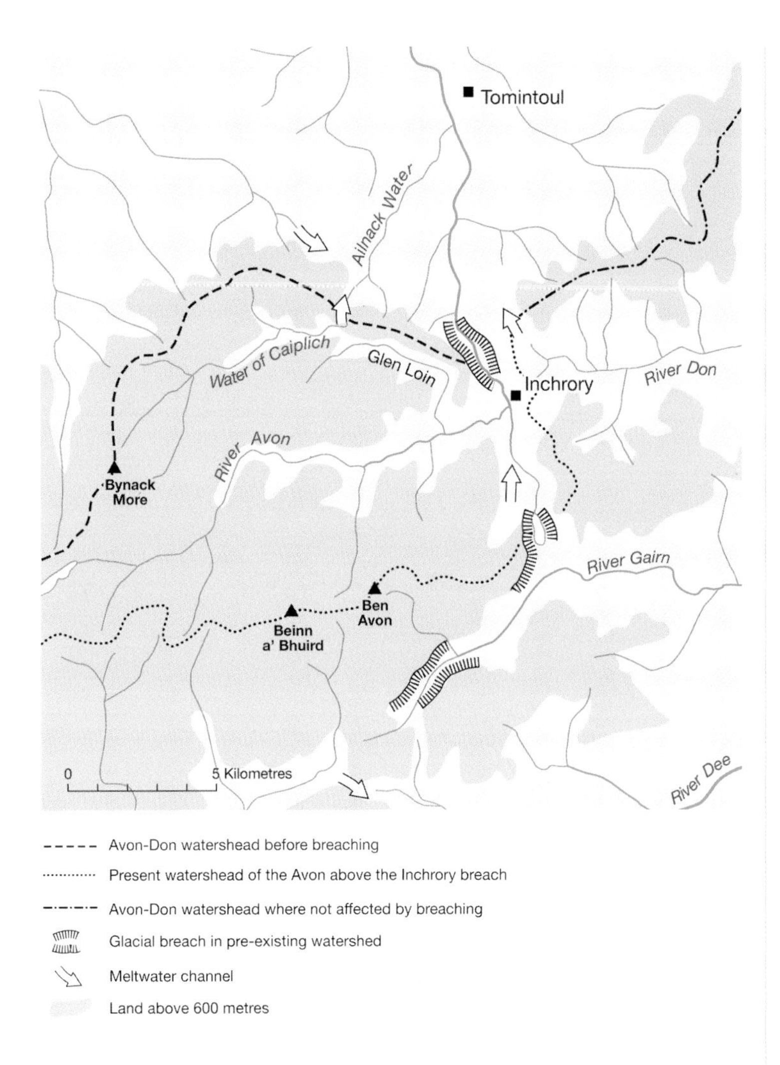

Glacial diversion of drainage in the eastern Cairngorms. The upper River Avon turns abruptly northwards through a glacial breach at Inchrory and flows towards the Spey. It was formerly part of the River Don, but glacial erosion lowered its northern watershed to such an extent that the postglacial course of the river was diverted. An abandoned valley indicates the eastwards continuation of the former course of the river towards the Don. The Water of Caiplich has also been diverted. It turns sharply north through a spectacular meltwater gorge to join the Water of Ailnack, its former outlet to the Avon along Glen Loin having been blocked by ice.

where, for example, Loch Maree, Loch Broom, Loch Laxford, Glen Shin and Glen Cassley all take their north-west to south-east alignment from the structural grain of the Lewisian rocks. The Great Glen is a large glacial trough excavated along the line of its namesake fault. In the Southern Uplands, the trough of Moffatdale also follows the line of a fault. In the Cairngorms, the major troughs of Gleann Einich, the Lairig Ghru and the Lairig an Laoigh all follow zones of hydrothermal alteration in the granite, which weakened the rock, and allowed it to be readily exploited by the pre-glacial rivers and then by the glaciers.

Where the glaciers had no existing glens to follow, they created new ones by carving through the pre-glacial watersheds. There are numerous examples of such glacial breaches, particularly in the north and west of Scotland, which now provide important transport corridors to the west coast – for example, along Glencoe, Glen Shiel and Glen Docherty. In the South-west Highlands a very striking radial pattern of glens and loch-filled rock basins indicates the movement of ice streams out from major ice accumulation centres over Rannoch Moor and the mountains to the south. Some of these ice streams breached the pre-glacial watersheds, which is how Loch Lomond was formed. In the Cairngorms, the watershed between the streams flowing to the Spey and those flowing to the Dee has been breached, but the rivers still flow in their original pre-glacial directions; the best examples are the Lairig Ghru and the Lairig an Laoigh. In some areas, however, the breaching of watersheds has completely altered the directions of river flow. For example, the present headwaters of the River Feshie formerly flowed east to join the Dee, but now turn sharply to the west to join the Spey; the Tarf Water was formerly a tributary of the Dee

Loch Lomond: glaciers create a new landscape

During successive glaciations, the western Highlands formed an important ice accumulation centre and this is reflected in the pattern of glacial troughs, breached watersheds and loch-filled rock basins eroded by the glaciers radiating out from the Rannoch Moor–Crianlarich area. Examples include Loch Treig to the north; Glencoe to the west; Glen Etive, Loch Awe and Loch Fyne to the south-west; Loch Lomond to the south; and Glen Dochart, Loch Tay, Loch Rannoch and Loch Ericht to the east and north-east. Loch Lomond illustrates particularly well the effectiveness of these glaciers in eroding the landscape and creating new glens by the process of watershed breaching.

The pre-glacial rivers of the Loch Lomond area flowed south-east from Glen Luss and Glen Douglas towards the Forth valley; east from Inveruglas towards the Teith valley via the through–valley now occupied by Loch Arklet; and north-east from Strath Dubh-uisge towards Glen Dochart. Over the course of successive glaciations, powerful ice streams flowed south from an ice dome over the Ben Lui area and excavated the trench that is now occupied by Loch Lomond. The glaciers carved through the pre-glacial watersheds, lowering them by as much as 600 metres and forming the narrow northern section of the Loch Lomond trough. They eroded down into their beds, forming deep rock basins and leaving the deepest part of the loch now some 180 metres below sea level. At the southern end of the loch, where the valley widens, the ice was able to spread out more and its power to erode was diminished. This explains the shallower depth of this part of the loch and the survival of the islands.

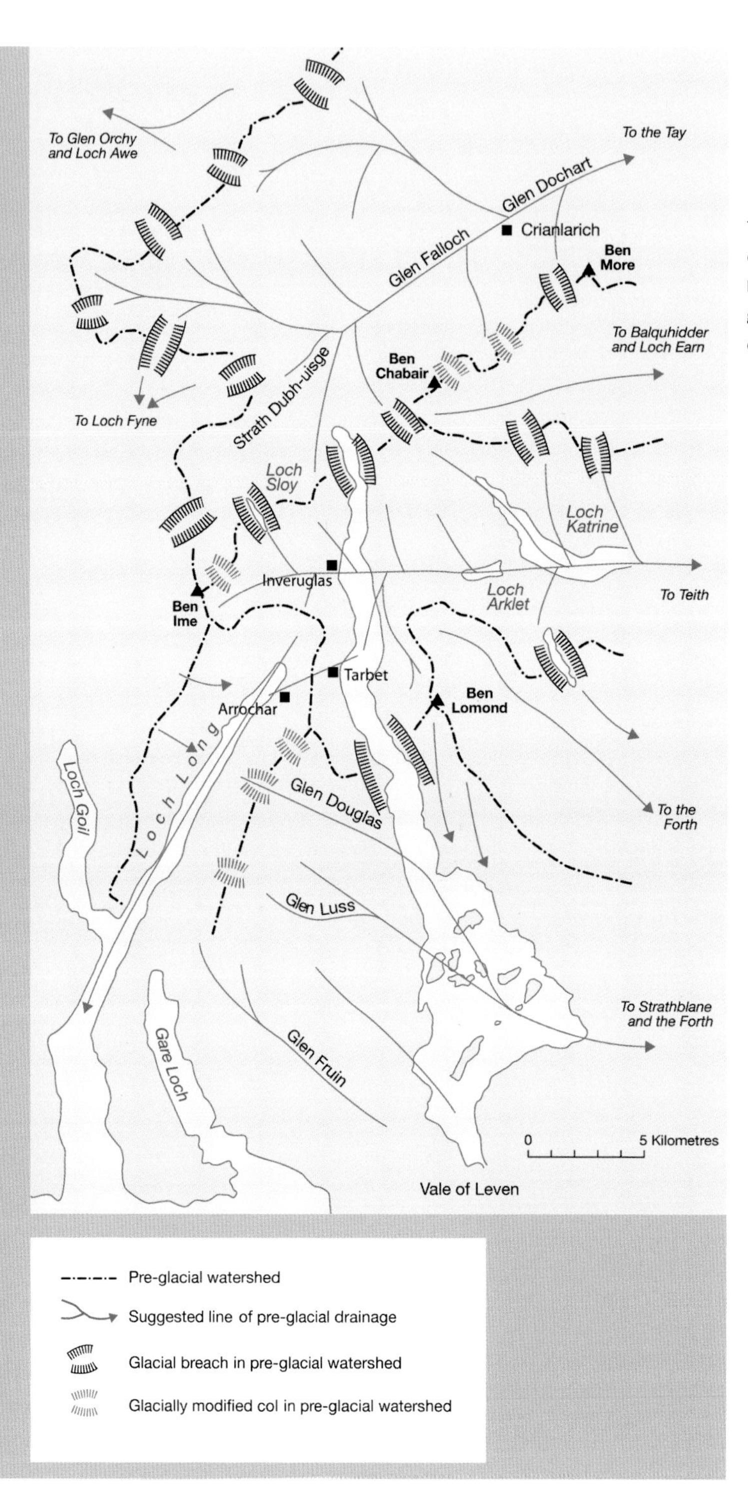

The pre-glacial drainage pattern of the Loch Lomond area has been significantly reshaped by glacial erosion and the breaching of the pre-glacial watersheds.

but has been diverted into the Tilt and now drains towards the Tay; and in the eastern Cairngorms, the River Avon has been diverted north from the Don to join the Spey.

Linear glacial erosion is not only a feature of the uplands. Powerful ice streams also deepened lowland valleys at a time when sea level was lower and the glaciers extended offshore. For example, deep basins were excavated in the floor of the Forth valley, reaching a depth of 206 metres below sea level east of Kincardine; and parts of the Tay and Clyde estuaries descend to 70 metres below sea level. The Beauly, Cromarty and Dornoch Firths have similar characteristics; for example, the Cromarty Firth has a new entrance carved by glaciers between the North and South Sutors, compared with its earlier course through Nigg Bay and out to sea in the Dornoch Firth. In effect, all these firths are glacier-deepened valleys later partially infilled by substantial accumulations of glacial and post-glacial sediments and then inundated by the sea. The glaciers also carved troughs and deep basins beyond the present coast. These were later flooded by the sea and now separate the islands of Skye, Mull, Islay, Jura and Bute from the adjacent mainland. In some of these troughs, the power of the ice weakened downstream and deep rock basins terminate in shallow rock bars at their seawards ends. Sometimes these are above present sea level, as at Loch Morar, but more often below, as at Loch Hourn and Loch Nevis.

Zones of landscape modification by glacial erosion. 1. Minimal glacial erosion. 2. Limited erosion, generally confined to minor ice moulding on lower ground and some interfluves. 3. Selective glacial erosion in the uplands, occasional corries, and ice-moulded lowlands with streamlined bedrock and crag-and-tail landforms. 4. Comprehensive transformation of valleys to troughs in the uplands and extensive ice moulding in the lowlands. 5. Landscapes dominated by glacial erosion with ice moulding and glacial roughening up to high summits; mountains and plateaux heavily dissected by glacial troughs, fjords and corries; and areal scouring on lowlands.

Glaciers shape the landscape

Glacial erosion has formed several characteristic types of landscape. The power of the ice was most effective where the glaciers were fast-flowing and warm-based (see chapter 2). These conditions occurred where steep, narrow glens channelled the ice flow, where the temperature and snowfall were relatively high and where the ice was thick. The intensity of landscape modification by glacial erosion was therefore greatest in the west and north and least in the south and east; this is reflected in the pattern of landforms and landscapes.

Extensive erosion by warm-based ice sheets produced landscapes of areal scouring. They are typified by the ice-scraped and ice-moulded surfaces of Lewisian gneiss in Sutherland and the Western Isles, especially in Harris and the south

Landscape of areal scouring. Suilven rises above an ancient platform of Lewisian gneiss that has been extensively scoured by warm-based ice sheets. The resulting mosaic of low rocky hills and lochans follows the structural grain of the bedrock.

of Lewis, which form a distinctive topography of low ice-scoured hills and lochans, sometimes very appropriately described as knock and lochan topography. The depth of erosion has probably not been great. Instead, the ice has emphasised the irregularities in the underlying bedrock, which are closely reflected in the topography.

In the northern and western Highlands, a combination of ice sheet and mountain glacier landforms dominates the landscape, including corries, glacial troughs, rock basins and fjords. The ice has heavily scoured much of the lower ground, and ice-roughened surfaces even extend up to the summits in Knoydart and Ardgour. The mountains are also heavily dissected by glacial breaches and troughs, and the pre-existing watersheds have been lowered by powerful ice streams. The glens are therefore highly interconnected, isolating the main mountain massifs and ridges. This is most evident in the mountains between Knoydart and Assynt. As glacial erosion was most intense in the western Highlands, the glens and straths there have been significantly steepened and over-deepened, and many have rock basins excavated in their floors. There is often a spectacular contrast between the more open glens and straths to the east, such as Glen Moriston and Strath Bran, and the deep, narrow

The Cairngorms: an old landscape modified by glaciers

The Cairngorm Mountains are a classic landscape of selective glacial erosion. The extensive plateau surfaces with tors have generally been little modified by glacial erosion and contrast sharply with the glacial trough of Glen Avon. This contrast is thought to reflect the former presence of thicker, faster flowing, warm-based glaciers in the glens, whereas the ice covering the plateaux was cold-based and relatively inactive.

The Cairngorms provide unique insights into long-term processes of mountain landscape evolution and environmental change over the last 400 million years of geological time. Broadly speaking, the massive form of the Cairngorms reflects the character of the granite mass intruded into the surrounding Dalradian rocks around 427 million years ago. The properties of the granite, and in particular the jointing and structural weaknesses, have strongly influenced the patterns of weathering, erosion and landscape evolution. The overall form of the current landscape, although modified by glacial erosion, is essentially an inherited one.

As the mass of Cairngorm granite was cooling within the Earth's crust, hot fluids moved along joints and fractures and chemically altered the adjacent rock, forming zones of weakness. The granite was exposed at the surface by erosion around 400 million to 390 million years ago and the major landscape features of the Cairngorms began to form soon after as processes of weathering and erosion selectively exploited the zones of weaker rock. Over time, the pre-glacial rivers developed the precursors of the present glens along these zones of weakness. During the Neogene, Palaeogene and early Quaternary, between about 65 million and 1 million years ago, episodes of uplift, deep weathering and stripping of the decomposed rock shaped the overall form of the relief into a series of plateau surfaces, with tor-capped summit domes where the granite is more massive and has fewer joints. Later, during successive phases of glaciation in the Quaternary, erosion by mountain glaciers and ice sheets selectively deepened and extended the pre-glacial glens, forming the glacial troughs of the Lairig Ghru, Gleann Einich and Glen Avon.

The relict landforms that originated before the Ice Age or during its early stages are unusual for their scale of development in a glaciated mountain area; they include tors, weathered bedrock and plateau surfaces. These features stand in sharp contrast with the cliffs of the adjacent corries, breached watersheds and deep glacial troughs. Together they form an outstanding example of a landscape of selective glacial erosion and show how the erosive effects of the Ice Age glaciers were very effective in particular areas but minimal in others according to whether the glaciers were warm-based or cold-based. For example, it has been estimated that the deepest glacial excavation amounted to about 500 metres in An Garbh Choire between Braeriach and Cairn Toul, whereas adjacent plateau areas may have been lowered by only a few metres or tens of metres. Such a landscape is exceptional in western Europe and is comparable with parts of Baffin Island, Canada.

glens like Glen Shiel and Glen Docherty that lead down to the west coast. The road to Ullapool also drops sharply down from the Dirrie More into the trough of Strath More leading to Loch Broom. Landscapes of alpine-type mountain glaciation are dominated by a high density of corries, arêtes and valley-glacier heads. They are spectacularly developed, for example, in the Black Cuillin of Skye, the northern mountains of Arran, the Rum Cuillin and the An Teallach area of Wester Ross.

In the central and eastern Highlands, the landforms of glacial erosion contrast sharply with the extensive older plateau surfaces into which they are cut. This type of landscape is described as one of selective glacial erosion. Such landscapes occur where powerful, warm-based ice streams excavated deep glacial troughs, but adjacent watersheds and plateau surfaces remained little modified under a cover of cold-based ice. These landscapes extend eastwards from Ben Alder and Creag Meagaidh to the Drumochter Hills, the Cairngorms and the Angus glens. The best example is the Cairngorms, where tors and deeply weathered rock on the plateau surfaces exist only a short distance from the deep glacial troughs of Glen Avon, Gleann Einich and Glen Geusachan.

Ice-moulded lowlands comprise streamlined hills and spurs and areas where the scouring action of the ice movement has emphasised the grain of the bedrock. Good examples occur throughout the Lothians and Fife, where the streamlined hills and ice-moulded topography are generally aligned west–east; for example at Dalmahoy Hill in West Lothian and the Garleton Hills in East Lothian. Often the more resistant volcanic plugs have protected weaker sedimentary rocks on their eastern lee sides from the force of the ice, forming crag-and-tail landforms. Edinburgh Castle Rock (crag) and the Royal Mile (tail) is a classic example. Others include North Berwick Law in East Lothian and the Bathgate Hills in West Lothian. Stirling Castle and the Wallace Monument sit on top of similar landforms developed on the resistant dolerite of the Midland Valley Sill. In the Borders, in the area north of Hawick, the ice moulding is boldly emphasised by the orientation of the grain of the Silurian bedrock, and again in Knapdale, Argyll, by the structural complexities of the Dalradian rocks.

In some places, the glaciers had very little effect. We can recognise such areas from the presence of relict features, such as deeply weathered bedrock and tors, which survived under cold-based ice. This type of landscape is particularly well displayed in Buchan, as described above.

Traces of past environments

The climate records from the ocean floor and ice cores indicate that Scotland experienced eight major glacial–interglacial cycles in the last 750,000 years. Successive glaciations added cumulatively to the shaping of Scotland's landscapes but also removed most, if not all, traces of earlier events on land. Therefore much of the evidence seen today relates to the last glacial period, and particularly its later part between about 33,000 and 11,500 years ago, and the period after the ice; relatively little is known about prior events. Nevertheless, there are a few sites in Scotland where earlier deposits have fortuitously survived. Typically they display organic layers of peat or soil, buried beneath later glacial deposits. Pollen, plant and insect remains preserved in these organic deposits provide a wealth of environmental information about past vegetation and climate. Sites with such records are largely confined to the areas that experienced limited glacial erosion in parts of North-east Scotland, the Central Lowlands and the Northern and Western Isles. In Buchan, a variety of deposits have provided some of the longest records of changing Ice Age environments in Scotland. One of the most important sites was discovered in a quarry at Kirkhill, north-west of Peterhead. Here, a unique set of deposits revealed remarkable evidence for three separate glaciations, two interglacials and at least four episodes of periglacial conditions. Sadly, the key sections were later buried by landfill and this crucial record can now only be accessed by boreholes in an adjacent field.

The last time that the climate was broadly comparable to that of today – it may even have been slightly warmer – was during the last interglacial, between about 128,000 and 116,000 years ago. For example, a peat deposit on the northern mainland of Shetland contains the remains of pine trees that were growing there at that time.

A series of climate fluctuations then followed, with birch and birch-pine woodland present at times, before a marked downturn in climate around 75,000 years

Resistant lowland hills

The landscape and landforms of the Central Lowlands reflect very clearly the juxtaposition of resistant volcanic rocks and weaker sedimentary rocks, and the action of differential weathering and erosion by frost, rivers and glaciers over many millions of years. Thick sequences of sediments accumulated during the Devonian and Carboniferous (416 million to 299 million years ago) and zones of weakness in the crust allowed magma to well up towards the surface at various times, giving rise to volcanoes, fissure eruptions and intrusions of igneous rocks. The more resistant igneous rocks, in the form of lavas, sills and volcanic plugs, form the conspicuous hill masses and individual hills, or laws, which characterise the Central Lowlands today. They include the Devonian lavas of the Pentland Hills, Sidlaw Hills and Ochil Hills and the Carboniferous lavas of the Garleton Hills, Bathgate Hills, Renfrewshire Hills, Kilpatrick Hills, Gargunnock Hills and the Campsie Fells. The spectacular volcanic vents of Arthur's Seat, Edinburgh Castle Rock, North Berwick Law and Dumbarton Rock also formed during the Carboniferous, as did other intrusions – the Midland Valley Sill, Salisbury Crags, Traprain Law and numerous vents and plugs in Fife, including West and East Lomond.

These igneous rocks were tilted later by movements in the Earth's crust and modified by erosion. The erosion had two effects: it removed the volcanic cones, leaving the more resistant feeder necks and sills as upstanding hills; and it lowered the neighbouring weaker sedimentary rocks to a greater extent than the harder lavas and sills. As a consequence, conspicuous scarp slopes formed by the more resistant volcanic rocks often rise sharply above the less resistant sedimentary rocks that underlie the lower ground. A particularly good example is the Midland Valley Sill, which extends over 1,600 square kilometres in central Scotland. Where it has been exposed by erosion of weaker overlying rocks, it now forms a striking landscape feature, as in the Lomond Hills and Stirling area.

Many of these volcanic hills have been moulded by the ice sheets, forming streamlined spurs and crag-and-tail landforms.

North Berwick Law, in East Lothian, is a good example of a crag-and-tail landform. The resistant volcanic plug has protected the weaker sedimentary rocks on its lee side from erosion, leaving a streamlined ridge or 'tail' seen to the left.

Reconstructing the past

Where sediments of different origins are deposited one on top of another, they provide a record of changing surface processes in the past; for example, river gravels overlain by glacial deposits may indicate a change from ice-free to glacial conditions. If the deposits are separated by a layer of peat or soil, this may reveal a significant time gap between them and, very importantly, the organic material may provide radiocarbon dates and information on the environmental conditions at the time of its accumulation.

If sediments and vegetation remains accumulate continuously over long periods of time in lochs and peat bogs, they build up a progressive record of past environmental changes. Cores of sediment can then be extracted from these environmental archives and analysed for the pollen, plant remains and insect remains preserved in the different layers. The ages of the different layers are commonly established using radiocarbon dating. Such archives tell us a great deal about past environmental changes and are examined in more detail later in this chapter.

An old gravel pit at Teindland in Moray shows a buried soil dating from the end of the last interglacial, around 120,000 years ago. The sloping black layer underlain by the pale layer in the centre of the photograph is a type of soil known as a podzol. Pollen grains preserved in the soil indicate the presence of hazel and alder woodland, followed later by pine woodland, heathland and finally sparse grassland. The upper part of the soil has been disturbed by intense freezing under tundra conditions, which, together with the vegetation changes, reveals a deteriorating climate and the onset of the last glacial period. The glacial deposits overlying the soil indicate the later presence of a glacier in the area.

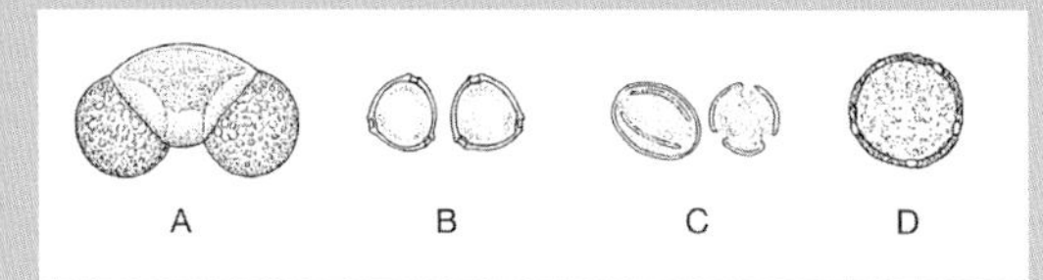

Pollen grains viewed under a microscope. Different types of plant can be identified from the characteristics of the pollen they produce. Pollen grains released by plants and preserved in lake sediments and bogs can be retrieved by coring and analysed to provide a record of vegetation changes over time. (A) Pine. (B) Birch. (C) Oak. (D) Elm.

Shetland's ancient woodland

Layers of buried peat at two important locations in Shetland give some clues about the climate and vegetation during the middle and later stages of the last interglacial. Such deposits are rare in Scotland and are of great significance to our understanding of environmental changes during the latter stages of the Ice Age. In particular, they show that around 120,000 and 100,000 years ago parts of Mainland Shetland were vegetated with no glaciers present. At Fugla Ness on the exposed west coast of Northmaven, the plant remains in a bed of peat buried beneath some 4 metres of later periglacial and glacial deposits suggest a less extreme climate than today around 120,000 years ago, with pine trees growing there, and shrub species present that are now found only on the European mainland or in the west of Ireland. At Sel Ayre, on the west coast of the Walls peninsula, a younger peat bed contains plant remains which show that heath and grassland were present there around 100,000 years ago. At both sites, we can tell from the thick accumulations of periglacial slope deposits which overlie the peat beds that the climate subsequently became colder; these deposits were later buried by till when the last ice sheet covered Shetland.

Glaciers shape the landscape

Glacial erosion has formed several characteristic types of landscape. The power of the ice was most effective where the glaciers were fast-flowing and warm-based (see chapter 2).

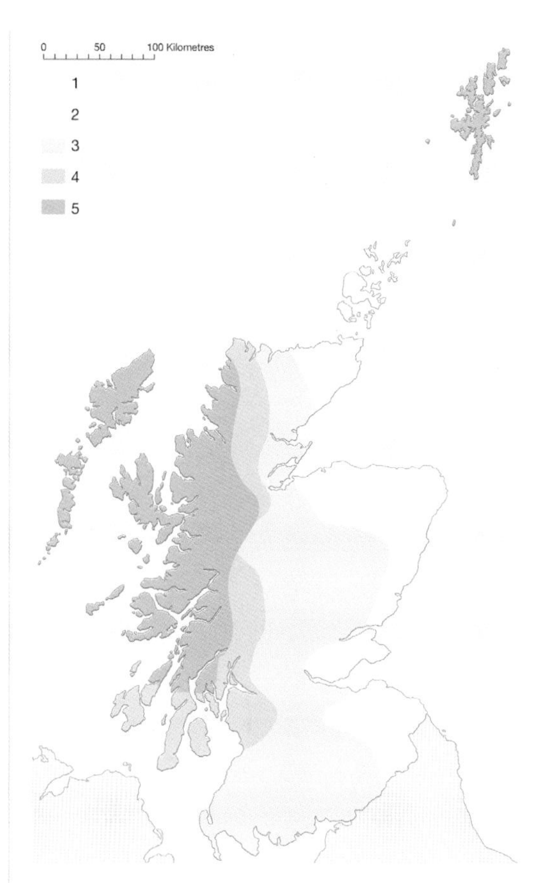

Landscape of areal scouring. Suilven rises above an ancient platform of Lewisian gneiss that has been extensively scoured by warm-based ice sheets. The resulting mosaic of low rocky hills and lochans follows the structural grain of the bedrock.

Above. Herds of woolly mammoth, reindeer and woolly rhinoceros roamed the tundra landscape of Lowland Scotland before the advance of the last ice sheet after about 33,000 years ago.

Opposite. Coastal erosion has exposed the remains of an ancient pinewood in a hollow at Fugla Ness in Shetland. The black layer (1) below the figure is a bed of peat with wood fragments from pine trees. These date from the last interglacial, 120,000 years ago, and grew on the surface of the deposits from an earlier glaciation. The peat is overlain by a pale brown layer (2) of debris washed down from the adjacent slopes under periglacial conditions. The stony, reddish-brown layer (3) at the top is glacial till deposited by the last ice sheet.

Ice Age bones

The limestones of the Inchnadamph area in Assynt contain Scotland's most important cave systems. These occur in the drainage basins of the Traligill River and the Allt nan Uamh. Both streams flow underground, except in very wet weather. Uamh an Claonite, the longest cave in Scotland, comprises 3 kilometres of underground passages. The caves are of particular interest for their calcite formations and animal remains. These provide a unique record of past environmental conditions extending back beyond the last glaciation, as well as the former presence of mammals now extinct in Scotland.

Water draining into the limestone has dissolved the rock along joints and fractures, forming cave passages. The oldest caves probably formed about a quarter of a million years ago. Glacial erosion later deepened the valleys, leaving these early caves abandoned at a higher level. New, lower cave passages formed in association with the lowered water tables. During warmer interglacial periods, water dripping into the cave passages formed stalactites and stalagmites. Dating of these formations, by a technique based on the rate of decay of uranium isotopes, indicates periods in the past when the area was ice-free.

Excavations in some of the caves in the Allt nan Uamh glen have uncovered a wealth of fossil bones. The remains from these Bone Caves indicate that herds of reindeer roamed across a cold, dry tundra landscape before the last ice sheet expanded after about 33,000 years ago. Brown bear, wolf, arctic fox, arctic lemming and northern vole were also present. These animals probably returned during the closing stages of the last glaciation when the glaciers were retreating. A particularly interesting discovery from the caves is a polar bear skull dating from the time of the last glaciation. Several species – brown bear, reindeer, northern lynx and wolf – were present at the start of the present interglacial 11,500 years ago but are now extinct in Scotland.

Today there is an attractive walk along the partly dry limestone glen of the Allt nan Uamh, leading up to the entrances of the Bone Caves, now perched above the glen floor. The Assynt caves, however, are only accessible to experienced and properly equipped cavers.

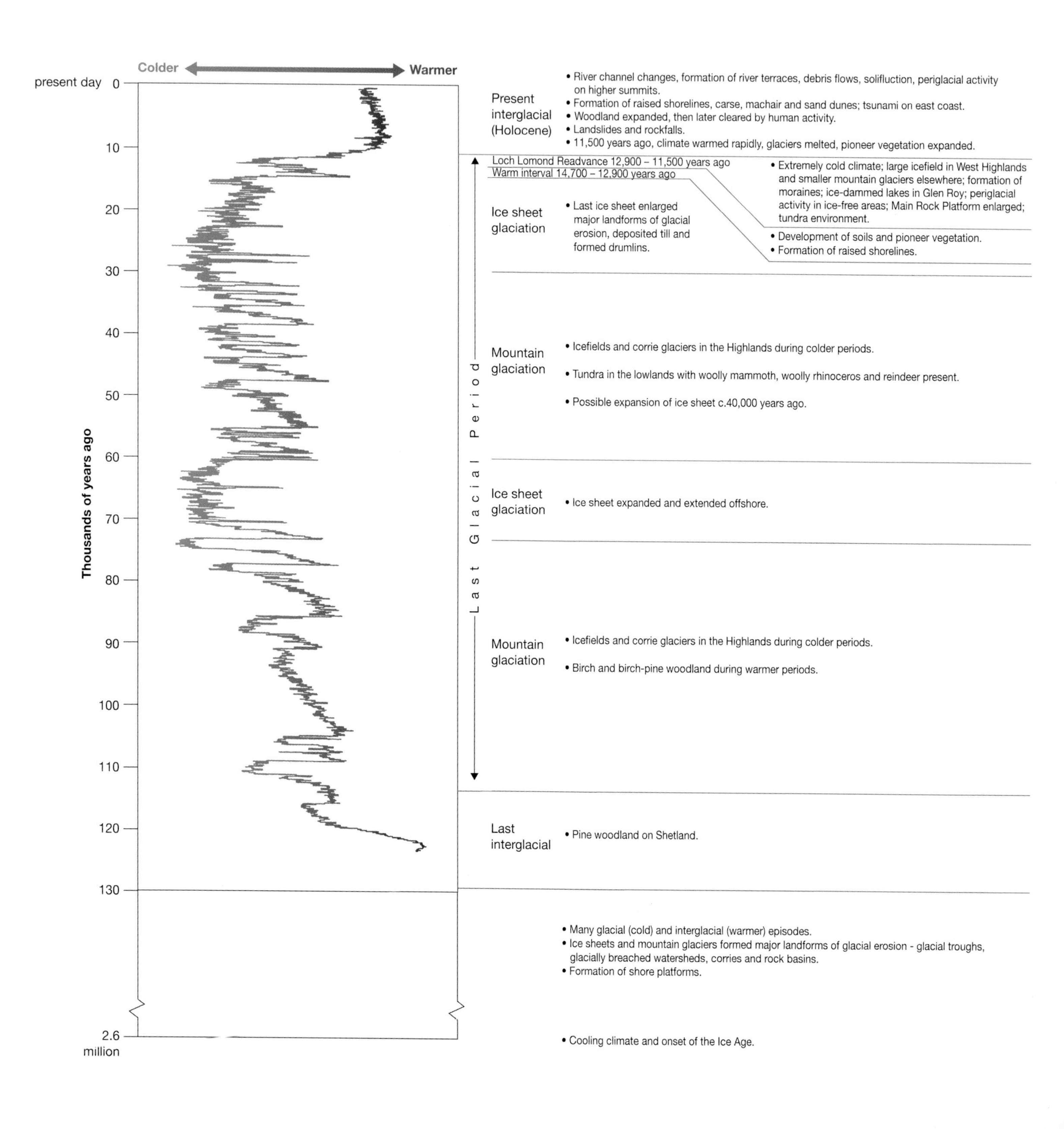
Colder
Warmer
present day
0
10
20
30
40
50
60
70
80
90
100
110
120
130
2.6 million
Thousands of years ago
Last Glacial Period
Present interglacial (Holocene)
• River channel changes, formation of river terraces, debris flows, solifluction, periglacial activity on higher summits.
• Formation of raised shorelines, carse, machair and sand dunes; tsunami on east coast.
• Woodland expanded, then later cleared by human activity.
• Landslides and rockfalls.
• 11,500 years ago, climate warmed rapidly, glaciers melted, pioneer vegetation expanded.
Loch Lomond Readvance 12,900 – 11,500 years ago
Warm interval 14,700 – 12,900 years ago
• Extremely cold climate; large icefield in West Highlands and smaller mountain glaciers elsewhere; formation of moraines; ice-dammed lakes in Glen Roy; periglacial activity in ice-free areas; Main Rock Platform enlarged; tundra environment.
• Development of soils and pioneer vegetation.
• Formation of raised shorelines.
Ice sheet glaciation
• Last ice sheet enlarged major landforms of glacial erosion, deposited till and formed drumlins.
Mountain glaciation
• Icefields and corrie glaciers in the Highlands during colder periods.
• Tundra in the lowlands with woolly mammoth, woolly rhinoceros and reindeer present.
• Possible expansion of ice sheet c.40,000 years ago.
Ice sheet glaciation
• Ice sheet expanded and extended offshore.
Mountain glaciation
• Icefields and corrie glaciers in the Highlands during colder periods.
• Birch and birch-pine woodland during warmer periods.
Last interglacial
• Pine woodland on Shetland.
• Many glacial (cold) and interglacial (warmer) episodes.
• Ice sheets and mountain glaciers formed major landforms of glacial erosion - glacial troughs, glacially breached watersheds, corries and rock basins.
• Formation of shore platforms.
• Cooling climate and onset of the Ice Age.

Opposite. A record of climate change for the last 130,000 years from the Greenland Ice Sheet and a summary of events in Scotland. In comparison with the relatively stable conditions of the present interglacial (the last 11,500 years), the climate in the past has fluctuated frequently and rapidly.

ago when an extensive ice cover developed over Scotland. Glaciers then remained, in one form or another, particularly in the Highlands, until 11,500 years ago. Much of Lowland Scotland was ice-free for a period of time before about 33,000 years ago, but harsh climate conditions and open tundra vegetation prevailed. Fossil mammal remains from this cold period have been recovered from a number of sites in the Central Lowlands, notably in parts of lowland Ayrshire and the lower Clyde valley. They include woolly mammoth and woolly rhinoceros, species that are now extinct, and reindeer. In North-west Scotland, remarkable accumulations of animal bones in some of the limestone caves at Inchnadamph have provided further revelations about the Ice Age fauna (see p. 175).

The summit of Ladhar Bheinn (1,020 metres) in Knoydart formed a nunatak rising above the surface of the last ice sheet. The upper level of the ice at around 950 metres is indicated by a periglacial trim line (arrowed). This marks the boundary between frost-weathered rock on the summit pyramid of the mountain and ice-scoured bedrock along the ridge below.

Landscape modifications during the last glaciation

The last ice sheet may have undergone two phases of expansion. During the earlier phase, between about 33,000 and 26,000 years ago, the Scottish ice probably merged with the Scandinavian ice sheet on the floor of the North Sea and extended out across the continental shelf west of the Western Isles. This could occur because world sea levels were about 125 metres lower than today as so much water was locked up in globally expanded ice sheets. The ice may then have retreated for a short time before advancing again around 22,000 years ago. During this later phase, the ice sheet was less extensive, the icefront lying some 30 kilometres off the present coast of eastern Scotland and some 80 kilometres to the west of the Uists.

The glaciers expanded outwards from major centres of ice accumulation in the western and northern Highlands, forming a vast icy dome that buried much of the landscape. The ice reached a thickness of over 1,300 metres in the western Highlands, but in the North-west Highlands and Hebrides the higher summits protruded above its surface as nunataks. Like a tidemark on a bath, the ice left behind a trace of its former surface level against the sides of these mountains. This takes the form of a periglacial trimline, so-called because the slopes above it experienced intense frost action and are covered by shattered bedrock and blockfields. Below the trimline, the frost-weathered debris was removed and the bedrock scoured bare by the ice. In the North-west Highlands, the trimline declines from over 900 metres in Wester Ross to 550 metres over the north coast, following the former gradient of the ice surface. It is well displayed on mountains such as Ben More Assynt, Conival, An Teallach and

Ladhar Bheinn and on Clisham in Harris. The higher hills of Skye, Rum and Mull were also nunataks. Possibly, the high tops of Ben Nevis, Bidean nam Bian and Ben Starav, which have large summit blockfields, also escaped an icy burial. However, even all of these summits may have been covered by earlier ice sheets.

Elsewhere, ice streams from the southern Highlands transported Highland erratics across the Central Lowlands. An ice centre also developed later in the western part of the Southern Uplands. Some of this ice moved northwards across the southern part of the Central Lowlands, merging with the Highland ice and deflecting it. This is reflected in the character of the glacial deposits in central and southern Ayrshire and in Midlothian where deposits with erratics from the Highlands are overlain by deposits derived from the Southern Uplands. Shetland supported its own ice cap but Orkney was covered by ice from the Scottish mainland moving north-west from the Moray Firth. Local ice centres also formed on the islands of Skye, Mull and Harris and along the west coast of the Uists, and merged with mainland ice flowing west and north-west across the Minch and to the north of Lewis. The directions of movement of the ice have been reconstructed from the patterns of striations and dispersal of erratics.

The last ice sheet left behind a variable cover of till and other deposits in the lowlands and in the Highland glens (see p. 182). In places, the till was moulded by the ice into low streamlined hills, known as drumlins. These are oval or elliptical landforms that may be up to 50 metres high and several kilometres long. They tend to occur in distinctive swarms. Generally, they are landforms of lowland areas and are best developed in Galloway, the Tweed valley, south-east of Loch Lomond and around Glasgow. The

The mountains of East Greenland reveal how the Highlands of Scotland might have appeared during the Ice Age. Powerful glaciers would have filled the glens and only the tops of the highest mountains would have risen above the ice surface as nunataks.

A reconstruction of the maximum extent of the last British Ice Sheet. Scottish glaciers were confluent with the Scandinavian Ice Sheet in the North Sea basin some time between about 33,000–26,000 years ago, but other sectors of the ice sheet may have reached their maximum extents slightly later (e.g. after about 22,000 years ago). The contemporary sea level may have been as much as 125 metres lower than it is today, allowing the ice to extend across the bed of the North Sea and also the continental shelf to the west. The inset map shows the possible limits of the Scottish Ice Sheet around 22,000 years ago, after it had separated from the Scandinavian ice, and also the extent of glaciers in Scotland during the Loch Lomond Readvance, between about 12,900–11,500 years ago.

0 50 100 Kilometres
Rock basins

Major rock basins eroded by the glaciers occur predominantly in the north and west of Scotland. In the West Highlands, there is a marked radial pattern associated with glacier flow outwards from an ice centre in the Rannoch Moor to Crianlarich area.

city itself is built on drumlins, and the hilly topography of the urban landscape is reflected in many of the local place names, such as Jordanhill, Maryhill, Hillhead and Drumchapel. Glasgow University occupies a prominent position on the crest of a drumlin, while curving street patterns in certain areas – Maryhill and Mosspark – follow the layout of the drumlin landforms. The drumlins are generally steeper on their western, up-ice sides and their orientations show how the flow of the glaciers fanned out eastwards north of the Clyde and south-eastwards south of the Clyde as they moved out from the Highlands.

Erratics

Those great blocks of granite so foreign to the place on which they stand, and so large as to seem to have been transported by some power unnatural to the place from whence they came.
James Hutton, *Theory of the Earth*, 1795

Erratics, ranging from small stones to large boulders, are a common feature of glaciated areas, as noted by James Hutton in the Alps. They differ from the surrounding bedrock and in some parts of the world have been transported hundreds of kilometres from their sources. Their distributions can be used to infer former ice movement directions. Erratics commonly occur in till, but in many parts of the North and West Highlands, they frequently appear as isolated boulders perched on ice-scoured rock surfaces. Good examples of Torridonian sandstone and Cambrian quartzite erratics are widespread in Torridon and Assynt. In the Drumochter Hills, granite erratics from Rannoch Moor are scattered over the hills up to the summits. In Galloway, blocks of Loch Doon granite occur on the greywackes (a type of sandstone) of the summit ridge of Merrick. Sometimes, erratics have become a feature of local folklore, such as Samson's Putting Stone on the lower slopes of Ben Ledi, near Callander. Often distinct 'boulder trains' can be traced away from their source outcrops; for example, boulders from an outcrop of the igneous rock, essexite, extend in a narrow zone eastwards from their source near Lennoxtown to the Firth of Forth.

Among the more intriguing erratics are two groups of rocks. The first are of Scandinavian origin and occur in till at various locations along the coast of North-east Scotland and inland between Aberdeen and Portsoy. They include fragments of rhomb porphyry and larvikite. In Shetland, there is a conspicuous boulder of tonsbergite, known as the Dalsetter erratic, which has been incorporated into a stone dyke in south Mainland. These erratics are thought to have been transported from southern Norway, their nearest known sources, during a glacial event pre-dating the last glaciation.

Erratic boulder of Torridonian sandstone resting on Cambrian quartzite bedrock at Beinn Eighe National Nature Reserve.

The second group are large 'rafts' of sedimentary rocks or marine sediments that have been transported onshore from the floor of the North Sea. One of the largest is a block of Lower Cretaceous sandstone which lies beneath the ground surface at Leavad, on the Latheron to Halkirk road in Caithness. It is nearly 8 metres thick and 800 metres long and was transported by the ice from at least 15 kilometres offshore, the nearest known occurrence of this rock type. Other examples are blocks of marine sediments containing shells at Clava east of Inverness, near Inverbervie in Kincardineshire, on the Kintyre peninsula and in Ayrshire. At one time these were thought to be in situ deposits and to represent sea levels much higher than at present.

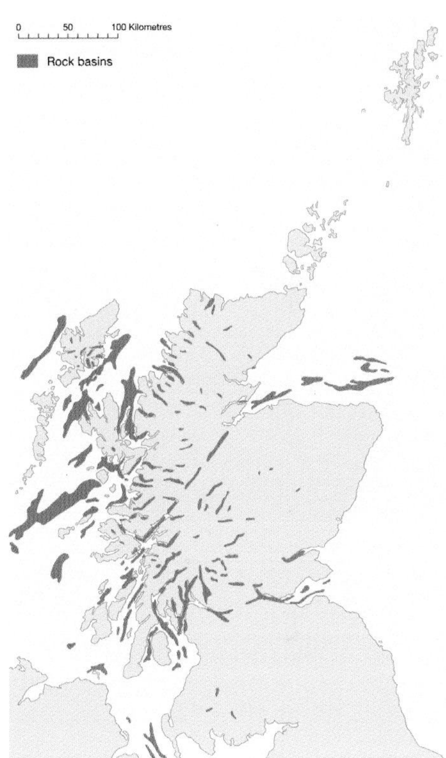

Left. Drumlins near Kirkintilloch, north of Glasgow. The former direction of ice flow was from right to left across the photograph.

Below. This modern glacial landscape in Iceland shows how Scotland might have looked during the melting of the last ice sheet. The land surface comprises unstable moraines, lakes impounded by the glacier and large outwash plains with rapidly shifting rivers building up thick deposits of sand and gravel.

The ice melts

Around 15,000 years ago, the climate warmed extremely rapidly in the space of a few hundred years from full glacial conditions to temperatures similar to those of today. Melting of the last ice sheet accelerated, releasing vast volumes of meltwater, which formed river channels along and underneath the ice margins. These meltwater channels are easily recognised today because they are quite unlike modern river courses (see p. 183): many are now dry and often run along hillsides at a shallow angle parallel to the former ice margins; some, such as the Poll Bhat and Clais Fhearnaig channels between Glen Derry and Glen Quoich on the Mar Lodge Estate, cut across spurs; and some even run uphill where the water was forced upslope by the pressure underneath the ice. The meltwater often formed major drainage routes which can be identified by systems of these channels, for example along the northern flanks of the Lammermuir Hills and Tinto Hill, the southern flanks of the Ochil Hills, the northern flanks of the Cairngorms, the southern margin of the Pentland Hills, in the Dinnet area of Deeside, and on the Struie between the Cromarty and Dornoch Firths.

The debris carried by the glacial rivers was washed and sorted then deposited as large accumulations of sand and gravel, often in the form of ridges and mounds. Good examples occur in a belt from south-west of Lanark to north-east of Carstairs, in the Teith valley, around Leuchars in North-east Fife, in Strathallan and along the Moray Firth coast between Inverness and Nairn. Sometimes the rivers flowed in confined tunnels beneath the glaciers. After the ice melted, their former courses were revealed as sinuous ridges of sand and gravel, known as eskers. Particularly good examples occur at

Glacial deposits

Of all the varieties of detritus left behind by the ice,
the most universal and characteristic is the till.
Archibald Geikie, *The Scenery of Scotland*, 1887

Debris eroded at the bed of a glacier may be transported either frozen into the basal layers of the ice itself or dragged along underneath. The fragments of rock are ground down where they come into contact with other particles and the bed itself. They lose their angular edges, and a fine rock flour is produced. Debris deposited directly by a glacier is called till, a word of Scottish derivation used to describe a coarse, stony soil. Till is poorly sorted material that typically consists of stones and boulders mixed with silt, sand and sometimes clay. It may be deposited underneath a glacier by melting out or by lodgement (the glacier is no longer able to drag the debris forward). Where a lot of water is present, the debris released from the ice may be moulded into streamlined landforms as the glacier continues to flow over it. Much of the till in lowland Scotland has been shaped in this way. At the front of a glacier, till may also be pushed up or bulldozed by the ice into mounds or ridges, known as moraines.

Till contains fragments of rocks picked up by the glacier from the ground it has flowed across. So the composition and colour of the till can give an indication of the direction of glacier flow. For example, the tills in Strathmore and northwards towards Aberdeen are a deep red colour, reflecting the Old Red Sandstone rocks from which they are derived; in the Glasgow area, the till is red where it contains Old Red Sandstone rocks and grey where it contains Carboniferous rocks. Till may also include far-travelled rock fragments, or erratics. In some coastal areas, such as parts of Caithness, Ayrshire and Aberdeenshire, and along the south coast of the Moray Firth, where the glaciers moved onshore after crossing the sea floor, it is even possible to find sea shells in the till.

Coastal erosion has revealed a thick deposit of glacial till south of Montrose. The red colour of the till reflects the derivation of its constituent materials from Old Red Sandstone rocks.

Meltwater channel on the northern flanks of the Lammermuir Hills, south-west of Dunbar.

An icy cauldron

The intensity and power of glacial meltwater rivers is difficult to imagine, unless the reader has ever attempted to cross one, for example in Iceland, or been deafened by the sound of boulders being carried along their beds. A unique and most spectacular feature formed by the action of a meltwater river is the Vat in the Howe of Cromar, near Dinnet in Aberdeenshire. Here a great cauldron has been carved in solid granite, forming a giant, nearly enclosed pothole over 15 metres deep and about 18 metres in diameter at its widest. The meltwater spilled from the main Dee valley over a low col between two hills and, armed with a bed-load of boulders, scoured out the cauldron. It is now partially filled with granite debris and the stream diminished to its present relatively small trickle. The Vat can be accessed easily from the National Nature Reserve visitor centre at Muir of Dinnet.

The Vat at Muir of Dinnet, near Aboyne. Only about half of the cauldron can be seen in the photograph; the rest is buried beneath the gravel deposits on its floor.

Eskers at Carstairs, near Lanark. The ridges of sand and gravel represent the former course of a glacial river system. The ridges are braided, or interlinked, with kettle holes between some of them.

Carstairs Kames

Carstairs Kames is a long-recognised classic example of an esker system and associated glacial meltwater deposits. It forms part of an extensive glacial drainage system extending in a belt from south-west of Lanark to north-east of Carstairs. Although called 'kames', after the historical usage of this term, the landforms are in fact eskers, consisting of a series of sinuous ridges and mounds of sand and gravel with intervening kettle holes. They have a long history of study dating back to the mid-nineteenth century, and have been frequently cited in textbooks, but their exact origin still remains a matter of debate, particularly whether they formed in tunnels under the ice or in open, ice-walled channels. A significant part of the landform complex has been removed by sand and gravel quarrying, leaving only a core area now protected as a Site of Special Scientific Interest. Because the landforms were formed by processes that are no longer active, a large part of the resource has been lost for ever. A small compensation is that while destroying a large part of the landform, the quarrying also provided information on the internal structure and composition of the deposits. There are now very few large esker systems in Scotland that remain largely intact, and while sand and gravel are important mineral resources, a significant part of Scotland's Earth heritage has been lost.

Carstairs, around Inverness and at Greenlaw Moor in the Borders.

Where the meltwater rivers emerged from subglacial tunnels at the glacier fronts, they were no longer confined by the ice and spread out in complex braided patterns, building up thick layers of sand and gravel in the form of outwash plains. There are good examples in Strathmore around Edzell and in Glen Roy at its junction with Glen Turret. Often these deposits were dissected by later river erosion to form suites of outwash terraces, now preserved along the margins of many glens, such as Glen Feshie. Where these deposits buried glacial ice, or where bodies of ice became cut off from the active icefronts, eventual melting of the ice produced a hummocky landscape of kames (mounds) and loch-filled kettle holes (hollows). Good examples occur in upper Strath Spey between Boat of Garten and Kincraig, with kettle hole lochs at Loch Garten, Loch Malachie, Loch Vaa and Loch Alvie, and in the Nith valley north of Dumfries. Often many of these landforms – meltwater channels, eskers, kames, kettle holes and outwash terraces – occur together as they are all associated with the melting of the ice sheet; particularly good examples are found at Dinnet on Deeside, in the Feugh basin near Banchory, between the western outskirts of Inverness and Nairn, at the Kippet Hills near Ellon, and in the Uig area of western Lewis.

The modern coast has also benefited enormously from meltwater deposits since the rivers transported vast volumes of sand and gravel from the melting glaciers to the coast and offshore. Later as postglacial sea level rose, the material was carried shorewards and incorporated into beaches and sand dunes.

A sting in the tail – the ice returns

By around 14,000 years ago, summer temperatures had risen rapidly to about 12°C or higher in southern Scotland, and much, if not all, of the last ice sheet had melted. Bare hillsides were rapidly modified by erosion and the unstable mineral soils washed downslope before vegetation became re-established. Pioneer open-tundra plant communities colonised the mineral soils and were replaced by closed grassland and heathland, with some scrub and birch woodland. The woodland was most extensive in the south and east, with scattered stands in particularly favourable places elsewhere. The climate, however, had an abrupt surprise in store.

After about 13,000 years ago, the climate cooled rapidly and glaciers once more expanded in the Highlands as the Gulf

On the esker trail

As the last ice sheet was melting, large volumes of meltwater were directed towards the Moray Firth from the Great Glen and along Strath Nairn. The water flowed alongside the glaciers and in tunnels underneath them. Traces of the subglacial rivers can still be seen today in the form of eskers – narrow, sinuous ridges of sand and gravel, left behind when the glaciers melted. Particularly good examples occur at Torvaine, west of Inverness, between Inverness and Nairn (the Kildrummie Kames) and at Littlemill in Strath Nairn.

At Littlemill, the glacial rivers formed several large (15 metres high and 2 kilometres long), parallel eskers. Here too, there are kettle holes, large depressions formed when masses of ice were cut off from the retreating glacier. These later filled with water to form small lochs and boggy hollows.

The Inverarnie Esker Trail at Littlemill has been developed by the Forestry Commission. This geomorphological trail takes the visitor on a journey around the landforms created by the glacial meltwaters and provides an opportunity to walk over and view some fine examples of eskers and kettle holes, which are protected as part of a Site of Special Scientific Interest.

Kame and kettle landscape near Lanark.

Stream and North Atlantic Drift shut down. This final glacial event is known as the Loch Lomond Readvance, named after the area where it was first described. The largest icefield built up in the West Highlands, stretching from Torridon to Loch Lomond, with smaller icefields on Skye and Mull (see inset map on p. 179). Outlet glaciers extended out from these centres and numerous small corrie glaciers and valley glaciers developed elsewhere. Mean July temperatures at sea level in Scotland were around 8–10°C; mean January temperatures, below -20°C or -25°C. Annual precipitation, including rain and snow, ranged from around 2,500 millimetres at sea level on Mull to possibly only 100 to 200 millimetres in North-east Scotland. As a result, the eastern Grampians had an arid tundra landscape, with only small glaciers in the Cairngorms, compared with the large icefield 50 kilometres to the west. In the uplands, frost-weathered debris on summits and slopes beyond the glacier limits was moved downslope by solifluction. Screes accumulated below steeper rock slopes. Frozen ground (permafrost) probably extended down to sea level. Trees and most tall shrubs disappeared and tundra vegetation was widespread. Rivers flowed in braided, unstable channels and filled the valley floors with sediments. Scotland would have looked like parts of the Arctic in northern Sweden or Svalbard today. This was a time of rapid climate change that had a significant impact on the landscape.

The Loch Lomond Readvance glaciers produced a variety of well-developed moraines that can often be traced over many kilometres. End moraines were formed by the dumping or pushing of debris at the glacier fronts and these often continued along the glacier margins as

Pollen and plant remains preserved in lake sediments and peat bogs can be used to reconstruct past environmental changes. This diagram shows changes in the relative amounts of different pollen types, and hence changes in vegetation, since the melting of the last ice sheet around 15,000 years ago at Abernethy Forest in Strath Spey.

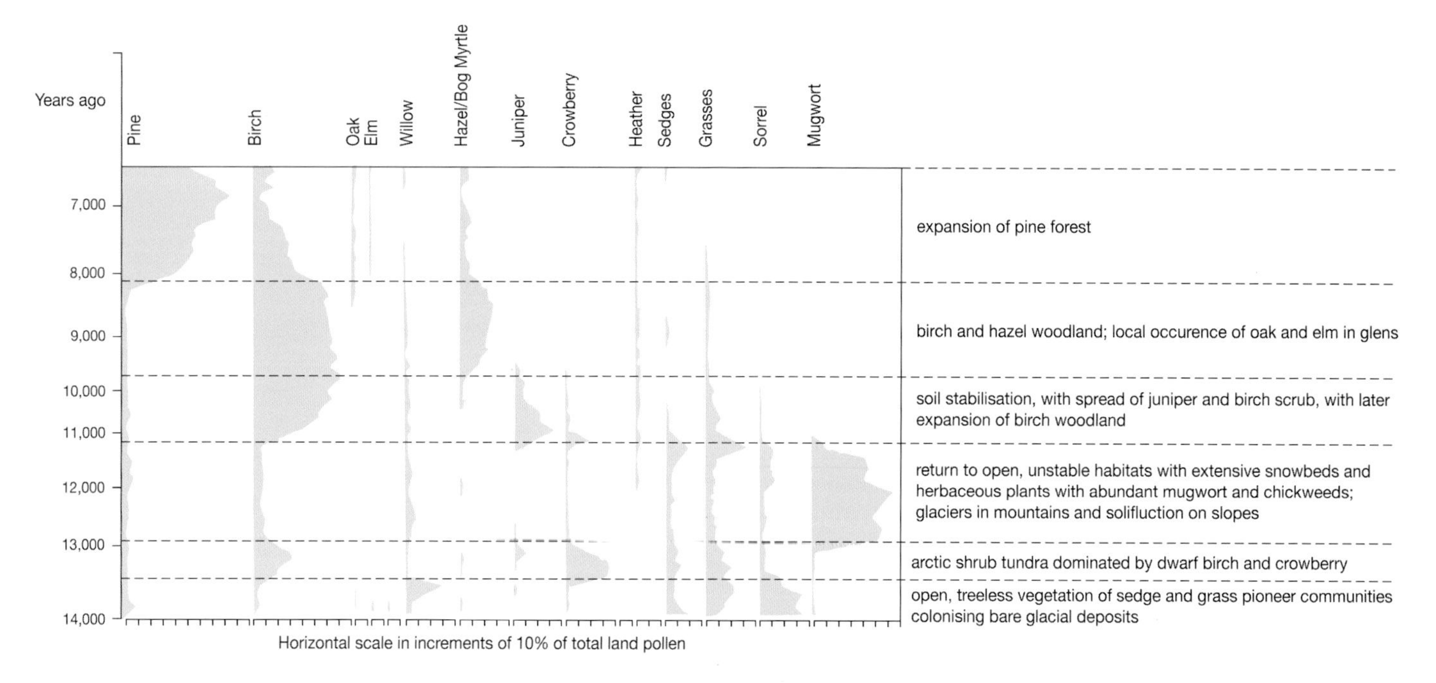

The Loch Lomond Readvance

The area at the southern end of Loch Lomond is particularly important for interpreting the final cold period at the end of the last glaciation. It is the 'type area' in Britain for the Loch Lomond Readvance, which occurred between about 12,900 and 11,500 years ago. This area was where the glacial readvance was first described in detail and its name is now enshrined in the scientific literature.

Several localities demonstrate the sequence of events and the landscape changes at this time. In the Blane and Endrick valleys, till deposited by the last ice sheet is overlain by the deposits of an ice-dammed lake, formed as the ice was melting sometime before about 15,000 years ago. As the ice retreated, the sea flooded from the west into Loch Lomond and the Blane and Endrick valleys, depositing estuarine silts with a boreal marine fauna. These can be seen in sections along the southern shore of the loch. In an abandoned railway cutting at Croftamie, a buried layer of plant remains directly overlies the last ice sheet till and contains species indicative of a tundra environment. In the nineteenth century a reindeer antler was found with these plant remains. The Loch Lomond Readvance glacier then pushed south, forming an end moraine that sweeps around the south end of the loch, from Conic Hill to Glen Fruin, burying the earlier deposits under a layer of till. East of Drymen, at Gartness, the glacier dammed a new lake (Lake Blane) in the Blane and Endrick valleys. Similar glaciers emerged from the Highland glens between Menteith and Buchlyvie and at Callander, where they formed large end moraines.

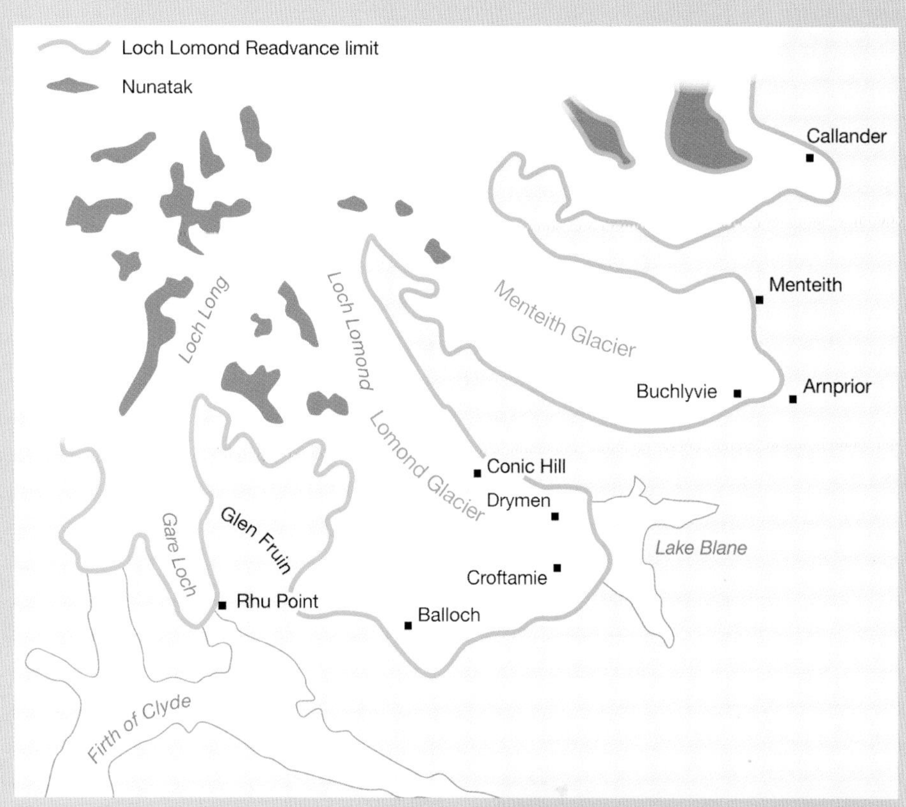

Above. During the Loch Lomond Readvance, glaciers from the icefield which formed in the West Highlands extended down the glens and spread out into the Central Lowlands. Their limits are marked by prominent end moraines. A large lake was dammed in the Blane valley in front of the Loch Lomond glacier.

Right. Deposits excavated in a former railway cutting through a drumlin at Croftamie, near Drymen, south-east of Loch Lomond, illustrate the sequence of events and landscape changes at the end of the last glaciation. The red till (1) at the bottom was deposited by the last ice sheet around 22,000 years ago. The thin dark band (2) immediately above contains the remains of tundra plants living in the area around 12,500 years ago. The blue layer (3) is mud that settled out on the floor of a lake formed in front of the advancing Loch Lomond glacier. At the top, the brown till (4) was deposited by the glacier when it advanced across the area about 12,000 years ago.

Left. Lateral and end moraines at An Teallach formed by a Loch Lomond Readvance glacier. The bouldery moraines contrast with the bare, ice-scoured bedrock on the right of the photograph.

Right. This modern glacier in Greenland shows how the landscape of An Teallach may have looked at the time the moraines were formed.

lateral moraines. There are many excellent examples in the Highlands and Southern Uplands, for example at An Teallach, Loch Glascarnoch and the north end of Loch Treig, north of Loch Rannoch, in Glen Tarbert in Ardgour, in several of the corries of the Cairngorms and Lochnagar, and at Loch Skeen, north-east of Moffat in the Borders. Distinctive hummocky moraine also occurs in many Highland glens and over large areas of Rannoch Moor. This appears as a chaotic form of mounds when seen on the ground, and is reflected in local names such as the 'Valley of the Hundred Hills' in Glen Torridon and the 'Valley of the Hundred Hills' at Glen Fuaron on the south side of Glen More on Mull. Other good examples occur on Skye near Sligachan, along the A9 in the Pass of Drumochter, and near Kingairloch in Ardgour. The mounds often have a clear alignment at an angle to the valley axis and are believed to have formed at the margins of the glaciers as they were retreating. Where the glaciers emerged from the uplands into the Central Lowlands, they formed large end moraines around the southern end of Loch Lomond, at the Lake of Menteith and near Callander.

Along the west coast, the outlet glaciers from the West Highland icefield terminated at sea level near the mouths of the glens and produced a variety of landforms, including end moraines and large outwash plains, some of which built out into the sea as deltas. One of the best examples is at Moss of Achnacree, north of Oban. Other good examples occur at Corran, Ballachulish and Loch Eil, where the glaciers formed flat-topped outwash deltas as they retreated. In several places, the glaciers advanced over the seabed and ploughed up marine deposits into their end moraines. This happened at Rhu Point on Gare Loch and at South Shian at the mouth of Loch Creran, and today it is possible to find the shells of marine organisms in these deposits. Radiocarbon dating of the shells has been used to establish the age of the glacier readvance.

Lakes formed at the margins of several of the Loch Lomond Readvance glaciers where they dammed tributary valleys. Fluctuating water levels produced 'staircases' of shorelines around

Valley of the Hundred Hills

Choire a' Cheud-chnoic, or 'the Valley of the Hundred Hills', is one of the most striking examples of hummocky moraine in Scotland. This has an apparently chaotic form of mounds, ridges and hollows when seen on the ground, but when viewed from the air or on aerial photographs, the mounds are often elongated and aligned in concentric arcuate loops or parallel rows. These landforms are typically associated with valley glaciers of the Loch Lomond Readvance and may have formed in a number of different ways. Where they occur as concentric arcuate loops, they are thought to have formed at the margins of actively retreating glaciers; although the glaciers were progressively retreating overall, they advanced forwards in winter, pushing up the ridges. In other cases, the moraines may have formed beneath the glaciers through the deformation and streamlining of previously deposited debris; such features are aligned parallel to the former ice-flow direction, rather than at an angle to it as in the case of the ice-marginal forms. Some examples of hummocky moraine elsewhere may have formed on top of the ice where rock debris covered the surfaces of the glaciers; such debris would have insulated the ice underneath and inhibited its melting. Variations in the thickness of the debris layer would have produced variations in the rate of melting and the development of a chaotic hummocky relief.

The landforms in Choire a'Cheud-chnoic in fact comprise two sets of features, which contribute to their apparent complexity. The first is a set of arcuate ridges and mounds aligned across the glen. These appear to be earlier features that formed at a retreating glacier margin during the melting of the last ice sheet. The second set is a series of parallel ridges, or flutings, which are aligned along the glen parallel to the former direction of ice movement. In places they are superimposed on the arcuate ridges, indicating that they formed later underneath the Loch Lomond Readvance glacier that occupied the glen. They were shaped and streamlined by the ice as it flowed over the earlier deposits. Thus in some situations, more than one generation of landforms may be present.

Hummocky moraine at Choire a' Cheud-chnoic, 'the Valley of the Hundred Hills' in Glen Torridon.

such lakes. The best example is at Glen Roy, where three prominent shorelines, known as the Parallel Roads, can still be seen clearly today. Deltas formed where meltwater rivers transported large volumes of sediment into these glacial lakes. Particularly fine examples occur at Achnasheen. These take the form of large, flat-topped terraces dissected by later erosion. Other lakes formed where the glaciers had receded from large end moraines, such as at Gartness in the Blane valley near Drymen.

Beyond the ice – periglacial processes

Although glaciers were the main agents that shaped Scotland's landscape during the Ice Age, there were long periods when periglacial processes dominated. These involved the action of frost and ice in the soil and rocks, and affected the ground that lay beyond the glaciers during the times of build-up and decay of the ice sheets and during periods of more restricted glaciation, such as the Loch Lomond Readvance. They also affected the mountain summits and slopes that rose above the ice surface as nunataks. The effects of periglacial processes are most apparent on mountain summits and slopes in the Highlands and Islands; although lowland soils were also extensively modified, the traces are now more difficult to see at the ground surface. Periglacial processes are still active on Scottish mountains today, although less intensively than in the past. They have produced a number of distinctive landforms.

Intense frost weathering of the bedrock has resulted in a widespread cover of debris on mountain summits and upper slopes. The character of this debris varies with the bedrock type. Resistant and well-jointed rocks, such as

Glen Roy

It is far the most remarkable area I ever examined . . . I can assure you Glen Roy has astonished me.
Charles Darwin, 1838, in a letter to Charles Lyell

The features which so astonished Charles Darwin are in fact a series of shorelines that formed along the margins of ice-dammed lakes in Glen Roy. Known as the Parallel Roads because of their distinctive form and appearance, they are one of the most striking landform landmarks in Britain and form three clear parallel lines running along both sides of the glen (see chapter 6).

During the Loch Lomond Readvance, glaciers from the West Highland icefield blocked the mouth of Glen Spean, impounding a lake which eventually filled up Glen Roy and extended eastwards through the present Laggan valley. The water reached a level of 260 metres above sea level, controlled by the height of an overspill col at its eastern end (Diagram A). As the glacier advanced, it then blocked off the entrance to Glen Roy. This allowed the lake level in the glen to rise until it was able to escape across a col at 325 metres above sea level (Diagram B). As the glacier advanced further up the glen, this exit too was blocked and the water rose to a third level, determined by a col at 350 metres above sea level at the head of the glen (Diagram C). At the same time, a lake was impounded in Glen Gloy at 355 metres above sea level. Later, as the ice retreated, the overflows were unblocked in the reverse order.

Periodically, as the lake levels rose, the ice dam failed and the waters drained in huge floods underneath the ice. The floodwaters travelled under the ice through the Spean gorge and into the Great Glen, depositing a large fan of sand and gravel where they emerged from the front of the glacier at Fort Augustus. The level of Loch Ness rose and the water flooded out along the Ness valley past the present site of Inverness and into the Beauly Firth. Such floods occur today in Iceland, where they are known as jökulhlaups or glacier bursts.

Each of the major lake levels in Glen Roy, Glen Spean and Glen Gloy is associated with the formation of a shoreline. These are best seen in Glen Roy as benches running along the hillsides. They were formed by a combination of intense frost weathering of the bedrock along the shoreline and wave erosion and deposition of the fractured rock.

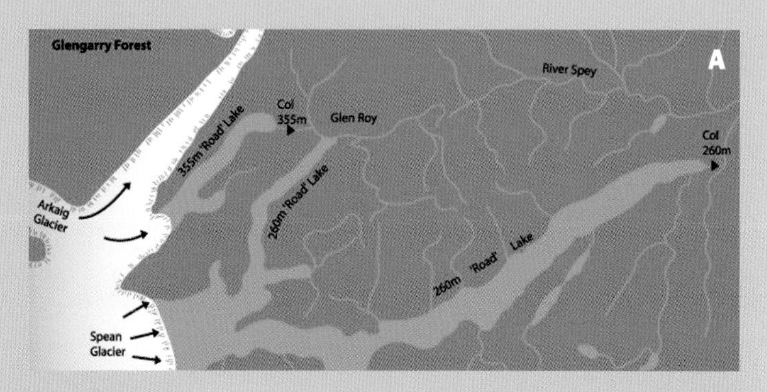

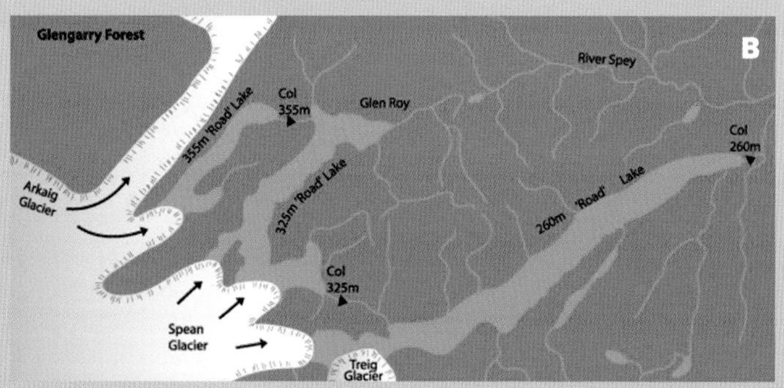

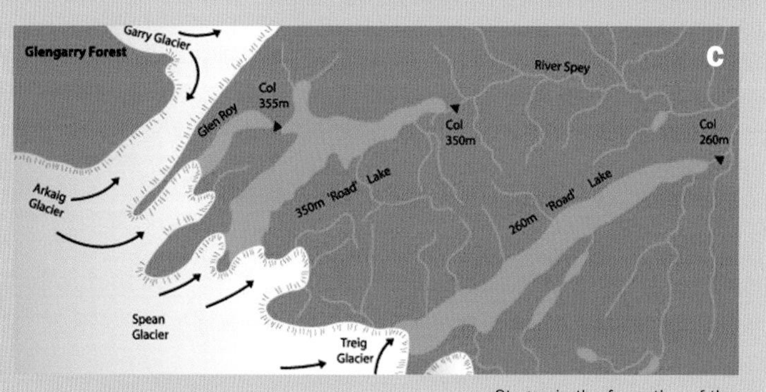

Stages in the formation of the ice-dammed lakes in Glen Roy, Glen Gloy and Glen Spean, showing the icefront positions and the locations of the overspill cols that controlled the different lake levels. The sequence of lake levels is explained in the text.

Blockfields on the summits of Ben More Assynt and Conival, in Assynt. The summits of these mountains lay above the surface of the last ice sheet and the quartzite bedrock was subjected to intensive frost shattering.

quartzite and microgranite, break down into angular lumps that form blockfields, for example, on the quartzites of the Paps of Jura, Ben More Assynt and Conival and locally on the Cairngorm granites as on Derry Cairngorm. Granites and sandstones usually weather through granular disintegration, producing a surface cover of stones mixed with coarse sand. Finer-grained rocks, such as schists, break down into a mixture of stones, silt and sand.

The frost-weathered debris has been modified by frost sorting in the soil and by solifluction. Frost sorting produces a range of forms that occur in distinctive patterns, hence the generic name, patterned ground. Finer-grained soils are more susceptible to the formation of ice lenses and therefore to frost heaving. This in turn aids the separation of the coarser from the finer material and leads to the formation of characteristic features such as sorted circles and stripes. Good examples of the former occur on Sròn an t-Saighdeir on Rum and Beinn Fhuaran near Ben More Assynt. The larger features, a few metres across, are now inactive, but smaller forms remain active. One of the best examples of active sorted stripes is on Tinto Hill.

Slow, seasonal downslope movement of the soil by solifluction processes has formed characteristic sheets, terraces and lobes of debris on many Scottish mountains. Particularly good examples occur in the Fannich Mountains. Where weathering of the bedrock has produced coarse debris, large boulder lobes have developed, as in the Cairngorms and on Lochnagar. Such features are now inactive. Soliflucted soils are widespread in the Southern Uplands and many valley floors have been partially infilled by soliflucted deposits, which have been subsequently dissected by the rivers.

During the cold episodes when glaciers were

Top left. Large inactive sorted circles on the summit of Sròn an t-Saighdeir, Western Hills of Rum. Freezing and thawing of the soil has sorted the debris, producing circles with coarse material at the margins and fine material in the centres.

Top right. Active sorted stone stripes on the slopes of Tinto Hill, south of Lanark. Freezing and thawing of the soil has sorted the debris into bands of coarse and fine material. The bands of coarse material are approximately 20 centimetres wide.

Bottom left. Slow downslope movement of the soil formed these large boulder lobes in Lurcher's Gully in the Cairngorms at the time of the Loch Lomond Readvance when the climate was intensely cold. Seasonal melting of the surface layers of the frozen ground allowed the saturated material to flow downhill.

Bottom right. Thick valley infill near Sourhope in the Cheviot Hills has been dissected by the Bowmont Water. Such infills are common in the Southern Uplands and comprise till, soliflucted till and soliflucted frost-shattered bedrock.

of limited extent, lowland soils would have been perennially frozen, with only a thin surface layer of the permafrost melting in summer. These soils would have exhibited polygonal-shaped surface patterns, like those that occur in the Arctic today. They are known as ice-wedge polygons, and form by the contraction of the permafrost in winter.

Lowland soils have also been extensively affected by solifluction, but this has generally left no clear surface expression. However, sometimes in old quarries or in exposures along forestry roads, the stones in the soil can be seen to have a clear downslope alignment due to the movement of the soil. Occasionally, the underlying bedrock has also been deformed.

Solifluction deposit at Pittodrie at the foot of Bennachie, near Inverurie. The slow downslope movement of the soil has deformed the upper layers of the weathered granite bedrock.

Frosty island hills

Periglacial landforms of various types occur on most of the higher mountains in the Highlands. They are also particularly well displayed at lower altitudes on a number of hills on the Scottish islands, notably on Ronas Hill in the northern part of Mainland Shetland, Ward Hill on Hoy in Orkney, and the Western Hills of Rum.

Ronas Hill (453 metres above sea level) has a superb suite of features formed by frost and wind action at the lowest altitude of anywhere in Britain. The northerly latitude and the highly exposed oceanic conditions have favoured their development over the past 15,000 years or more, right through to the present day. The landforms include a relict summit blockfield in the granite bedrock, and active terraces with turf-banked fronts formed by the gradual movement of debris down the slope (solifluction). The effects of the wind have produced narrow stripes of vegetation and intervening areas of bare sand and gravel. Together, the action of solifluction and wind have combined to form a striking pattern of terraces and vegetated stripes.

Ward Hill (479 metres above sea level) is exceptional for features formed by wind action and solifluction. The effects of the wind are clearly seen in the presence of active vegetated stripes and deposits of blown sand. These sand accumulations had previously been stabilised and covered with vegetation, but are now being actively eroded. It is unclear whether this recent erosion has been initiated by natural processes or through grazing pressure. In a number of places, there are alternating layers of sand and soil, showing that in the past there have been periods of erosion, represented by the sand layers, and periods of stability, represented by the layers of soil. There are also many fine examples of solifluction terraces on Ward Hill, formed by the slow, downslope movement of the weathered rock debris.

The Western Hills of Rum rise to less than 600 metres above sea level but display some excellent examples of periglacial landforms. The list of features present on the three hills, Ard Nev, Orval and Sròn an t-Saighdeir, reads like a classic textbook selection: blockfields, scree, boulder sheets and lobes, turf-banked terraces, ploughing boulders and, most superb of all, sorted patterned ground forms. The different types of bedrock have strongly influenced the distribution of the landforms. On Sròn an t-Saighdeir, weathering of the microgranite has produced an extensive, relict blockfield and slopes with large boulders. Locally this material is arranged into large, partly vegetated sorted circles 2 to 3 metres in diameter on the flat summit, and into sorted stripes where the slope steepens. Adjacent slopes are covered extensively in boulder lobes and terraces. On Orval and Ard Nev, the basalt and granophyre rocks have weathered to produce a combination of fine and coarse material. The finer grained material is susceptible to frost heaving, and small sorted circles 50 centimetres in diameter and sorted stripes 20 centimetres wide are currently active.

Turf-banked terraces formed by solifluction and the effects of wind on vegetation growth and cover, on Ward Hill, Hoy, Orkney. The surfaces of the terraces are maintained largely free of vegetation by the wind.

THE HOLOCENE – LANDSCAPE CHANGES AFTER THE LAST GLACIATION

THE LAST 11,500 YEARS

Around 11,500 years ago, the climate warmed rapidly again and the last glaciers disappeared from Scotland as temperatures rose to levels probably slightly above those of today. This marked the beginning of the present Holocene interglacial, commonly known as the postglacial period. The initial part of the Holocene was characterised by rapid changes as plants recolonised and the landscape readjusted to non-glacial conditions. Rockfalls modified unstable slopes, rivers carried more sediment, and there were dramatic changes at the coast as the relative level of the land and the sea varied.

Plants return

As the glaciers melted and frozen soils thawed, pioneer vegetation of herbs and shrubs colonised the bare, unstable soils. Later, a succession of tree species spread into Scotland. By 8,500 years ago, birch and Scots pine forest had expanded over a large area from Wester Ross to Strath Spey. In the south of Scotland, oak, hazel and elm were dominant. Along the west coast from Kintyre to Skye, and in the north-east lowlands, the woodland mainly comprised birch, hazel and oak. Further north and in the Northern and Western Isles, birch and hazel formed significant woodlands in contrast to the largely treeless landscapes of today. Mesolithic hunter-gatherer activity from about 10,500 years ago onwards probably only had localised and temporary impacts, so that the landscape around 6,000 years ago would have been extensively wooded.

Relatively little is known about the climate during the early part of the Holocene, but generally it may have been slightly warmer, drier and less variable than today. After about 6,500 years ago, the climate appears to have become more variable, with the onset of abrupt shifts to wetter conditions and longer periods of higher precipitation and cooler temperatures, interspersed with shorter, drier periods. During one of these wet shifts about 4,000 years ago, pine forest growing on the blanket bog in the far north of mainland Scotland and on Lewis contracted rapidly. The wetter episodes probably saw the land surface becoming less stable and more debris flow activity.

Following the development of settled farming communities at the beginning of the Neolithic

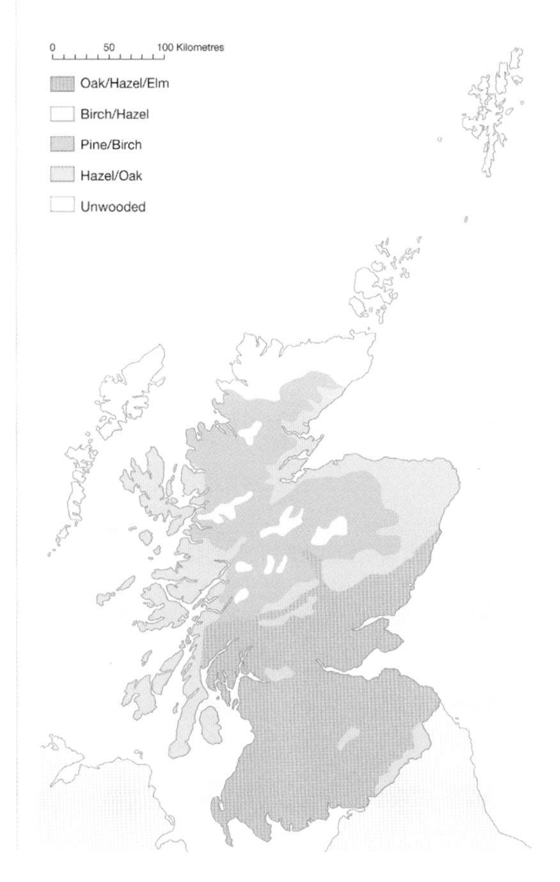

Distribution of the dominant woodland types in Scotland around 6,000 years ago, before significant human clearance. Only some of the unwooded summit areas are shown.

around 6,000 years ago, human activity increasingly interacted with natural processes to bring about changes in the vegetation cover and soils of Scotland. Pollen and lake sediment records indicate phases of woodland clearance, expansion of agricultural activities and accelerated soil erosion during the Neolithic, and later during the Bronze and Iron Ages. Evidence of landscape disturbance from prehistoric farming activity in the uplands reveals that such areas were once more productive and attractive for settlement. For example, the sediments at Braeroddach Loch, near Aboyne in Aberdeenshire, record that soil erosion increased after 6,000 years ago as the woodland was opened up for pastoralism and accelerated around 4,000 years ago when cereal cultivation began. The reasons for the later abandonment of such areas probably involve a complex interplay of climate shifts, soil deterioration, and social and economic factors.

Although it is often thought that peat formation is a relatively recent phenomenon of the last few thousand years, the real picture is more complex. Peat began to form at different times in different places, often much earlier. The first peat growth was probably in the early Holocene, and then it expanded during later wetter phases. Loss of woodland cover is believed to have contributed to increased leaching and surface wetness in the uplands, particularly after 4,500 years ago. Wet heathland replaced the woodland and contributed to podzolisation (leaching of organic material and soluble minerals from the upper soil horizons and their deposition below, sometimes forming an iron pan) of soils and development of blanket bog and

Kinloch Glen, Rum, was the site of one of the earliest Mesolithic settlements in Scotland, dating from around 10,000 years ago. The landscape was wooded, with alder, hazel and willow scrub present. About 5,000 years ago, human activity intensified and the woodland was increasingly cleared for cultivation; this was accompanied by soil erosion. The peat cover on the island also expanded as the climate became wetter and the soils deteriorated.

peat, which became extensive in the uplands and on lower ground in the north and west.

The last few hundred years have seen an apparent increase in slope instability and soil erosion in the uplands. Climate change and extreme rainfall associated with the so-called 'Little Ice Age' of the mid-fifteenth to mid-nineteenth centuries may have been the causes, but human activity and especially land-use changes may have contributed.

Unstable slopes

Rock slopes eroded by the glaciers were often left steep and unstable when the ice melted. Many of these slopes subsequently collapsed following the release of stresses in the glacially steepened bedrock because the glaciers were no longer present to buttress the valley sides. Over 600 examples of such rock slope failures have been recognised in the Highlands. The most

The Little Ice Age

The skating minister. Sir Henry Raeburn's painting, *The Rev. Dr Robert Walker Skating on Duddingston Loch* (c.1795).

The period from the middle of the fifteenth century to the middle of the nineteenth century is known as the 'Little Ice Age', a time of climate deterioration. This period saw extremes of weather, including the coldest conditions since the end of the last glaciation, severe snowy winters, storms, heavy rainfall and summer droughts. There are records of flooding, shipwrecks, loss of life and livestock, and famines. The late seventeenth century was particularly cold and a time of great hardship. However, it also provided opportunities for some; for example, the world's first skating club was formed in Edinburgh in 1742. As a test for admission, aspiring members had to jump successfully over three hats and perform a circle while skating on one foot. As captured in Sir Henry Raeburn's painting, the Rev. Dr Robert Walker skating gracefully on Duddingston Loch would surely have passed.

Farther north, in the Cairngorms, travellers noted late-lying snow in the mountains, and it is likely that there were many more snowbeds surviving throughout the year than today. Snow may even have covered the Cairngorm plateau for decades at a time. John Taylor, writing in 1618 in *The Pennyless Pilgrimage*, noted: "I saw Mount Benawne [Ben Avon], with a furr'd mist upon his snowie head instead of a nightcap, (for you must understand, that the oldest man alive never saw but the snow was on top of divers of those hilles, both in Summer, as well as in Winter)." Later, Thomas Pennant, in *A Tour in Scotland 1769*, described a contemporary view of the hills from Deeside, as follows: "naked summits of a surprising height succeed, many of them topped with perpetual snow." In the *Statistical Account of Scotland* for 1793, the Rev. Charles M'Hardy, minister of the parishes of Crathie and Braemar, referring to Lochnagar, Beinn a'Bhuird and Ben Macdui, wrote that: "Upon these mountains, and others connected with them, there is snow to be found all the year round; and their appearance is extremely romantic, and truly alpine." However, despite some speculation, there is no evidence that glaciers re-formed in Scotland during the Little Ice Age. Today, the late-lying snowbeds are much more restricted and ephemeral. In 2003, all the snowbeds in Scotland had disappeared by late August, the earliest that this occurred of all recorded years in which no snow survived: 1933, 1959, 1996 and 2003.

impressive features include the large-scale landslides at Trotternish on Skye, notably at The Storr and Quiraing, and on Mull and Raasay, where thick beds of Tertiary basalt lavas overlie weaker Mesozoic sedimentary rocks. Locally in the Highlands, rock slopes have failed completely, producing tongues and spreads of boulders on the valley floors below. Particularly good examples are in the Lost Valley in Glencoe and on Beinn Alligin in Torridon. The limited dating evidence available suggests that some of the large rock slope failures occurred several thousands of years after the glaciers disappeared; for example, those at The Storr and Beinn Alligin are dated at about 6,500 years ago and 4,000 years ago, respectively.

The most spectacular rock avalanche in Scotland is on Beinn Alligin. A large part of the rock face collapsed below the ice-steepened summit of Sgùrr Mhór, leaving a scar that can be seen from the south shore of Loch Torridon. The rock avalanche generated sufficient energy to form a tongue of bouldery debris extending over 1 kilometre down Coire Toll a' Mhadaidh Mór. Ian Mitchell, in *Scotland's Mountains before the Mountaineers* (1999), wrote: "Just below the summit of Beinn Alligin is a deep gash noted and avoided by all walkers. This is called Eag Dhubh na h-Eigheachd, the deep gash of the crying or wailing. The local legend has it that shepherds used to hear a wailing or cry of despair from this gash, which only ended when one of their number, descending to investigate, fell to his death. The spirit of the mountain appears to have been appeased by this unintended sacrifice, and the wailing ceased."

Much more typical of the metamorphic rocks of the Highlands, however, are rock slope failures where the rock mass has moved only a short distance downslope and has remained coherent. A variety of distinctive features include open fractures in the bedrock, fissures along summit ridges and rock scarps facing uphill. Some particularly good examples occur in Kintail. Here, on Beinn Fhada above Gleann Lichd, much of the slope below the south-west side of the summit plateau is the site of one of the largest areas of rock slope deformation in Scotland. The slope has not failed completely, however, and there is no spread of boulders or debris on the floor of the glen below. In Arrochar and the Trossachs, good examples of similar slope failures occur on Beinn an Lochain, Ben Donich and Benvane. The Cobbler owes its remarkable triple peaks to an extensive slope failure on its south-west slope. Similarly, the craggy summit of Ben Lomond, viewed from the north-west at Tarbet or Inveruglas, has been shaped by a rock slope failure. A particularly striking example in the eastern

The Beinn Alligin rock avalanche. The source area of the slope failure is clearly visible as the fault-bounded scar on the corrie headwall on the right, below the summit of Sgùrr Mhór.

Slopes on the move

Schematic diagram of the Trotternish escarpment, showing the geology and the pattern of successive landslides. The weaker sedimentary rocks were unable to support the overlying rocks after the last ice sheet had melted. The main blocks partly rotated as they collapsed and slid downwards.

The Trotternish escarpment is the site of the largest continuous area of landslides in Britain, with two of the most spectacular at The Storr and Quiraing (see chapter 6). The development of these landforms is closely related to the underlying geology. At the time of the opening of the North Atlantic Ocean, fissure eruptions produced a 300-metre-thick series of basalt lava flows that buried the Jurassic sedimentary rocks of the Trotternish peninsula. In addition, dolerite sills were intruded into these sedimentary rocks. Later faulting, tilting to the west and erosion of the rock strata produced an escarpment extending along the eastern side of the peninsula. This escarpment was probably steepened by glaciers as they moved northwards along the Sound of Raasay during the Ice Age. When the glaciers melted, the support the ice provided for the escarpment was removed. As a result, joints slowly opened up due to stress release in the glacially steepened cliffs, weakening the rock and leading to slope failure. The weaker sedimentary rocks resting on the stronger dolerite sills could not support the weight of the overlying lavas, and blocks of the escarpment repeatedly collapsed.

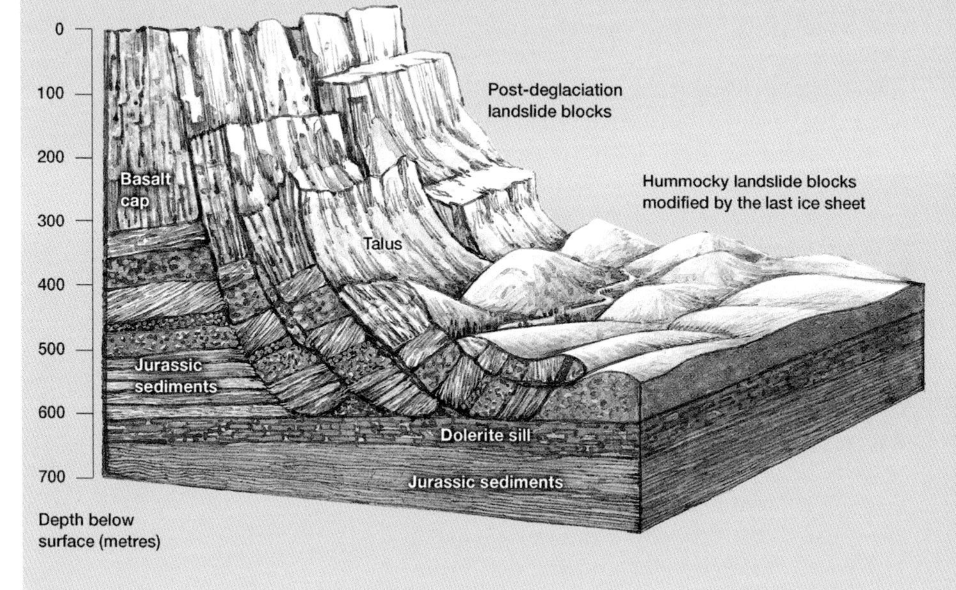

The landslides form two zones. The inner zone is a spectacular maze of tabular and tilted slipped blocks, weathered pinnacles, buttresses and remarkable rock architecture. These features formed during the postglacial. The most recent landslide at The Storr has been dated to about 6,500 years ago, about 10,000 years after deglaciation, suggesting a period of progressive weakening of the rocks. The outer zone consists of subdued, hummocky ground formed by earlier landslides later moulded by the glaciers. These outer slipped blocks predate the last glaciation and may have occurred following earlier episodes of deglaciation. Overall, the scarp face has retreated about 600 metres inland through a combination of glacial erosion and landsliding.

Rocks fallen from the cliffs above have accumulated on the scree slopes below the escarpment. From the volume of material, it has been calculated that the cliff face has retreated by 5 to 6 metres since deglaciation. This retreat is still continuing, but more slowly than in the past, at a rate of about 1 centimetre per 100 years. The scree slopes are also extensively cut by deep gullies formed by debris flows during periods of prolonged or intense rainfall. Old soils buried in the scree show that there have been periods of slope stability in the past and that debris flows have been intermittently active for at least the last 6,000 years.

Left. Large rock slope failure on the flank of Corrie Brandy in Glen Clova, Angus. The headscarp is about 40 metres high.

Opposite left. Active debris flows on the slopes of the Lairig Ghru in the Cairngorms.

Opposite right. Accelerated soil erosion on Ben Mór Coigach, north of Ullapool, has led to the loss of montane habitat and its replacement by a stony lag surface. Buried soils in the remaining islands of sand deposits indicate periods of erosion and stability in the past, but the entire surface may now soon be stripped.

Highlands occurs at Corrie Brandy, in Glen Clova, where a large area of the corrie headwall has partially slipped downslope.

In the Central Lowlands, the steep slopes along the edges of the lava plateau of the Campsie Fells are notable for a number of large landslides and rock avalanches. The glacially steepened slopes have failed sometime after the disappearance of the ice. Good examples occur along the north-facing scarps of the Fintry and Gargunnock Hills and also above the villages of Strathblane and Lennoxtown. The source areas of the rock debris are clearly visible in the cliffs. In the Kilpatrick Hills, the curious rock formations of the Whangie are also a form of slope failure. Large slabs of basalt have become detached from the main rock face; one of these remains standing while others have collapsed. Legend has it that chasms between the blocks were cut by the tail of the Devil, lashing out as he flew over the area.

Scree slopes are a common feature below glacially steepened cliffs. As a result of intensive frost weathering and rockfalls, many scree slopes accumulated in ice-free areas during the cold conditions at the time of the Loch Lomond Readvance. Generally, these are large and stable compared with screes formed during the Holocene, which are often active today as debris continues to build up from intermittent rockfalls. Scree slopes in many parts of the Highlands have been modified by debris flows. These are often triggered by intense rainstorms and comprise rapid flows of saturated debris. In some places, they originate in rock gullies; in others, as bowl-shaped scars where the slopes failed. Their tracks appear as long narrow gashes on many hillsides, formed as the debris flowed downslope. Particularly good examples occur along the flanks of the Lairig Ghru, Glen Docherty and the Pass of Drumochter. Over time, repeated flows from the same source areas have built up cones of debris at the base of the slopes, with some of the largest below the Chancellor on Aonach Eagach in Glencoe. There is some evidence that debris flows may have been more frequent in the last

300 years. Possible causes include more intense rainstorms, woodland clearance and increased animal grazing affecting slope stability. In the eighteenth century, the settlement of Achtriochtan in Glencoe was swept away by debris flows. Such flows are still active today and sometimes block the A82 road.

Wind activity has been important in the production of accumulations of sand on Scottish mountains, particularly on the Torridonian sandstone hills (for example, An Teallach, Ben Mór Coigach and Cùl Mór), on the Trotternish ridge and on Ward Hill on Hoy in Orkney. Today, these sand accumulations are being actively eroded by the wind where the surface vegetation cover has been broken and the underlying soil exposed. These deflation surfaces typically occur on exposed plateaux, cols and ridges, with spectacular examples on An Teallach and Ben Mór Coigach. The wind-blown sand is deposited in sheltered lee sites. Wind activity today is also associated with a variety of vegetated patterned ground forms on exposed mountain slopes. Wind stripes comprise bands of vegetation and bare soil and are particularly well developed in parts of the Cairngorms and on Ronas Hill in north Mainland in Shetland.

RIVERS

As described earlier in this chapter, glaciers have had a major influence on present-day river valleys, deepening them, infilling parts of them with large quantities of till and outwash deposits, and forming rock steps. Consequently, Scottish rivers display a great deal of variability along their courses: lower gradient alluvial reaches alternate with steeper rock sections, and stable reaches alternate with unstable ones where the rivers shift backwards and forwards across their floodplains.

The contrast between the rivers in the west and those in the east, north of the Highland Boundary Fault, is very noticeable. Except where altered by glacial breaching, the main watershed was established before the Ice Age and lies well to the west. In the west, precipitation is higher, river gradients are steeper and there is higher run-off. To the east, precipitation is lower, gradients are generally lower and there are more glacial deposits that have been reworked into terraces.

Scotland's rivers display a number of different forms of channel. These can be straight, meandering or braided. Straight channels are relatively rare over any distance since bars of sand and gravel develop in the channel, giving rise to sequences of pools and shallows and the development of a sinuous flow pattern. Erosion is then focused on alternate banks downstream and a meandering channel pattern develops. Meandering channels usually occur on floodplains with low gradients. They change naturally over time through lateral and downstream migration of the meander bends and the cutting off of meander necks. Good examples of meandering rivers include the River Endrick, flowing into Loch Lomond, the River Glass in Inverness-shire and the River Clyde at its junction with the Medwin Water.

Braided rivers display a network of interconnected channels separated by gravel bars. Typically they occur in high-energy environments where channel gradients are steep, sediment supply is abundant and banks are easily eroded. Some rivers show characteristics of both meandering and braiding and are known as 'wandering gravel-bed rivers'. They are highly dynamic, switch courses frequently and leave abandoned channels on their floodplains. Good examples are the River Feshie and the lower River Spey.

During flood periods when sediment supply is plentiful, rivers tend to build up their floodplains. However, if sediment supply is then reduced or the amount of water increases, then the river will cut down into its floodplain, forming a new lower floodplain and leaving the former one abandoned at a higher level as a river terrace. Particularly good examples of river terraces occur in Glen Feshie and along the River Findhorn. Where a tributary river joins a main valley, it often builds up a fan-shaped area of sediment known as an alluvial fan. Good examples occur in Glen Feshie and Glencoe and at the junction of the Quoich with the Dee, upstream from Braemar.

The River Quoich has formed an alluvial fan at its junction with the River Dee, near Braemar.

Winding rivers

Major changes in river channel position occur from time to time, depending on the frequency and patterns of flooding and the nature of the bed materials, or in response to significant land use changes in the river catchments. Many lowland meandering rivers in Scotland, however, have been artificially stabilised by bank protection works and flood embankments to prevent the loss of land and to mitigate flooding.

The River Clyde at its junction with the Medwin Water is a particularly good example of an actively meandering river that has not been subject to major engineering works. Over the last few hundred years, the river channel has shifted position as individual meander loops have grown increasingly sinuous and then been cut off across their narrowing necks. Old maps and aerial photographs provide a detailed record of these changes. Former positions of the river are also revealed in the presence of well-preserved abandoned channels and meander cut-offs, or ox-bow lakes, on the floodplain. The sediments within these channels record the past flood history of the river, providing an important baseline for comparing the frequency and magnitude of future floods, which are generally projected to increase as a result of climate change.

This aerial view of the River Clyde, near Carnwath, illustrates many of the characteristics of meandering rivers. As the degree of channel sinuosity increases, meander necks have been cut off, leaving abandoned channels or oxbows on the floodplain. In the centre of the photograph, the meandering Medwin Water joins the Clyde.

Going with the flow

The River Feshie is one of the best sites in Britain displaying river processes and landforms associated with a highly active, gravel-bed river with a mountainous catchment. It is a steep, powerful river prone to large floods and rapid channel changes. Floods can occur at any time of the year, during heavy rainfall associated with weather fronts, convective storms in summer or following rapid snowmelt. The Feshie is a typically 'flashy' river since water levels can rise and fall very quickly. Large amounts of sediment are available to the river and are readily moved during floods. Several reaches of the river have braided channels, which are wide and shallow and characterised by rapid and frequent changes in position. The river is reworking its floodplain, and rates of bank erosion can exceed 10 metres per year. It is one of the most active gravel-bed rivers in Britain.

One of the best examples of an active alluvial fan (a fan-shaped accumulation of sediment) in upland Britain occurs at the junction of the River Feshie with the River Spey. This, too, is characterised by rapid riverbank erosion and channel switching during floods. During a flood in 1990, the position of the main channel shifted from the centre of the fan into a previously abandoned channel to the south. To reduce flooding on agricultural land, it was later artificially moved back into the centre of the fan by means of channel straightening and bulldozed gravel banks. In 1996, the river again changed naturally to a more northerly course where it has remained.

The dynamic nature of the River Feshie highlights the problems in managing active rivers and attempting to control flooding within a very narrow zone. Traditional methods of confining such highly dynamic rivers to single engineered channels and protecting riverbanks from erosion are not only detrimental to the natural environment, but also will almost invariably fail. Alternative approaches give rivers the freedom to move within broader flow corridors on their active floodplains (see chapter 5). This allows the natural processes to continue to operate and helps to reduce the flood hazard downstream in the catchment by letting floodplains act as water storage areas during floods.

The River Feshie is a highly dynamic, gravel-bed river noted for rapid channel changes during floods. This braided reach is the most active of any gravel-bed river in Britain.

Bedrock channels are often highly confined within gorges, sometimes called linns in Scotland, which have stepped long profiles, waterfalls, potholes and plunge pools. Good examples include the Linn of Dee, the North Esk near Edzell, the Black Rock of Novar at Evanton, Corrieshalloch Gorge, and Randolph's Leap on the Findhorn. Many of these gorges were probably formed by meltwater rivers as successive ice sheets melted. However, at the end of the last glaciation, some rivers did not return to their previous courses where these had been infilled with thick glacial deposits, but instead excavated new valleys, leaving their old courses buried under the glacial deposits. A good example is the Clyde upstream from New Lanark. Here, the river has cut a new course through a narrow 7-kilometre-long gorge with two spectacular waterfalls at Bonnington Linn and Cora Linn.

Although there is a great deal of variability, some generalisations can be made about the pattern of river landscapes from the mountains to the sea. Typically in the upper reaches of the Highland glens, steep boulder-bed torrents are interspersed between bedrock reaches with gorges and waterfalls. These streams may be able to shift their bouldery bed material only during extreme floods, but can produce dramatic effects when they do so, as in the case of the Allt Mór, which swept away part of the access road to the ski slopes on Cairn Gorm during a flash flood in August 1978.

Stepped long profiles continue downstream, reflecting patterns of glacial erosion and bedrock outcrops; bedrock channel reaches separate wider alluvial reaches with meandering channels developed on infilled glacial valley floors and basins. The rock sections often display impressive gorges, sometimes following lines of geological weakness. Many such gorges have probably been formed by glacial meltwaters during

The Muckle Spate

There are few such well-documented floods as the Muckle Spate of 1829. Sir Thomas Dick Lauder recorded the event and gave a graphic description of its effects in his book, *An Account of the Great Floods of August, 1829 in the Province of Moray and Adjoining Districts*, published the following year. He had reason to do so as most of his land on the lower reaches of the River Findhorn at Relugas was devastated. Quite simply, there was a tremendous storm over the Cairngorm and Monadliath Mountains, which gave rise to the most extreme historic flood ever recorded in Scotland. The amount of rain that fell over the area at the centre of the storm is not known, but 95 millimetres was recorded over a 24-hour period at Huntly, many miles away. The whole of the area from Inverness in the north-west to Montrose in the south-east was affected by the flood. Such was the damage and devastation caused to land, roads and bridges, and to houses and other buildings, that a national emergency relief fund was established.

One of the best places to gauge the size of the flood is at Randolph's Leap on the lower River Findhorn, about 10 kilometres upstream from Forres. Here, the river course has been cut by glacial meltwater into a gorge around 400 metres long, 8 metres wide and 18 metres deep. To give some idea of the intensity of the flood during the storm, at the narrow point of the gorge where it is only 3 metres wide, the river rose by about 15 metres and dumped huge boulders 1 to 2 metres in diameter.

Elsewhere, the depth of water at the height of the flood is now rather difficult to comprehend, with depths of 6 metres being recorded on the floodplain of the River Dee where Mar Lodge now stands. The flood carried with it large quantities of debris eroded from the surrounding hillsides and valley bottoms, and the alluvial fans at the junctions of side glens with the main straths were considerably extended.

There are many other manifestations of this great flood, particularly the debris slides and debris flows on the hillsides in many of the glens in the Cairngorms, most notably the Lairig Ghru. Such features form where the soil and weathered debris become so wet that they lose their coherence and flow and slide down the hillside. Many new slides were initiated by the Muckle Spate and have also been active during later intense rainfall events, for example in 1956.

The River Findhorn at Randolph's Leap as sketched by Sir Thomas Dick Lauder. The river rose about 15 metres above its normal level during the Muckle Spate of 4 August 1829.

successive periods of deglaciation. Corrieshalloch is a spectacular example of such a slot gorge cut along a line of geological weakness.

In the lower-gradient alluvial reaches, wandering gravel-bed rivers are typical. They may have single or braided channels, which have migrated backwards and forwards across their floodplains, forming suites of river terraces cut into glacial sediment infilling the glens. Modern floodplains are often bounded by bedrock or the bluffs of older terraces near the valley sides. Erosion and reworking of such deposits provides a continuing supply of sediment for the rivers to form temporary, shifting bars. Studies of old maps have shown that braiding was more extensive in the past on gravel-bed rivers, for example on parts of the Tay. However, widespread construction of flood embankments during the nineteenth and twentieth centuries engineered a change from braided to more stable single channels. Over the last 250 years, channel instability in gravel-bed rivers has been pronounced during floods, and the rivers have extensively reworked their valley floors through rapid rates of lateral movement. Alluvial fans commonly occur where tributary steams join the main rivers. In many glens, lochs are another legacy of glaciation and act as sediment traps with the upstream ends infilled by delta deposits. Loch Lubnaig north-west of Callander is a good example.

Lowland reaches of rivers are relatively limited in extent along the coastal plain of the Moray Firth and east coast and it is only really in the Central Lowlands that the rivers take on a lowland character. Generally they have low gradients and a relatively stable meandering form with sand rather than gravel beds. Their stability is often maintained by bank protection and floodbanks. The lower courses of the Rivers Forth, Earn and Cree extend across the raised estuarine sediments, or carselands, of the Forth, Tay and Solway estuaries, respectively.

Opposite. Meanders are well developed on the lower River Cree between Newton Stewart and its mouth at Wigton Bay on the Solway Firth.

THE CHANGING COAST

Sea level changes

The relative level of the sea and the land has changed throughout geological time. The most recent changes during the Ice Age have left a strong imprint on the form of the coast, its position and the type of features present, including shorelines and beaches that are now raised above present sea level. Broadly speaking, during the last glaciation between 23,000 and 19,000 years ago, global sea level was around 125–135 metres lower than today since the expanded ice sheets had locked up vast amounts of the world's fresh water. Sea level then rose in steps as the glaciers melted and released their water back into the sea: to -70 metres by about 13,000 years ago, -10 metres around 7,000 years ago, and roughly to its present level around 4,500 years ago. However, this general pattern of global sea-level (glacioeustatic) change (see chapter 2) is complicated by a wide variation in *relative* sea level around Scotland over time because the level of the land has also moved up and down due to glacioisostatic changes – the crust sinking under the weight of the growing Scottish ice sheet and then slowly rising up again when it melted.

The last ice sheet was thickest over the western Highlands, but thinned radially away from this area towards North-east Scotland, the Northern and Western Isles and Galloway. Because the crust was pushed down further where the ice was thicker, areas near the centre of the ice sheet, centred on an area between Rannoch Moor and Crianlarich, have had further to rebound and are still continuing to rise today. In contrast, in those areas near the periphery of the ice sheet, most of the uplift was completed thousands of years ago, so that relative sea level has been rising there. To these effects are now added the rise in sea level resulting from global climate change, and in the future, most parts of the Scottish coast will experience a relative sea-level

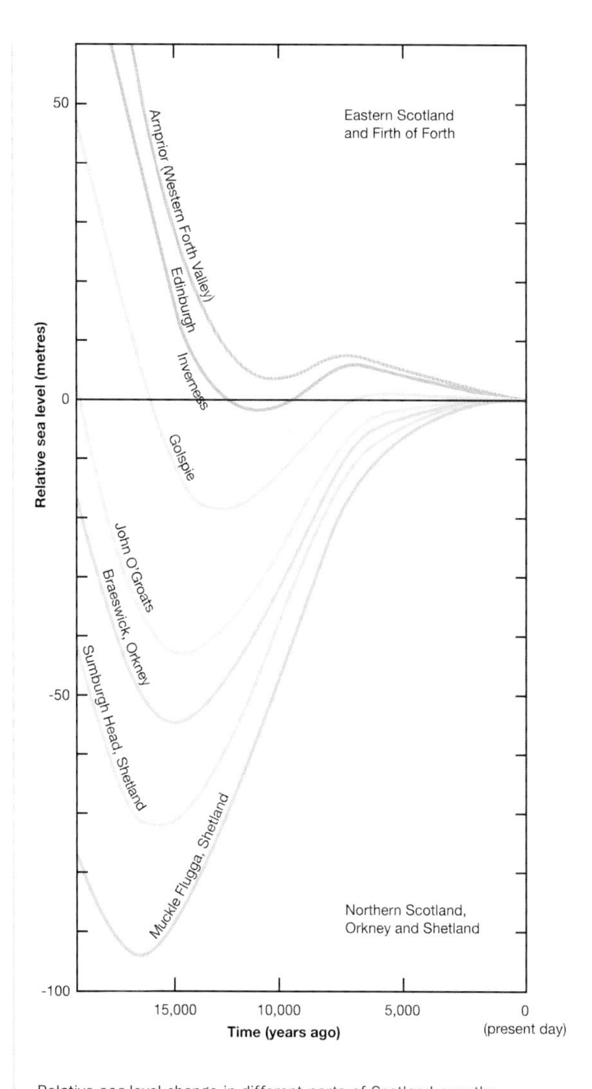

Relative sea-level change in different parts of Scotland over the last 19,000 years. The curves are based on a model which takes into account geographical variations in the former thickness of the Scottish ice sheet and the properties of the Earth's crust. The rise in relative sea level began much earlier and has been much greater in the Northern Isles than in central Scotland, reflecting differences in the ice thickness and therefore the amount of rebound of the land.

rise. The form and position of the present coastline of Scotland are therefore an amalgam of the effects of all these different changes in land and sea levels, in the past and at present.

The effects of changing land and sea levels are clearly seen in the landforms around the coast of Scotland and in the shifting position of the coastline. 'Staircases' of raised shorelines occur extensively in those areas where uplift of the land has outpaced the global sea-level rise at different times in the past. Formerly, such shorelines were commonly named according to their apparent altitude above sea level, as the '25-foot' or '50-foot' raised beaches, for example. However, detailed surveys have revealed that the shorelines are not horizontal but are tilted as a result of the variations in the amount of isostatic uplift. The altitudes of individual shorelines typically decrease away from the centre of isostatic uplift in the Rannoch Moor to Crianlarich area. Consequently, individual shorelines cannot be named according to their altitude.

Relative sea level around much of the mainland was at its highest before 15,000 years ago when the last Scottish ice sheet was melting and the land was still pushed down. Shorelines formed at sea level at that time were then raised by later uplift of the land and now occur as much as 40 metres above present sea level. Particularly good examples occur on the west coast of Jura where there are magnificent raised beaches consisting of quartzite gravel (see chapter 6). Other notable examples are in eastern Fife, north of Fife Ness, and at Earlsferry. Elsewhere, raised shorelines are cut into the seaward edges of outwash terraces and deltas at the mouths of some west coast glens. Good examples are in the lower reaches of Glen Euchar around Kilninver, south of Oban, and in Kilmartin Glen, near Crinan.

As the last glaciers retreated, the sea flooded the lower-lying land and significant thicknesses of silts, sands and clays accumulated in the Forth, Clyde and Tay estuaries. These deposits contain the shells of marine faunas found at much higher latitudes today. In the Glasgow area, they extend up to 35–40 metres above present sea level, indicating that the sea must have entered the lower Clyde and Kelvin valleys and formed a large embayment in the Paisley–Linwood area. At this time, too, the sea flooded into Loch Lomond from the Clyde, creating what must have been a most impressive sea loch. In the Forth valley, the sea extended westwards to the area of Menteith,

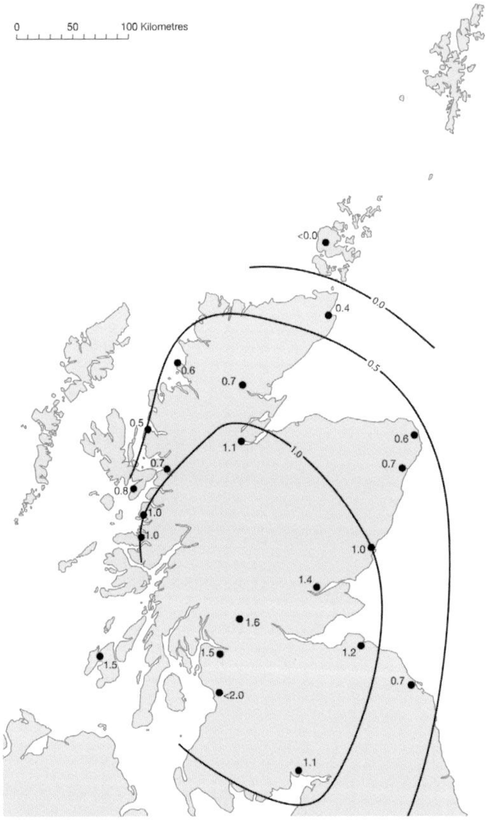

Recent changes in relative land and sea level. The figures are in millimetres per year. Values above zero indicate a relative rise in the level of the land, and vice versa. Note that, although global sea level is rising, the rate of uplift of the land is greater over most of mainland Scotland because of isostatic uplift. The pattern of uplift reflects greater postglacial rebound of the land in the South-west Highlands, where the last ice sheet was thickest.

If projections of future global sea-level rise are borne out, then the whole of Scotland will experience a net sea-level rise by the end of the present century (see chapter 5).

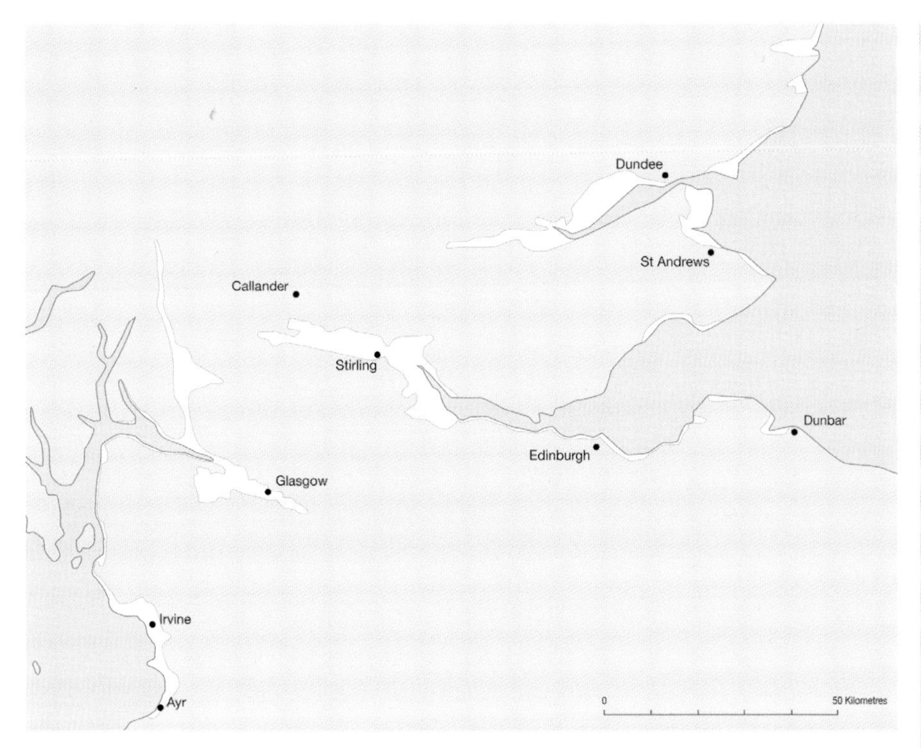

Areas flooded by the sea in central Scotland at the maximum of the postglacial rise in sea level around 7,000 years ago. Earlier, as the last ice sheet was melting, an even greater area was flooded.

forming a series of beaches that were later buried by the postglacial carse deposits. Only a narrow land connection lay to the west, so that the geography of central Scotland was rather different from that of today. Along the Tay estuary, the sea continued west along Strathearn and the Methven valley, nearly to Crieff. Further north, the sea also flooded the land at the head of the Beauly Firth and along the Cromarty Firth.

For several thousand years, relative sea level then fell in central Scotland as local uplift of the land (due to glacioisostatic rebound) outpaced the rate of global sea-level rise. Later, however, between about 9,500 and 6,500 years ago, relative sea level rose again, following the final melting of the North American and Scandinavian ice sheets, and large areas of the coastline were once more drowned, particularly along the lower ground of the Forth, Clyde and Tay estuaries. In the Forth valley, the sea extended westwards to Flanders Moss. In the west, the postglacial sea flooded into the lower Clyde valley, and again into Loch Lomond. Once more, northern Scotland nearly became a separate island from the rest of Britain. In the Irvine area in Ayrshire, the sea extended 4 to 5 kilometres inland from the present-day coast, forming a large embayment and depositing beach sands and gravels. Thick deposits of silts and clays accumulated in the extended estuaries, particularly in eastern Scotland along the Forth and Tay, the Beauly Firth and Montrose Basin, and also along the Solway Firth coast. Later as relative sea level fell, all these areas re-emerged from the sea and now form extensive areas of flat carseland. Shells and even entire skeletons of whales have been found buried in these raised beach and carse deposits. Today, the western Forth valley contains one of the most detailed records of relative sea-level change in Scotland.

Extensive raised beaches are also associated with these higher postglacial sea levels and occur at heights up to about 14 metres above present sea level; some of the best examples are on the west coast of Jura, at Rhunahaorine Point in Kintyre, along the west coast of Arran, and at Spey Bay. The most extensive feature is known as the Main Postglacial Shoreline. Together with its cliffline, it forms a prominent feature of the coastal landscape of eastern and western Scotland. Large coastal forelands formed where sediment supply was particularly abundant, such as Morrich More at the mouth of the Dornoch Firth, and at Spey Bay. These postglacial raised beaches provide flat building land for settlements; for example at Arbroath, Montrose, Leith and Grangemouth. They also provide sites for some

of our most famous links golf courses: St Andrews, Carnoustie and Troon.

In the Northern and Western Isles, the last ice sheet was thinner, so that the land there rebounded less and was overtaken by the rising sea level. Here, in contrast to much of central mainland Scotland, relative sea level has been rising for the last few thousand years, creating a drowned coastline. Submerged peat deposits, sometimes with tree remains, have been recorded off the coast of these areas, at the Ayres of Swinister in Shetland, on Sanday in Orkney and on the Uists and Benbecula in the Western Isles, for example. Dating of these deposits indicates that relative sea level has risen by at least 5 metres in the last 6,000 years or so in the Western Isles, and by about 3 metres in the last 7,500 years on Sanday. This is in marked contrast to most areas on the Scottish mainland where it has fallen by more than that amount over the same period.

One particularly remarkable event recorded in the coastal sediments of eastern Scotland is a tsunami that occurred about 8,100 years ago. It is believed to have been created by a huge submarine slide on the Norwegian continental shelf and to have flooded the coast from Shetland to the Borders.

Another distinctive feature of sea-level change

Above. Raised shorelines form a series of 'steps' at Kincraig Point on the south coast of Fife.

Opposite. The carselands of the Forth valley are bounded to the north by the fault scarp of the Ochil Hills. This area was inundated around 9,000–7,000 years ago during the postglacial rise in sea level, when the estuary extended as much as 25 kilometres west of Stirling. Stirling Castle and the Wallace Monument stand on glacially eroded remnants of the Midland Valley Sill. The Castle Rock would have been an island in the middle of the estuary.

The carselands of the Forth: buried beaches, bogs and stranded whales

The fertile carselands of the Forth valley west of Stirling contain a remarkable archive of past environmental changes and human activities. Buried beaches, peat bogs, whale skeletons and archaeological remains point towards dramatic changes in postglacial sea level and the local flora and fauna, as well as human impacts on the landscape.

Carse, literally meaning an extensive stretch of low alluvial land bordering a river or estuary, is the name given to the raised estuarine mudflats bordering many Scottish estuaries and extending inland along the river valleys at their heads. The most extensive carselands are those of the Firth of Forth, stretching some 50 kilometres from Grangemouth to near the Lake of Menteith.

As the last ice sheet melted, the land was still depressed by the weight of the ice and the sea flooded westwards along the Forth valley, forming an estuary which extended nearly to Aberfoyle. However, as the land recovered and glacioisostatic uplift outpaced the rising sea level, much of the area emerged during the early part of the Holocene. A cover of peat developed over the poorly drained surface of the emergent estuarine deposits. Later, around 9,000 years ago, the rising sea level for a time exceeded the uplift of the land, and the Forth valley once more became an extended estuary. The peat was buried by the estuarine mudflats. Locally, rock islands, like that on which Stirling Castle stands, and peat islands in parts of Flanders Moss remained above the sea. By about 7,000 years ago, relative sea level was again falling in the upper Forth valley, and the carse surface began to be covered in oak woodland on the drier areas and extensive raised bog elsewhere. In the late eighteenth century, large stumps of oak with axe marks were discovered beneath the peat, suggesting that the woodland was cut down by humans; alternatively, it has been suggested that the cuts could be beaver teeth marks. The remains of a wooden construction dating from around 5,200 years ago may have been a hunting or fishing platform or pier used by Neolithic people to exploit the natural resources of the mudflats and creeks.

Other remarkable finds in the carse deposits include beds of mussel and oyster shells and the skeletons of seals and whales. At high tides, the surface of the carse was covered by the sea over an area 25 kilometres long by 5 to 6 kilometres wide. At low tide, whales were left stranded from time to time on the extensive mudflats. Thomas Jamieson, in 1865, noted the discovery of a whale skeleton at Blair Drummond and another was found at Abbey Craig. Altogether, at least fifteen whale skeletons have been reported in the carse deposits west of Stirling, some up to 21 metres long. Bone tools found alongside some of the skeletons suggest that opportunistic Mesolithic people made use of this resource.

From the eighteenth century onwards, large areas of the surface peat were cleared as part of agricultural improvements in order to access the fertile carse deposits beneath. The peat, up to 6 metres thick in places, was stripped by hand and floated down the River Forth. The first extensive clearance was initiated by judge, philosopher and agricultural improver, Henry Home, Lord Kames, in 1767. His wife inherited the Blairdrummond Estate, and he leased out land on condition that tenants, the so-called 'moss lairds', cleared the peat. The practice was finally stopped in 1865 by an Act of Parliament because the peat was causing problems for the salmon and oyster fisheries downstream. Today, East and West Flanders Moss are important remnants of the once extensive lowland raised bogs of the Forth valley and are highly valued for their wetland ecosystems. They represent one of the most vulnerable types of habitat worldwide and are the largest remaining area of lowland raised bog in Europe.

Shetland: a drowned landscape

Shetland is a good location to see the effects of submergence on the landscape. Relative sea level probably began to rise here around 16,000 years ago and since then could have risen by as much as 90 metres. At several locations, deposits of peat lie submerged below present sea level. A radiocarbon date from one such deposit at Symbister harbour on Whalsay suggests sea level has risen by at least 9 metres in the last 6,000 to 7,000 years.

There are a number of other tell-tale signs of submergence and these are best seen along the inner shores of the numerous inlets. Many of these distinctive features have been given local names. The inlets formed where the lower reaches of the valleys have been submerged by the sea are known as voes. Unlike the fjords of the western Scottish mainland, the form of these voes generally owes less to glacial deepening than to drowning by the sea; good examples are Dales, Basta and Aith voes. At the shoreline, there is usually a bar of sand and/or gravel gently fashioned by the waves, known locally as an ayre. Unlike the sand and gravel bars on the Scottish mainland, which are formed of materials that were combed up from the sea bed as the sea rose after the glaciers had melted, the Shetland ayres appear to be made of material eroded from the valley floors and sides and then carried to the coast by rivers; the gravel has not been smoothed and rounded to any great extent by the waves. Immediately inland from the ayres there is often a lagoon where the salinity varies with the level of the tide and the storminess of the sea. These are known locally as oyces and are often named The Houb. In some places, the oyces are tidal, and in other places they contain brackish water. Further inland still, is marshy ground with a cover of peat. Occasionally, more than one ayre has developed, as at the Ayres of Swinister. An equally distinctive landform is where a bar runs between the mainland and an offshore island, forming a tombolo – a name derived from the Italian word for sand-dune. The best example connects St Ninian's Isle with the Mainland at Bigton in the south of Shetland.

The range of features associated with the drowning of the coastline in Shetland is well seen at the Ayres of Swinister where three ayres or bars have formed. They enclose a tidal basin called The Houb.

Scotland's tsunami

At the end of 2004 tsunamis were the centre of attention worldwide after an earthquake off the coast of north-west Sumatra triggered a massive tsunami causing appalling loss of life and the devastation of coastal communities and infrastructure around the Indian Ocean.

A tsunami is a series of waves generated in a body of water by a major disturbance such as an earthquake, volcanic eruption or landslide. Tsunamis often occur following earthquakes below the sea bed, particularly in subduction zones where the Earth's plates are grinding past each other (see chapter 2). The motion of the plates is intermittent, so that stress builds up to a point where it is released by a sudden movement. The movement can cause the sea bed to be raised or lowered by many metres, sometimes accompanied by huge underwater slides of rock and unconsolidated sediments on steeper slopes. The resulting displacement of the overlying water creates a series of waves at the sea surface. In the open ocean, these are relatively far apart, over 100 kilometres, and are called long waves. They have very wide fronts, often up to 500 kilometres, and are not very high, maybe less than a metre, but they travel at very high speeds of up to 800 kilometres per hour. When one of these waves reaches shallow water near the coast, a tremendous transformation occurs: the front of the wave slows down and increases dramatically in height as the rest of the wave, still travelling fast, piles up behind it. The resulting tsunami hits the coast with great ferocity and its height can be anything up to 10–15 metres: a wall of water that has a huge destructive power.

Although Scotland lies in a geologically stable area not generally associated with major geological hazards, a tsunami did in fact strike the east coast in the relatively recent past, about 8,100 years ago. It was caused by a huge submarine slide on the Norwegian continental slope off the west coast of Norway at a location called Storegga, north-west of Trondheim, in water 500 metres deep. An earthquake may have triggered the slide, possibly accompanied by release of methane from gas hydrates. Evidence for the tsunami takes the form of a buried layer of sand that is preserved in the coastal deposits of eastern Scotland from Shetland to the Borders; the event has also been recorded along the west coast of Norway, in the Faeroe Islands and in North-west Iceland. In Shetland, the tsunami deposit occurs at several locations as a buried sand layer within coastal peat and lake sediments. The best place to see it is at the viewpoint lay-by on the south shore of Sullom Voe where it lies sandwiched in the roadside peat and contains sand and lumps of peat ripped up by the waves. Elsewhere, the sand layer can sometimes be seen at Montrose Basin near the Scottish Wildlife Trust visitor centre, if a section has been excavated.

The tsunami that affected the coast of Scotland around 8,100 years ago is represented by a 30- to 40-centimetre-thick layer of sand and ripped-up blocks of peat at Sullom Voe in Shetland. The height of the run up of the tsunami here was big – about 22 metres above the sea level at the time.

Based on the altitude above the contemporary sea level reached by the sand layer, the estimated run up height of the tsunami may have exceeded 5 metres along the coast of mainland Scotland and reached an astonishing 25 metres in some places in Shetland. It is possible that the waves themselves may have been 10 metres high. While the event was extensive, its human impact was probably small, because of the relatively low population at the time. However, at least one Mesolithic settlement at Inverness was inundated and others may have been flooded at Broughty Ferry and in North-east Fife. Two further tsunamis are known to have struck the coast of Shetland in the last 5,500 years. This indicates that such events are not unique and that Scotland is not exempt from tsunami hazards over longer timescales.

has been the formation of shore platforms with fossil clifflines. Some occur in the present intertidal area, whereas others have been raised as much as 45 metres above present sea level. The ages of formation of these platforms are uncertain but several appear to pre-date the last glacial maximum as they are ice-moulded or overlain by till and may have been occupied by the sea on several occasions. These platforms are particularly well developed in the Inner Hebrides where three main generations have been identified (see chapter 6: Islay and Jura). The High Rock Platform, ranging in height from 18 to 45 metres above present sea level, is best seen on the north-west coast of Islay, the west coast of Jura, the west and south coasts of Mull, and on Colonsay, Oronsay and Rum. On Bac Mór in the Treshnish Isles, it appears as the distinctive platform which forms the 'brim' of the Dutchman's Cap. The High Rock Platform was probably formed by a combination of wave, sea-ice and frost action during successive glacial periods when the margin of the Scottish ice sheet lay to the east of the islands. On Coll and Tiree, it is represented by rock surfaces several kilometres wide.

The Main Rock Platform is a tilted shoreline. It slopes southward from around 10 metres above present sea level on Lismore to a point below present sea level at Machrihanish and on the coast south of Ayr, reflecting the greater uplift of the land in the north of the area where the ice was thicker. The platform is best seen around the south-east coast of Mull from Salen to Carsaig, around Kerrera and on Lismore, where it is up to 150 metres wide and backed by a 15-metre-high cliff. Duart Castle, one of the sights on the ferry crossing from Oban to Craignure on Mull, is built on this platform. Further south, it forms a

The intertidal shore platform at Dunbar has been planed across a series of dipping beds of Carboniferous sedimentary and volcanic rocks.

conspicuous rock feature along much of the coast of the Firth of Clyde and in the Clyde estuary. It is particularly well displayed between Largs and Inverkip, between West Kilbride and Ardrossan, at Heads of Ayr and around the Ardmore peninsula near Cardross. On Great Cumbrae it forms a raised bench running the whole way round the island. Throughout the area, the Main Rock Platform is backed by a prominent cliff and is overlain by postglacial beach deposits. It was partly cut during the cold climatic conditions at the time of the Loch Lomond Readvance glaciation through a combination of intense frost weathering and sea-ice and wave action, under conditions similar to those experienced on the coast of Arctic Canada today. However, it is probably in part an older feature that was reoccupied by the sea. Like the High Rock Platform, the Main Rock Platform is often cut in resistant rock and it is unlikely that wave energy alone could have formed it.

The Low Rock Platform occurs as an intertidal feature at heights between 0 and 2 metres on Jura, Scarba, Islay, Colonsay and Oronsay. As it is not tilted and shows evidence of having been glaciated, it is believed to have formed during one or more interglacial periods before the last glaciation. Intertidal platforms also occur extensively along the east coast of Scotland, one of the best examples being at Dunbar. Again these are likely to be old rather than modern features. Offshore, extensive marine erosion surfaces with backing clifflines and submerged caves, such as those around St Kilda, have been identified on the continental shelf to depths of over 100 metres. Probably these features formed during glacial stages when relative sea levels were much lower. Because they occur some distance from the area of maximum crustal depression by the glaciers, they have not been raised above present sea level.

The coast today: cliffs and beaches

Scotland today has a very long coast – 10,192 kilometres. It is also very varied: high and steep cliffs with no beaches, abandoned cliffs with beaches at their foot, long sandy beaches backed with dunes, extensive sand plains, both deep and shallow estuaries, small bays of sand backed with cliffs and many other variations. This diversity reflects a combination of varied rock types, the glacial legacy and different wave-energy conditions.

The coastline is the point where the land and sea meet. Its position is a reflection of the types of rocks and geological structures, the effects of waves and currents, changes in sea level and whether the land is rising or sinking. Where the land is steep, then the position of the coast does not vary very much even though the relative height of the sea may change considerably, as along the west coast of Orkney and Shetland, and in the fjords along the west coast of the mainland. Elsewhere, it can change considerably over centuries as the land either builds out into the sea, as on the north and south sides of the Tay estuary, or erodes rapidly, as along the west coast of South Uist. The shape of the coast is also influenced by the existence of old river valleys, like those of the Forth and Tay estuaries and the fjords of the west coast, all of which have been deepened by the glaciers.

The Scottish coast can be broadly divided into rocky and soft types: in the former category, hard rock is the predominant material and the key element is cliffs, and in the latter, gravel and sand are the predominant materials and the key elements are beaches and dunes.

Rocky coasts

Most of the coastline of Scotland consists of rock. Changes along rocky coasts depend on the resistance of the rock itself, which is determined by the presence of joints, faults and other fractures. Geological structures are also very important in influencing the shape of the coast; for example, the coast of the Black Isle northwards to Caithness follows the line of the Great Glen Fault. Although rocky coasts change slowly, there is nevertheless a wide variety of landforms.

Scotland has a number of spectacular stretches of rocky coast with plunging cliffs that extend well below sea level. They occur in the Northern Isles, especially on Foula, at St John's Head on Hoy, and on St Kilda, where Conachair at 430 metres is the highest cliff in Britain. Other good examples are at Cape Wrath and St Abbs Head. Current wave action makes little discernible impact on these cliffs which are vertical or nearly vertical and cut in resistant rocks.

Other rocky coasts have retreating cliffs, resulting in a fascinating array of landforms:

Waves and tides

The rocks and sediments along the coast are fashioned by water movements. The level of energy in the water, the degree of resistance of the material forming the coast, the availability of sediment and whether relative sea level is rising or falling are the main factors determining whether the coast is being eroded, is static or is building outwards.

Waves are by far the most significant element in shaping the coast. They are generated by wind, which is usually associated with atmospheric pressure gradients offshore. Sometimes waves generated thousands of kilometres away reach the coast in the form of swell waves. These are generally constructive and carry sediment towards the beach. More frequently, locally generated wind waves affect the shore. During storms, when these winds are strong, the waves can be extremely destructive and carry sediment away from the beach. Equally, if their height and frequency are lower, they can be constructive by carrying sediment from offshore onto the beach and therefore enabling it to better withstand wave attack in stormier conditions. The angle of wave approach is very important. The more it is at an oblique angle to the coast then the greater chance there is of sediment being carried along the coast to form spits and bars at and around river mouths and estuaries. Often there is a pattern of erosion and deposition in the same coastal unit, termed a 'sediment cell'. Erosion may occur in one section largely by wave action, perhaps aided by a river mouth lowering the beach surface and making the coast more vulnerable to wave attack. Further along the coast in the direction of the movement of the water, there may be an accumulation of sediment. Good examples of such patterns occur between St Andrews and the mouth of the Tay and at Montrose Bay. Where the water immediately offshore is relatively shallow, such as in large bays on the east coast, then waves lose much of their energy before reaching the coastline itself and are therefore generally less destructive. On the other hand, where the water remains deep up to the coastline, usually where there are cliffs whose base is well below the water level, then the energy at the point of wave impact on the coastline is very high, leading to erosion of even the hardest rocks and the formation of caves, stacks and other features. The cliffs around Esha Ness, on the west coast of Shetland, have these characteristics.

Tides are significant around the Scottish coast as they are continuously moving seawater around. There is a twice-daily rise and fall and a lunar monthly pattern of fortnightly high (spring) tides and low (neap) tides. Generally, the spring tide periods tend to produce the more dramatic effects of erosion along soft coasts, especially if accompanied by low pressure and strong winds. The tide rises and falls at different times around the coast and therefore creates some lateral movement, for example southwards along the east coast. In confined channels and sounds, such as the Pentland Firth and between the islands of the Inner Hebrides, the tidal race is swift and can carry a lot of sediment away from the coast.

Generalised map showing the many different materials that form the coastline of Scotland.

caves, arches, stacks, rockfalls and shore platforms. Sometimes, in the most exposed areas there are modern boulder and gravel beaches on the cliff tops, thrown up by large waves during storms. The joints in the cliff faces are penetrated by sea, rain and groundwater and are further weakened by the pressure from the impact of the waves. These coasts change over hundreds of years so that, for example, headlands with caves become stacks, which are eventually reduced to piles of boulders. Notable examples are: the west coast of Hoy, Orkney (see chapter 6); Esha Ness, west Shetland (see chapter 6); Papa Stour, Shetland; north-west Sutherland; Duncansby Head, Caithness; and the Bullars of Buchan, south of Peterhead.

There are also many examples of abandoned cliffs partly or wholly covered with scree and vegetation. Frequently, there is a narrow raised beach or shore platform in front of the cliffs, representing higher sea levels in the past. Sometimes abandoned caves and sea stacks are also present. Good examples occur along parts of the coast of Kincardineshire, Ayrshire, Arran and Kintyre.

Drowned coasts are areas of land where the surface has been shaped by ice, water and waves but are now all or partly below sea level because the sea has risen relative to the land. Examples are the fjords of the west coast from the Firth of Clyde to Cape Wrath, the deep estuaries of the east coast, especially the Cromarty Firth and Firth of Forth, the archipelago of North Uist and the Sound of Harris, and the voes of Shetland.

Soft coasts

Coasts made of sand and gravel occur throughout Scotland. The critical features in their

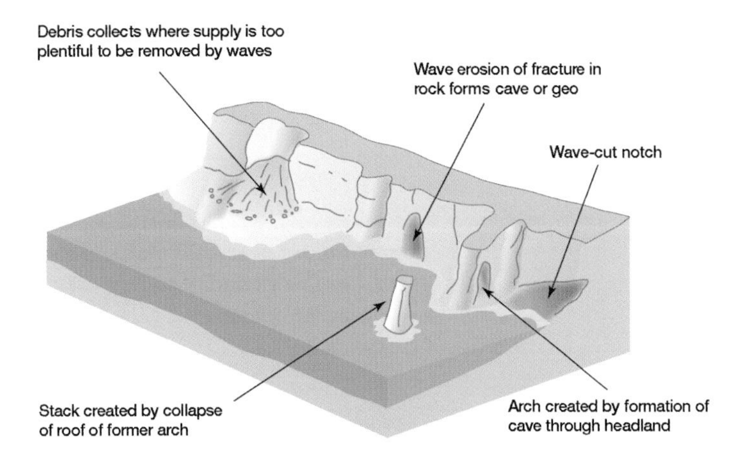

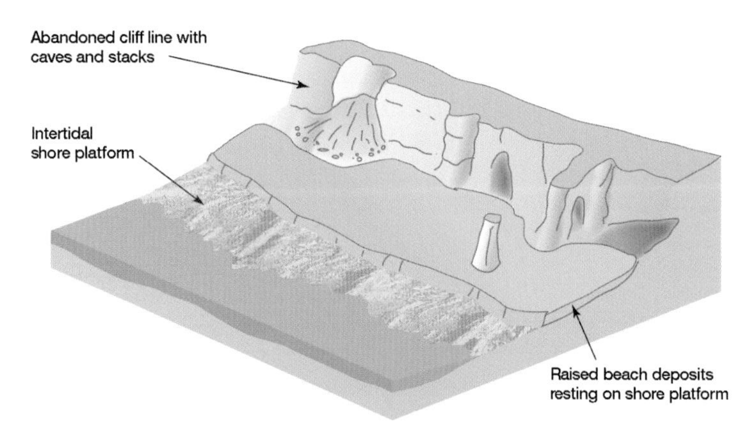

Rocky coasts display a variety of landforms. Many features along a cliff coast show the gradual evolution and disappearance of landforms as the waves remorselessly attack the cliff face. A sequence can be traced from a notch at sea level to a cave, to a geo, to a stack, and finally to a pile of rubble.

development are the supply of sediment to build beaches and dunes, the pattern of waves and currents, and to a lesser extent the strength and direction of the wind.

Although cliffs are still being eroded and sand and silt are being carried to the coast by rivers, the amounts of material are minor. The most significant source of material for constructing Scotland's soft coasts has been the debris carried to the coast and offshore by glaciers and meltwater during the Ice Age glaciations. This debris has been reworked by the waves and the wind to form sand and gravel beaches and sand dunes as sea level varied after the last glaciation.

Beaches and dunes are regarded by many as the most pleasing and interesting of the soft coastal landscapes. They are certainly the most frequently visited. Scotland has a number of very large sandy beach and dune systems. Some of the best examples are the Sands of Forvie, Aberdeenshire; Barry Links and Tentsmuir at the mouth of the Tay; Culbin Sands, Moray (see chapter 6); Morrich More, Dornoch Firth; north of Wick and at Dunnet Bay in Caithness; Killinallan at Loch Gruinart, Islay; and Torrs Warren and Luce Sands, Galloway. Many of the sand-dune systems are founded on a base of gravel ridges deposited as the sea fell to its present level several thousand years ago. There are also extensive spreads of raised gravel beaches at Spey Bay and Culbin. These are fronted by active beaches, and, in the case of Culbin, by a superb gravel spit.

Where there has been an abundant supply of sediment, the coast has built out seawards over the last few thousand years, forming extensive promontories termed 'forelands'. Good examples are Morrich More, on the south side of the Dornoch Firth, and Tentsmuir, south of the Tay estuary, where successions of offshore sand bars

Shetland: attacked by the sea

The power of the waves to throw pebbles and boulders well above sea level is clearly seen on the cliff top at Esha Ness, north-west Mainland, Shetland.

Shetland has a very long coastline dominated by cliffs. Most notable are the steep, plunging cliffs of Bressay and Noss in the east, cut in sandstones, and Esha Ness and the Villians of Hamnavoe in the west, cut in volcanic lavas and tuffs. The energy of wave action on the west coast of Shetland is higher than anywhere else in Britain. The cliffs plunge well below sea level and there is nothing to stop or reduce the full force of the Atlantic waves hitting the cliff faces and washing over the land on the cliff tops. Waves caused by wind over the sea surface and swell generated by bad weather anywhere in the North Atlantic reach these cliffs. Waves up to 20 metres high have stripped the vegetation and soil off the cliff tops, breaking up the bedrock, piling it into impressive ridges up to 3 metres high, and scattering pebbles and other debris up to 100 metres inland. This is not the result of some freak past events: it is happening at the present time.

Other features are also characteristic of the intense energy affecting these cliff faces. The pressure from water trapped in the joints and fissures in the rock exploits the slightest weakness, creating and enlarging cavities in the rocks. The cliffs are punctuated with caves, which, when their roofs collapse, form blowholes known locally as gloups. During storms, water spouts onto the cliff tops and emerges through other clefts in the cliffs. With further erosion, the cliffs retreat inland leaving some of their more resistant parts isolated as headlands with arches and stacks. In turn, these too eventually succumb to wave attack and collapse into the sea.

At the edge of the world

St Kilda forms a dramatic landscape of cliffed islands – Hirta, Soay, Boreray, Dun and Levenish – and sea stacks – Stac Lee, Stac an Armin, Stac Soay and Stac Biorach. The cliffs plunge well below sea level and it is likely that the stacks have been detached from the larger islands by the very high energy in the waves that constantly attack the shores of the archipelago. It is exposed to the swell waves and wind waves from the whole North Atlantic and, because the water is very deep right up to the base of the cliffs, none of the energy is dissipated until the waves reach the cliffs themselves. The effects of the waves can be felt as deep as 70 metres below, and up to 100 metres above, present sea level; a remarkable range not experienced anywhere else in Britain.

Evidence of substantial changes in sea level is found well below present sea level around St Kilda. Three remarkable submerged rock platforms at –120, –80 and –40 metres were formed when sea level was much lower during one or more episodes of glaciation and are testimony to the effectiveness of frost shattering, sea-ice processes and wave erosion at this incredibly exposed location. The platform at –40 metres is backed by a submarine cliff with spectacular caves, stacks and arches. The rising postglacial sea submerged these platforms and progressively isolated the islands.

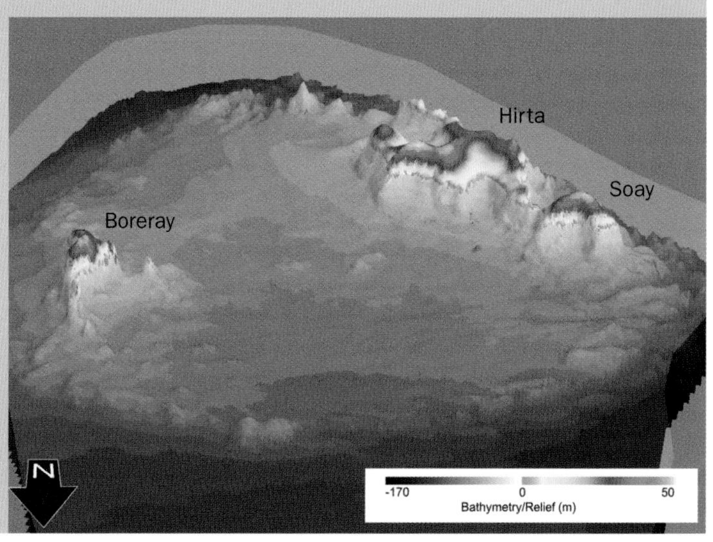

Perspective view of St Kilda from the north-west, based on a 3-D model of the water depth and surface topography. The cliffed coastline of the islands and stacks plunges to 40 metres below present sea level. Two extensive submarine platforms are present at 120 metres and 40 metres below present sea level.

Building the coastline

The coastline of North-east Scotland is remarkable for a number of large gravel and sand systems which are highly dynamic and internationally significant: Morrich More at the southern entrance to the Dornoch Firth, the coast between Ardersier and Findhorn including Whiteness Head and Culbin Sands and Bar, the area around the mouth of the Spey between Lossiemouth in the west and Portgordon in the east, the Strathbeg area south-east of Fraserburgh, and the Sands of Forvie to the north of the Ythan estuary. The material for the formation of these and other features was carried to the coast and offshore by meltwater rivers as the last ice sheet was shrinking. The principal rivers were those flowing into the Dornoch, Cromarty and Beauly firths; the Nairn, Findhorn and Spey in Moray; the Ythan, Don and Dee in Aberdeenshire; and the North and South Esk at Montrose. During the last 15,000 years this material has been moved and fashioned into a tremendous variety of sand and gravel landforms. During the postglacial sea-level rise between about 9,500 and 6,500 years ago, the debris was combed up to a new coastline. The cliff formed at this time is associated with the Main Postglacial Shoreline and is a clear feature at many locations. Relative sea level has then fallen in North-east Scotland, leading to the formation of extensive gravel bars at successively lower levels between the old cliff and present sea level. These are best seen in the 'staircase' of gravel ridges west of the mouth of the Spey, and in the bar that ultimately cut off what has become the Loch of Strathbeg from the sea. Similar gravel bars also lie underneath most of the other large sand systems in the area.

In recent centuries, gravel bars have continued to extend at very rapid rates, for example 34 metres per annum west of the Spey where there has been an exceptional supply of new gravel from the river and from coastal erosion further east. The rate of growth is now much reduced due to a combination of coastal protection works and a natural reduction in the supply of material.

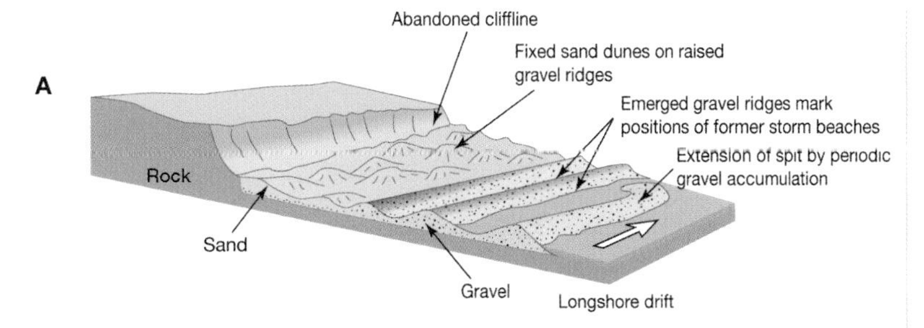

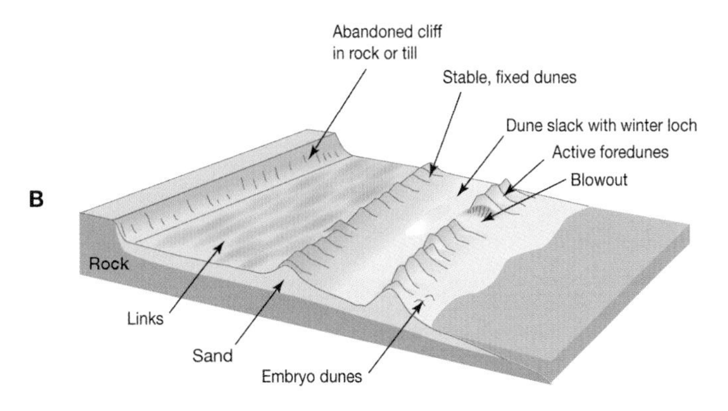

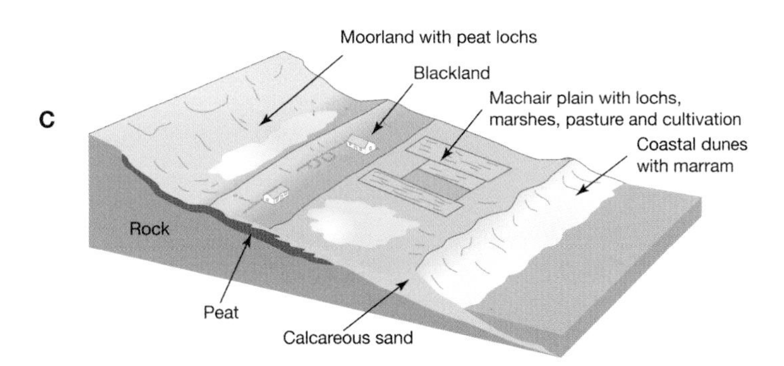

The landforms of soft coasts comprise a range of beach, dune and gravel features. **A.** Gravel beach and fixed dune coast characteristic of the Moray Firth (e.g. Culbin and Spey Bay). **B.** Where there is a supply of sand, as along parts of the east coast, the modern gravel beach is replaced by a sandy beach backed by a line of foredunes (e.g. Strathbeg). **C.** Machair coast typical of the Western Isles and parts of the Inner Hebrides, the north-west mainland and Orkney and Shetland. The dunes and marram are often absent on the Uists.

have become attached to the coast. Around the mouth of the Tay, the forelands at Barry Links on the north shore and Tentsmuir on the south shore are amongst the most active and most rapidly growing in Scotland. They have grown out 3–4 kilometres from the postglacial shoreline in the last 5,000–6,000 years. At a much smaller scale, pocket beaches are attractive features on otherwise rocky coasts. Inevitably, they are popular recreational spots. They occur all round the Scottish coast, with excellent examples on the Ross of Mull and in Wester Ross.

Quartz predominates in the sands of Scotland's beaches and dunes because it is more resistant to weathering than most other minerals and it is also a commonly occurring mineral. There is one notable exception: the sands of the Western Isles and parts of the Inner Hebrides, the north-west mainland and the islands of Orkney and Shetland. These are formed from shells of marine molluscs and other animals living on the shallow shelf immediately offshore. Once the living creatures have died, currents move the shells inshore where they are broken up by the waves and the fragments carried landwards. In exposed locations on the Atlantic coasts, this shell sand is readily blown inland to form sand plains or machair.

The machair is a particularly distinctive feature of the Scottish coast, and is unique to the areas noted above and to the west of Ireland. It is supremely developed along the west coast of the Western Isles (see chapter 6). The shell sands may comprise over 80–90 per cent shell fragments. The shells are made of calcium carbonate and give rise to lime-rich soils. Machair began to form mainly during the middle and late Holocene, at different times in different places, as the rate of sea-level rise slowed, allowing large volumes of sand to be transported from the shelf offshore towards the

Blowing sand everywhere

The sandy beach and dune systems of the Moray Firth and Aberdeenshire coasts are among the most spectacular in the country. They include the Sands of Forvie, a very extensive area of sand dunes on the north side of the Ythan estuary in Aberdeenshire. Just north of the river, there is a large amount of blown sand and huge unstable dunes with little or no vegetation. Many dune fields have distinct patterns, which clearly display how they have been formed and their current direction of movement. However, this is not the case at Forvie where the dunes have no clear pattern.

The Sands of Forvie comprise three distinct elements. In the south, a vast area of bare sand is continually on the move. The sand circulates as if it were on a conveyor belt between the river, the beach and the dunes. Immediately to the north of this bare sand area, vast curved dune ridges 25 metres high and 1 kilometre long are moving northwards. The northernmost section of the Sands of Forvie has a very different appearance, akin to the dune fields elsewhere around the Scottish coast, with most of the dunes relatively stable. However, in addition, there are exceptional parabolic-shaped dunes, nine in all, with steep faces at the front and gentler faces at the back, reflecting the effect of the wind, which is moving them along.

The instability of this dynamic coast has also had a human impact. The northward movement of the sand has buried the site of a prehistoric village, and in 1413 the Old Kirk of Forvie and adjacent cultivation rigs were abandoned due to encroachment of the sand.

Morrich More

Morrich More is a unique example in Scotland of a massive strandplain, consisting predominantly of a series of sand ridges that have built outwards from the postglacial cliff (arrowed) by 8 kilometres in the last 7,500 years or so. The sand came originally from the Dornoch Firth and also the Cromarty Firth when its outlet was through Nigg Bay and into the Dornoch Firth. During recent millennia the foreland has also moved and grown eastwards, supplied with sand from its western side as this has been eroded by the waves. In all, some fifty sand ridges represent five distinct phases in the evolution of the foreland. The low ground between each ridge has developed into salt marsh where the sea still encroaches periodically. On top of the ridges, small sand dunes have developed. A small area of Morrich More has a pock-marked appearance as a result of the area being used as a target range for low-flying military aircraft, but much is otherwise untouched and natural.

coast. This material was then built into beaches by the waves and blown inland to form dunes and machair. However, the amount of sand offshore was finite and after about 7,400 years ago, a dwindling supply, combined with continuing sea-level rise, led to erosion and recycling of sand from the fronting beaches and dunes, which helped to maintain the machair. Today, a combination of storms and rising sea level is continuing this pattern and producing widespread erosion at the front of the dunes and machair.

The amount of erosion and deposition of material on the soft coasts is dramatically influenced by the strength of the waves, and the

Machair

The meaning of the Gaelic word machair is 'an extensive low-lying fertile plain'. But this strict translation fails to give anything of an impression of the natural diversity within the suite of landforms comprising the machair system, or the degree of natural change that occurs between the seasons and over the years, or of the complex interaction between natural processes and human use. All of these characterise the surface expression of the machair in the Western Isles. The machair is therefore best described as a gently sloping sand-plain formed largely by wind-blown shell sand and used for generations for traditional agriculture.

A typical suite of landforms occurs inland from the coast (see diagram on p. 221). Just behind the beach, there is occasionally a dune ridge with a great deal of bare sand and marram grass. Where the dune surface has been disturbed, there are hollows and gaps where the sand is blown inland onto the flatter ground to form a broad sand plain. The shape of the sand plain reflects the presence of rocks and the water table, so it is often undulating, with rocks protruding through the sand and water at the surface during the winter. The wetter areas also form salty lagoons, freshwater lochs and fens, which provide important habitats for plants and birds. The drier areas are cultivated or used for grazing. Peat occurs extensively inland of the sand plain, with a transitional area in between forming the so-called 'blackland', where sand blown onto the surface of the peat forms a rich soil of organic material from the peat and lime from the sand. Sometimes the sand is blown even further inland onto the moorlands and up onto the western faces of the hills running down the eastern side of the islands.

Machair Newton and Lingay Strand, North Uist

pattern and strength of currents. The more exposed parts of these coasts are far more vulnerable to erosion and the loss of land than more sheltered locations. Wholesale retreat appears to be a common feature, for example, of the soft coasts along the Atlantic shores of the Western Isles. In contrast, both erosion and growth sometimes occur along any one stretch of the soft, sheltered coasts of eastern Scotland, for example between the mouth of the Tay and St Andrews and along the shores of the Dornoch Firth. Tentsmuir is the most actively advancing area on the Scottish coastline, having moved forward 100 metres in the last fifty years, this advance being partly fed by erosion of its southern end. By contrast, an average of 4 metres of coastal retreat has occurred annually in the dunes to the north of Montrose, representing one of the most rapidly eroding areas of coastline in Scotland. The sediment eroded from this area has moved north and contributed to the coastal accretion at St Cyrus (see p. 6).

Generally, there is little new material being supplied to the coast at present, either from offshore or from rivers. The waves, currents and winds are therefore largely redistributing existing material in the coastal zone. If coast defences reduce the availability of this material and its movement along the coast, or if it is lost offshore because of rising sea level, then it is not surprising that the adjacent shorelines are retreating. So, interrupting the natural recycling of sediment will exacerbate an already fragile situation on Scotland's soft coasts, particularly in those areas where relative sea level is currently rising and at a time when rising sea level is likely to outpace land uplift over an increasingly larger geographic area in the future. This pattern is echoed around the globe and raises important management issues discussed in the next chapter.

SOILS

Soils are an integral part of the landscape. They provide a medium for plant growth, support habitat diversity and control the flow of water. Soils are a product not only of natural processes but also of human activities. Soils in Scotland began to develop as the glaciers retreated. Pioneer plants quickly colonised the bare glacial deposits, and weathering released nutrients to support further plant growth and soil formation. Where micro-organisms decomposed the plant remains, organic matter accumulated at the surface, which helped to improve the structure of the soil and its ability to retain moisture. Today, the soils of Scotland principally reflect variations in the properties of their parent materials, climate and land use. For the last 5,000 years or more, human activity has increasingly modified soils through changes to the vegetation and through land use for agriculture and forestry. Very broadly, four main groups of soils occur in Scotland: peat and other organic soils, gleys, podzols and brown forest soils. In addition, shallow montane soils occur on the higher hills and mountains. However, there are many variations representing transitions between the different soil groups.

Peat and other organic soils form where wet and cool climate conditions contribute to surface waterlogging and suppress or slow down the biological decomposition of dead plants. The dead plant material accumulates and over time forms peat which is extensively developed in the north and west of Scotland, on higher ground in the east, and in the Southern Uplands. Peat and other organic soils cover about 66 per cent of Scotland and represent a hugely important carbon store; for example, it has been estimated that Scottish soils contain over 50 per cent of the UK's soil carbon stocks. Many areas of peat are undergoing severe erosion in the Grampian Highlands, in Shetland and elsewhere due to grazing and trampling pressures by domestic and wild animals and the effects of the harsh climate. This is a cause for concern, since erosion and oxidation of peat release greenhouse gases and there is an impact on the quality of freshwaters downstream through increasing concentrations of dissolved organic carbon. The blanket peats of Scotland, represented par excellence by the Flow Country of Caithness and Sutherland, are also internationally important as wetland habitats and support important bird populations.

Gleys develop in areas where the drainage is poor and the ground becomes waterlogged. They are recognisable by their mottled grey, blue-green and rusty colours as iron and other compounds are changed chemically from their normal orange-brown forms to forms that are grey-green in colour. The waterlogged conditions impede decomposition of dead plants, and a shallow organic layer may accumulate at the surface. Gleys are common in central and southern Scotland where they are often used for grassland. Peaty gleys, where the surface organic layer is thicker, are frequent in the western Highlands and islands where they tend to support wet heathland and grassland plant communities.

In the drier east, podzols are a common soil type, particularly on well-drained, acidic parent materials. Organic material accumulates at the surface because decomposition of dead plants is slow under the acidic conditions. The soils are well drained, so that organic and mineral compounds are leached from the surface and deposited lower down the profile. These processes produce a very distinctive soil profile, comprising an organic surface layer, a pale leached horizon with an ash-like colour, then a deeper dark brown layer where the leached minerals accumulate. Sometimes an iron-pan (a thin indurated layer cemented by iron oxide) is

Peat everywhere

The landscape of much of Caithness and part of Sutherland, covering about 400,000 hectares, is dominated by blanket peat – the Flow Country. The peatlands may appear uniform at first glance, but this is far from the case. They display many variations in their vegetation and in their surface patterns, including systems of linear ridges and elongated pools lying at right angles to the slope. The pools, locally termed 'dubh lochain' (black lochans), are steep-sided and can be up to 4 metres deep in level or gently sloping areas. They are likely to have formed where increased surface wetness and local waterlogging, perhaps due to climate change and/or changes in local hydrology, reduced the growth of peat in natural depressions in the bog surface, while adjacent hummocks continued to grow. Over time, the differences in peat accumulation became accentuated and the pools expanded and coalesced.

High humidity, persistent cloud cover, over 1,200 mm of rain a year and regular rainfall throughout the year, impeded drainage, low nutrient status and naturally acidic conditions are ideal for the growth of *Sphagnum*. This plant is the basic building block of the peatlands, a natural sponge that can absorb around ten times its weight of water, and does not require many nutrients to survive. The lower parts of the plant die but the upper parts continue to grow. In the acidic and waterlogged conditions, the dead plants do not decay but form a dense mat of material. This is slowly transformed into peat, with the help of the weight of the accumulating plants on top. It is a very slow process, perhaps progressing at a rate of about 1 millimetre per year, but often more slowly. As the peat accumulates over many millennia, it forms a blanket over the ground surface; hence the name 'blanket bog'. There are two layers: a thin upper one with living plants and flowing water and a much thicker lower one with dead plants and limited water flow.

Peat formation began at different times in different places. The earliest accumulations occurred locally in wetter gullies and depressions around 10,500 years ago. Although the Flow Country was never extensively forested, an open woodland with birch, hazel and willow scrub probably developed in drier, sheltered areas, with grassland elsewhere, until about 7,000 years ago. After that, the soil condition deteriorated as nutrients were progressively leached and heathland species expanded, followed by peatland species. The cause of the peat expansion was probably a deterioration in climate to wetter and cooler conditions during the second half of the Holocene, ideal for increased waterlogging and peat growth. Locally, early settlers may also have burned the woodland and peat replaced it. Around 4,500 years ago, there was a brief drier interval of a few hundred years when pine woodland was growing on the blanket bog; the remains of this woodland can be seen today in the tree stumps preserved in the

Patterned blanket bog in eastern Caithness, showing the array of pools, or dubh lochains, of different shapes and sizes.

peat in many areas. However, this woodland expansion was short-lived and was probably terminated by wetter conditions around 4,000 years ago. The development of the ridge and pool patterns on the surface of the peat appears to be a relatively recent feature of the last few thousand years. The peat is continuing to accumulate today and thicknesses of up to 6 metres have been measured.

The peatlands of the Flow Country are a vitally important environment and one of the most extensive intact mire systems in the world. They are home to many rare breeding birds, including waders, divers and raptors, and to a limited range of plants that are adapted to a very special and harsh environment. The latter include many types of bog moss (*Sphagnum*), sundews, cotton grass, bog myrtle and wet heath species. The peatlands also contain significant stores of carbon. Some of this is released naturally as methane. Much more is released as carbon dioxide when a bog is drained or eroded or the peat removed for horticulture or fuel. Methane and carbon dioxide are both important greenhouse gases. It is estimated that a metre depth of peat over a hectare in area contains 550 tonnes of carbon and that the world's peat bogs store many times more carbon than the tropical rain forests.

present. Podzols are common in the uplands and on lower ground in North-east Scotland where they support heather moorland and native pinewoods. They are low in nutrients, but in North-east Scotland they have been improved for agriculture.

Brown forest soils also occur on well-drained parent materials. They formed originally under deciduous woodland, where the plant litter was less acidic and well decomposed. This gives them a good structure and a brown colour, becoming lighter with depth as the organic content decreases. Brown forest soils are deep and fertile and occur mainly in the east and south of Scotland where they provide some of our most productive agricultural land. They also support important semi-natural woodland habitats.

Because of the extreme weather and thin vegetation cover, montane soils are usually shallow and poorly developed. Growth of surface vegetation and decomposition of organic matter are slow, leading to the formation of a very thin organic layer at the surface. These soils, which commonly include alpine podzols, are very fragile and vulnerable to erosion.

An alpine podzol in the Cairngorm Mountains with a distinctive pale leached horizon below the surface organic layer. These thin montane soils are fragile and vulnerable to disturbance from trampling and natural processes.

Peat erosion on the Ladder Hills, east of Tomintoul.

HUMANS SHAPE THE LANDSCAPE

Human activity is now a widespread agent in shaping the landscape. For example, globally, it has been estimated that the amount of rock and soil transported annually as a result of human activities – for building cities, motorways and dams and through mineral extraction and soil erosion from agricultural land – is now ten times that transported by all natural processes together. In the last few hundred years the spread of urban areas, together with industrial and commercial development and infrastructure, has greatly modified the surface landforms and soils of Scotland. The effects of this activity are clearly evident in a variety of environments as described in chapter 1.

Many of these effects are likely to be exacerbated in the future through climate change, as parts of the country become wetter or have more intense rainfall and as sea level rises. This theme is explored further in the next chapter.

This chapter has revealed how various different processes have shaped the surface landforms, or geomorphology, of Scotland today. These land forms do not occur in isolation and it is their particular combination, together with the underlying geological framework, that has endowed Scotland with its distinctive landscapes and remarkable diversity of scenery that are such highly valued assets today. A key theme has been the dynamic nature of the landscape. It is not static and will continue to change in the future. The next chapter looks forward to what some of the changes might entail and how the face of Scotland might look in the future.

An example of large-scale landscape modification by human activity – opencast coal mining near Kelty in Fife.

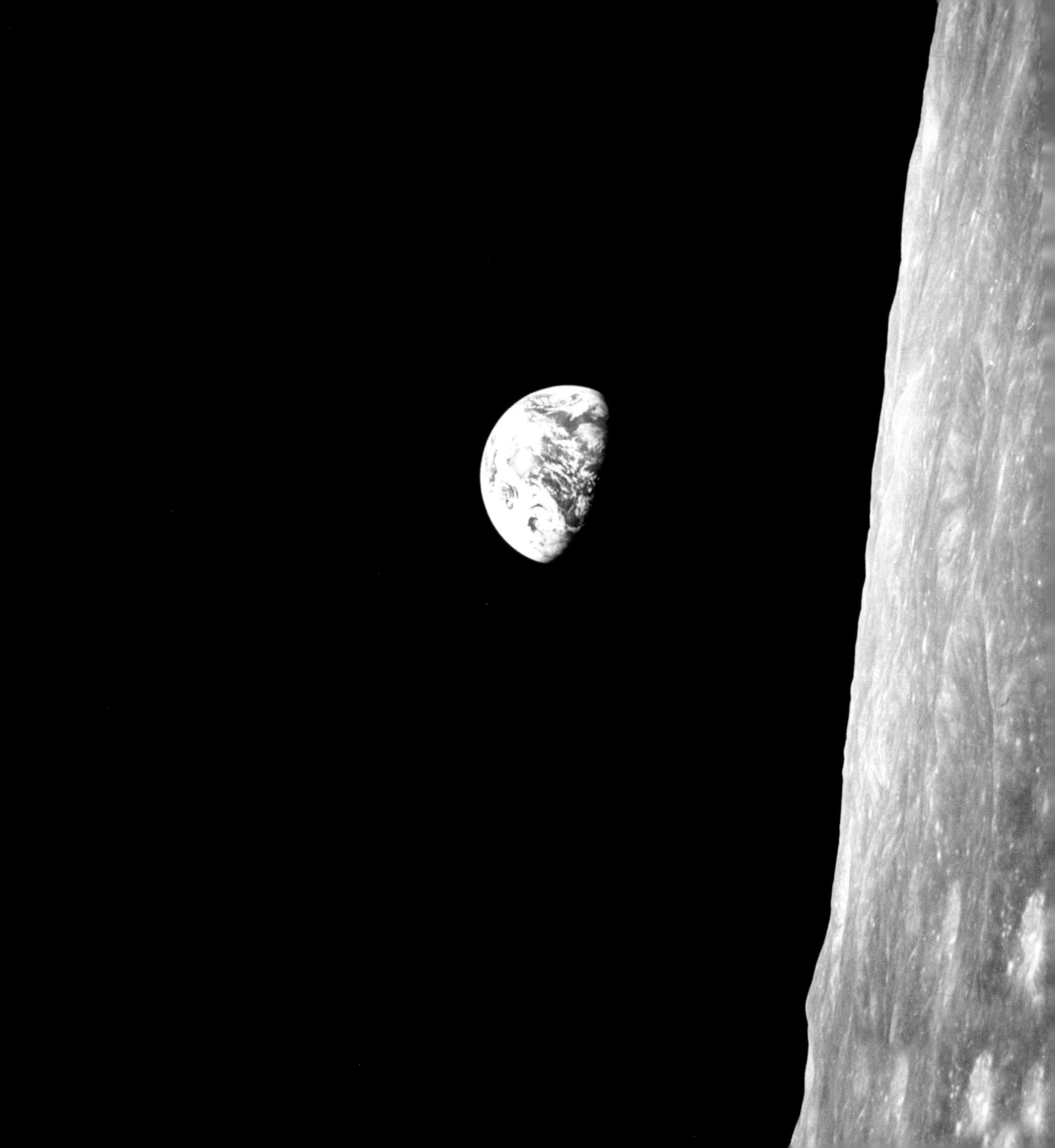

CHAPTER 5

FUTURE LANDSCAPES: SCOTLAND IN A GLOBAL CONTEXT

What happens to us
Is irrelevant to the world's geology
But what happens to the world's geology
Is not irrelevant to us.
We must reconcile ourselves to the stones,
Not the stones to us.

Hugh MacDiarmid, *On a Raised Beach*, 1934

Opposite. The geological history and future of Scotland are closely bound up with global and planetary processes and events. The Earth is exceptional in the solar system as a life-supporting planet. Its geological, oceanic, atmospheric, climatic and living systems are all highly interconnected and self-regulating. We are therefore part of a much larger, but finite, system, and our future is dependent on that system. In this symbolic picture of the Earth, taken by Apollo 8 astronaut William A. Anders in December 1968, the planet appears as a small, delicate blue orb rising above the barren surface of the Moon and set against the black void of space.

Previous chapters have revealed how the landscape of Scotland has been shaped over different timescales as a result of geological processes and climate change. While our world in Scotland may seem to be relatively stable on a day-to-day and year-to-year basis, apart from the occasional storm, flood or landslide, the message from the past is that we live on a highly dynamic planet where changes are inevitable over both short and long timescales. Some processes are small, slow and ongoing, like the weathering of rocks or the widening of the Atlantic Ocean, but over long timescales they produce large changes, lowering mountain ranges or rearranging the continents. Other processes are large and sudden but happen only occasionally – like major floods or volcanic eruptions – and produce substantial changes in a short time. It is also important to appreciate that Scotland does not exist in isolation from the rest of the world. The natural systems of our planet are highly interconnected through plate movements and the dynamics of the atmosphere and the oceans, which drive our weather systems and climate. This means that events in one part of the world can have global impacts; for example, large volcanic eruptions in the Far East affect the climate in Europe, and melting glaciers in Antarctica contribute to sea-level rise around our coast. Equally, the effects of global processes, such as climate change, are felt in Scotland.

In this chapter, we consider some changes that might affect Scotland over three different timescales in the future. First, we examine the possible effects of climate change and rising sea level during the remainder of the present century and beyond. We then look thousands of years further ahead to when the present interglacial might end. Finally, we touch briefly on possible changes over millions of years as Scotland continues its onward journey across the surface of the globe, driven by the movements of the Earth's plates. In addition, we identify some possible cataclysmic events that could affect the Earth at any time.

CHANGES IN THE TWENTY-FIRST CENTURY AND BEYOND

Climate change

As noted in chapter 2, the ice core records from Antarctica have revealed that temperature and greenhouse gases have changed very much in parallel over the last three-quarters of a million years. Carbon dioxide concentrations in the atmosphere have typically been about 180–190 parts per million by volume during glacial periods and 280–300 parts per million by volume or less during interglacials and have been maintained between these limits by natural processes. However, as a result of greenhouse gas emissions from human activities, the concentration has now risen from a pre-industrial level of 280 parts per million by volume to 394 parts per million by volume in 2012. Measurements from the Dome C ice core in Antarctica indicate that this level is unprecedented in the last 800,000 years and that the present rise is about 200 times faster than any other rise during this time. Similarly, the concentration of methane, also a powerful greenhouse gas, is now around 1,800 parts per billion by volume, considerably greater than pre-industrial levels which did not exceeded 800 parts per billion by volume over the same time period. Over the last hundred years, the average surface air temperature of the Earth has risen by about 0.8°C and three-quarters of this rise has occurred since the 1970s. In the period of instrumental records from 1850 to 2012, the ten warmest years have all been since 1998. The 2000s were the warmest decade in that period, and the late twentieth century onwards has been the warmest period in at least the last 1,300 years. There is now a general consensus among climate scientists, as reported by the Intergovernmental Panel on Climate Change (IPCC) in 2007, that most of the warming observed over the last fifty years is "very likely" to have been due to the increase in concentrations of greenhouse gases, such as carbon dioxide and methane, produced by human activities.

There are many uncertainties in making projections about future global climate change – uncertainties about future emissions of greenhouse gases, how the climate will respond to these emissions and the natural variability of the climate. The IPCC considered a number of different scenarios of future global emissions of greenhouse gases. Under a high emissions scenario, assuming that economic growth is rapid, market mechanisms dominate and there is social and economic convergence between different regions of the world, then global temperatures could increase by 4.0°C by 2100; with lower emissions, the rate of warming would still be significant compared with anything previously experienced by the human race in the last 10,000 years. Even if greenhouse gas concentrations were stabilised now, which would require global emissions to be cut by 60–70 per cent, further climate change is inevitable because of the gases already in the atmosphere. Consequently, we are already committed to a further warming of 0.5–1.0°C from emissions in the past. The full extent of climate change in the second half of this century and beyond will be influenced by the volume of greenhouse gases emitted during the next few decades. This, in turn, depends on decisions made now and in the near future at national and global levels on reductions in the use of fossil fuels, development of new technologies and more efficient use of energy. The longer the delay in acting, the greater the changes that are likely to be in store. The question, then, is not about whether the climate will get warmer, but rather by how much and what the effects will be.

The report from an international conference organised by the UK government in 2005 concluded that limiting global warming to about 2°C would probably be necessary to avoid serious risk of 'dangerous' climate changes, giving rise to irreversible melting of the Greenland and West Antarctic ice sheets and other effects. This would mean stabilising carbon dioxide at a concentration of 350–400 parts per million by volume or less. (However, the inclusion of other greenhouse gasses and aerosols complicates the picture.) Whether this is practical or not is another matter, and some experts believe that 450 parts per million may be the most realistic target that can be achieved. Such a level seems likely to be reached in the middle of this century if present emission trends continue. The projected temperature rise for such an increase is about 3°C and implies greater risk of adverse impacts. These figures may seem quite small and insignificant, given the much larger daily and seasonal variations in temperature across the planet. However, the average temperature of the Earth does not change greatly. For example, variations of only a few degrees in the global average temperature in the past have been accompanied by extreme changes in climate. At the last glacial maximum, when glaciers covered much of northern Europe and North America, the global climate was only

Flooding on the River Earn, east of Kinkell Bridge, Perthshire, in January 2005. The January average rainfall was exceeded in the first six days of the month.

about 5°C colder than today. The last time that global temperatures were 3°C higher was 3-5 million years ago during the Pliocene and sea levels then were up to 25 metres higher. If the geological past is a guide to the future, there may not be a large margin before the onset of significant climate and other changes.

It is also worth noting that climate change may not be steady because it will be superimposed on an underlying level of natural climatic variability. Further, it is important to appreciate that the often-used term 'global warming' does not adequately describe the projected changes, which may include greater variability, increased frequency and intensity of extreme events, and, due to our incomplete understanding of the complexities of the Earth's systems, possibly unanticipated, abrupt changes or 'climate surprises'. At a global scale, a warmer world means greater instability in the climate and significant changes in weather patterns. It means more frequent and intense droughts, floods, storms and heat waves, as well as rising sea level. This matters because of the likely impacts on human activities and societies, with attendant economic and social costs for mitigation, repairs and population displacements.

What might this mean for Scotland in the remainder of the twenty-first century? Accepting that there are uncertainties in down-scaling the global climate models to produce climate change projections at a regional level, the following scenario emerges. Natural variability from year to year will continue to be a feature of the climate, but overall, by the end of the present century, Scotland as a whole is likely to be warmer and wetter. Projections based on high emissions of carbon dioxide show a possible overall annual warming of up to 3.5°C, and up to 4°C in the summer in the south and east. Winter precipitation may increase by over 25 per cent in the east of Scotland. While the increase may be lower in the west, the west will still remain significantly wetter than the east. There are likely to be more frequent intense rainfall events, particularly in the west in winter. Summers may be drier, particularly in the east. Snowfall amounts may reduce by 60–90 per cent. Storminess and the frequency of severe gales may increase if projections of a stronger westerly circulation are borne out, although there is less confidence about this.

A number of effects on the landscape in Scotland can be envisaged from such changes in climate and weather patterns. There are likely to be more landslides and other slope failures, especially where slopes are prone to waterlogging during intense or prolonged rainfall. We are likely to have more and bigger floods, especially in the east, and some rivers could shift parts of their courses outside their currently active floodplains. Big flood events, which previously had a probability of occurring only once a century, could become twice as common and increase in scale by 20–30 per cent by the 2080s. The effects are likely to be exacerbated because of alterations to natural river systems through drainage of land, more impermeable surfaces in built-up areas, development on floodplains and entrainment of river channels by flood embankments. As a result of such modifica-tions in the past, many floodplains no longer fulfil their natural function of providing temporary storage of floodwaters, so that the flows are accelerated downstream. Increased flood risk will have knock-on effects by increasing demand for flood protection and restrictions on the use of floodplains for building developments and agriculture, as well as creating greater risks of soil erosion. In particular, a serious rethink will be needed on how floodplains are used, addressing demands to restore their natural function as flood buffers. In mountain areas, higher spring temperatures may produce earlier and rapid snow melting, with increased flood risk if accompanied by high rainfall. Reduced snow-fall may also change the appearance of the winter landscapes so favoured in calendar photographs. Greater and more intense winter rainfall could lead to enhanced soil erosion on agricultural land where fields are bare or sown with winter crops. Conversely, in summer, warmer drier conditions in the eastern High-lands could lead to the drying out of peat bogs and erosion where the surface vegetation cover is broken.

Debris flows blocked the A84 in Glen Ogle, north of Lochearnhead, following intense summer rainfall in August 2004. Such occurrences may become more common in the future.

Working with natural processes: river management

For centuries people have sought to protect their land from flooding as agriculture, industrial development, housing and infrastructure have increasingly encroached on river floodplains. Rivers have been straightened and canalised with the result that they now flow faster. Flood embankments have been created and progressively extended and raised in height, disconnecting rivers from their floodplains. Floodplain wetlands have been drained for agriculture. Roads, housing schemes, industrial estates and other developments have been built with hard impermeable surfaces, speeding run-off into drains and ultimately into rivers. All of these activities, as well as periodically greater amounts and intensity of precipitation, increase the chances of flooding and will continue to do so.

Like the coast, rivers are natural systems and should be managed in their entirety, rather than in small sections related to the ownership of adjacent land. Local flood protection works may provide a local solution but pass on bigger problems downstream. Ironically, many floodbanks on the River Tay failed during the floods in 1990 and 1993, so that by accident rather than design, the floodplains performed their natural function as temporary floodwater storage areas. This undoubtedly took some of the pressure off Perth, particularly in 1993, when the Tay upstream at Dunkeld inundated the whole floodplain. The most effective form of flood management would involve planning for whole river catchments from river sources to mouths. Such integrated catchment management plans are presently being considered in Scotland.

As part of such planning, river managers are now looking at developing more natural solutions. Increasingly, this will be driven by policies to implement 'sustainable flood management' under the Water Environment and Water Services (Scotland) Act 2003, the legislation which implements the EU Water Framework Directive in Scotland. One possible approach is to develop the concept of 'creating room for rivers' – analogous to managed realignment at the coast. It would involve repositioning floodbanks back from the active channel margins, allowing rivers to flood over their floodplains, and increasing the temporary storage capacity upstream in their catchments. This would help to reduce flood peaks downstream where there is high capital investment at risk. The effectiveness of such an approach could be enhanced, and have additional environmental as well as flood mitigation benefits, if it were combined with restoration of natural habitats such as wetlands on the floodplains. Other measures to reduce the speed of run-off and attenuate flood peaks may need to address land use in the uplands, blocking drains, replacing impermeable with permeable surfaces in urbanised areas, and, particularly, reviewing planning policy on floodplain development. As with coastal management, many economic, social, planning and land management issues will need to be resolved in a co-ordinated way. The choice ultimately is between engineering large parts of river catchments at great cost or allowing some land to be periodically inundated in a planned way in order to protect more valued properties and infrastructure downstream. So, in a few decades, the character and appearance of many upland river floodplains in Scotland could change significantly.

The Insh Marshes are a natural wetland on the floodplain of the River Spey between Kingussie and Kincraig. As well as being an important wildlife habitat, the marshes act as a natural floodwater storage area for the river.

Rising sea level

As a result of the thermal expansion of the oceans in a warming world and freshwater inputs from melting of small glaciers and ice caps, global sea level has risen over the last century by 10–20 centimetres. This rise is projected to continue during the twenty-first century by as much as 1m or more, depending on future levels of greenhouse gas emissions. It is important to emphasise that, even if the levels of greenhouse gases in the atmosphere and global temperatures are stabilised, sea level will continue to rise for at least several centuries, if not millennia, afterwards. It has been estimated that due to past emissions of greenhouse gases we are already committed to a sea-level rise of about 1 metre over the next few centuries. This is because the warming of the surface waters of the oceans will slowly penetrate down into the deeper waters, causing their gradual warming and expansion, and ice sheets will continue to melt. For the present century, melting of the world's glaciers will contribute significantly to sea-level rise. In the longer term, there has been speculation about the possibility of extreme high sea levels associated with the collapse of the great ice sheets in Antarctica and Greenland. For example, total melting of the Antarctic Ice Sheet would raise global sea level by about 57 metres – 52 metres from the East Antarctic Ice Sheet and 5 metres from the West Antarctic Ice Sheet – and the Greenland Ice Sheet by about a further 7 metres. Fortunately, sudden collapse of these ice sheets is unlikely in the foreseeable future, although in the longer term, there are uncertainties about the inherent stability of the West Antarctic Ice Sheet and the response of the Greenland Ice Sheet to climate change.

The record from the past also carries an

Melting ice sheets?

For the foreseeable future, the world's glaciers will continue to melt progressively and contribute to sea-level rise. Global warming is likely to result in the disappearance of smaller mountain glaciers, most of which are in retreat throughout the world; for example, about three quarters of the glaciers in the Swiss Alps could disappear by 2050. Of greater concern, because of their size and the potential for extreme sea-level rise if they melt or collapse, is the future of the Antarctic and Greenland ice sheets. For example, total melting of both would raise global sea level by about 64 metres. Fortunately, this is unlikely in the foreseeable future, although there are uncertainties in the longer term about the stability of part of the Antarctic Ice Sheet and the future of the Greenland Ice Sheet in a warmer world.

The Antarctic Ice Sheet is in two parts, the East Antarctic Ice Sheet and the West Antarctic Ice Sheet. The much larger East Antarctic Ice Sheet contains a volume of ice equivalent to a sea-level rise of 52 metres if it all melted. However, this ice sheet is land-based, extremely cold and considered to be stable, and it may even increase in size because of higher snowfall in a warmer world. Particular concern, however, has centred on the stability of the West Antarctic Ice Sheet and its ice streams, which if they collapsed would raise world sea level by about 5 metres, destroying coastal settlements, infrastructure and ecosystems. What distinguishes this ice sheet is that much of its bed lies below present sea level. Hence it might be particularly vulnerable to rapid recession and collapse through accelerated calving (masses of ice breaking off into the sea), and there is evidence that it has been much smaller during past interglacials. While the jury is still out on its longer-term stability, the consensus view has generally been that it is unlikely to collapse drastically in the next few centuries and that the large ice shelves which help to buttress it would remain stable in a warming world. However, it may continue to shrink further, either as part of its slow response to the climate warming at the start of the present interglacial, or if the present interglacial is significantly prolonged either naturally or by global warming. That said, concerns about its stability have been renewed, linked to recent changes in the state of the ice shelves on the Antarctic Peninsula.

Ice shelves are thick floating masses of ice, formed through the extension of glaciers from the land into the sea and by snow accumulation on their surfaces. There has been wide media coverage of the spectacular disintegration of several of the more northern ice shelves on the Antarctic Peninsula. Some of the most dramatic changes have affected the Larsen Ice Shelf. Large parts of it disintegrated completely in 1995 and 2002 within a few days, sending thousands of small icebergs into the Weddell Sea. These changes have been portrayed by some as a direct effect of global warming. Others have suggested that they are of local significance

Future melting of the world's ice sheets and glaciers as a result of climate change will lead to higher sea levels.

only since the effects are confined to a small part of the continent, and there is no evidence that similar changes are imminent elsewhere. In fact, the climate warming on the Antarctic Peninsula is not happening elsewhere in Antarctica. In any case, the loss of the ice shelves on the Antarctic Peninsula will not contribute to a rise in sea level because they were already floating.

However, increased concern has arisen following a number of recent studies. First, although some of the ice shelves may have disappeared during the warmer climatic conditions that began around 5,000 years ago, and then re-formed at some time within the last few thousand years, the collapse of the Larsen Ice Shelf is unprecedented during the last 11,500 years. Second, the land-based glaciers which formerly flowed into the Larsen Ice Shelf have accelerated following its collapse; this has raised serious questions about whether the West Antarctic Ice Sheet would behave in a similar fashion if its ice shelves were thinned or removed, so that its ice streams speeded up and hastened its collapse. Third, ice shelves respond rapidly to climate change and they may not survive as long in a warmer world as previously thought. It is still unclear what critical temperature rise would be needed to trigger the collapse of the bigger and more southern ice shelves which buttress large parts of the West Antarctic Ice Sheet. Finally, the effects of warming ocean water flowing underneath the floating ice shelves are also unknown, along with how, and over what timescale the ice sheet might respond.

While much attention has focused on the stability of the West Antarctic Ice Sheet, the Greenland Ice Sheet should not be overlooked. It was probably much smaller during the last interglacial, which was slightly warmer than at present, and its shrinkage contributed significantly to a sea level 4–6 metres higher than that today. Because its bed is above present sea level, unlike the West Antarctic Ice Sheet, its response to global warming is likely to be less drastic than that of the West Antarctic Ice Sheet, which could disintegrate rapidly. Nevertheless, its complete melting would raise world sea level by some 7 metres and have a huge effect on coastal geography and societies across the globe. Climate models suggest that global warming above 3°C, which is possible by the end of this century or earlier, could be sufficient to cross a threshold that would set in train the irreversible, long-term (i.e. over a period of several thousands of years) melting of the ice sheet, even if the temperature then stabilised. Although snow accumulation might increase at higher elevations on the ice sheet in a warmer world, this would likely be offset by greater melting at lower levels. Hence on the basis of current climate and ice sheet models, the Greenland Ice Sheet may substantially disappear, with a consequent sea-level rise of 5–6 metres over a timescale of centuries to thousands of years, depending on the ice sheet dynamics and how the climate evolves. In recent years, the lower parts of the ice sheet have undergone record levels of melting. Some of this meltwater appears to have penetrated to the bed of the ice sheet, speeding up the flow of the glaciers and leading to further loss of ice.

Although the response times of the ice sheets are generally thought to be in thousands of years, these observations suggest that their dynamic reactions in a warmer world could be faster, in terms of decades or less. In particular, warming of ocean waters, melting of ice shelves, increased amounts of meltwater penetrating to the glacier beds and accelerated flow of ice streams in Greenland and West Antarctica give cause for concern. Also, by the end of this century, temperatures seem likely to be sufficiently high to initiate melting of the Greenland Ice Sheet sufficient to raise sea level by several metres over the next few centuries, and may even set in train its irreversible melting. The uncertainties are still high, but so are the potential risks.

important warning. During the warmest part of the last interglacial, 120,000 years ago, when average global temperatures were slightly higher than today, the sea was several metres above its present level. A substantial part of this rise may have come from significant melting of the Greenland Ice Sheet.

In Scotland, future changes in relative sea level will reflect a combination of two effects: the global sea-level rise due to thermal expansion of the oceans and glacier melting, and the extent to which the land is currently rising in response to the unloading of the crust at the end of the last glaciation. Because the latter varies geographically depending on the variations in the former thickness of the last ice sheet (see chapter 4), the amount of sea-level rise will vary around the coastline. According to one best-estimate scenario, broadly equivalent to an intermediate level of greenhouse gas emissions, by 2100 the projected net sea-level rise could be as much as 60 centimetres in Shetland and exceed 40 centimetres in Orkney and the Western Isles, but be below 29 centimetres in the Inverness area, the Forth and Tay estuaries and along the west coast between Fort William and Stranraer, and be between 29 and 34 centimetres elsewhere.

The effects of rising sea level will be exacerbated if changes in weather patterns result in more frequent storms and higher waves, especially in northern Scotland. Extremes of sea level, which occur during storm surges caused by low atmospheric pressure and strong winds and when storms coincide with high tides, are of particular concern and inflict most damage. A best-estimate scenario for 2100, combining sea-level rise and storm-surge elevations for a one-in a-hundred year event, gives figures in a range from 3 metres to over 6 metres at different points around the coast. This means that land below

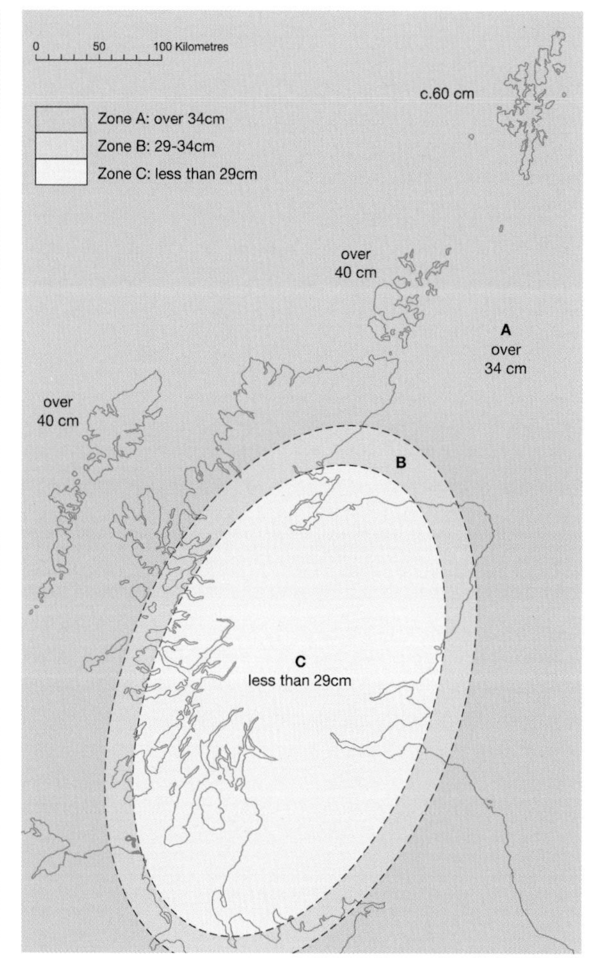

One best-estimate scenario of possible net sea-level rise for Scotland by 2100. This takes account of the projected future rise in global sea level and the regional pattern of ongoing crustal rebound following the last glaciation. Three zones (A, B and C) are identified according to the variations in the current rate of crustal rebound. The elliptical pattern of these zones reflects the variation in the former thickness of the last ice sheet, which was greatest in the South-west Highlands. Accordingly, the rise in sea level could vary from less than 29 centimetres in central Scotland to 60 centimetres in Shetland. The effects of such sea-level rise could be significantly enhanced by the occurrence of storm surges. During such conditions, sea-level rise could temporarily exceed 3 to 4 metres throughout Scotland.

Waves batter the seafront at Kirkcaldy. Pressure on coastal defences is likely to increase through a combination of rising sea level and storm surges elevating the water surface.

these elevations would be vulnerable to extreme sea levels. Reductions in the return periods of events of a particular magnitude are also likely. For example, by the 2050s it is estimated that the hundred-year high-water level will become on average a twenty- to forty-year event, which will increase the flood risk. However, there will be geographical variations, with shorter return periods in the north and longer ones in Argyll and Ayrshire. Add to this the continuing net deficit in sediment supply to the coast and it is inevitable that soft, exposed coasts will retreat inland where not protected by natural or man-made defences. Conversely, sheltered and rocky coasts may show little discernible change.

Low-lying soft coasts, including dunes, machair, beaches and estuaries, face the greatest increased risk of flooding, coastal erosion and loss of coastal habitats and recreation facilities, such as links golf courses. This is already happening. As the coast retreats, beaches, dunes and machair and their associated habitats will shift landwards or move laterally along the coast in response to changing wave energy patterns. Where they come up against hard barriers inland, such as natural rock outcrops or human infrastructure, settlements and industrial areas, they will become squeezed out and disappear. The loss of these natural forms of coast defence will place additional pressure on existing sea defences and necessitate a significant degree of adaptation elsewhere. Rising sea level is also likely to lead to estuary margins moving inland, with increased flooding and permanent inundation at the heads of estuaries. Outer estuaries may be particularly vulnerable to loss of their natural bars and barriers, which act as a natural form of sea defence, leading to increased risk of flooding and erosion.

In concluding this section, it is important to emphasise that the changes outlined above are not predictions for what will happen, but are possible scenarios based on current understanding of the climate system and the complex links between the ice sheets, oceans and atmosphere. Basically we are undertaking an uncontrolled experiment with the Earth's natural systems at a time when there are still large gaps in our understanding of the key processes and the feedbacks that could amplify or dampen global changes. The risk is that if emissions of greenhouse gasses are not reduced as a precautionary approach while these gaps are being addressed, thresholds will be crossed which result in abrupt and unexpected changes. Even with significant reductions in future emissions, adaptation is still going to be required to address the effects of climate change and sea-level rise which we are already locked into because of past and present emissions. If the changes are slow and progressive, then planned adaptation should be possible but may be costly. If, however, change is abrupt, or, more likely, accompanied by a greater frequency of extreme events, then options for adaptation will be more limited.

Coastal erosion at Montrose has led to the loss of part of the links golf course. An intermediate solution to coast protection between 'do nothing' and the high-cost installation of rock armouring along the full length of the affected coastline has involved the creation of a series of artificial headlands (indicated by arrows) using blocks of heavy rock. The underlying purpose is to create a more stable coastal configuration of headlands and bays. While not halting erosion, it should slow down the overall rate of coastal retreat and allow natural processes to continue operating. The design, however, has not been entirely successful at this particular location.

Working with natural processes: coastal management

Similar issues and dilemmas as for river management have also come to the fore in relation to coastal management. For many decades, solid structures of concrete or rock have been built along the coast in an attempt to stop the sea encroaching onto the land and to prevent the loss of valuable property, infrastructure and recreational facilities, such as golf courses. Such an approach, however, of 'holding the line' everywhere is becoming no longer practical or financially realistic in a situation of rising sea level and increased frequency and height of extreme sea levels. Increasingly, land will be lost to 'coastal squeeze', which is the progressive erosion, or narrowing, of intertidal areas in front of hard coast defences or other hard structures inland, such as cliffs, so that the natural buffering of these areas is lost. Moreover, many existing sea walls have had unintended effects such as increasing erosion of beaches, which are a natural form of coast protection, or shifting the erosion elsewhere along the coast beyond the area protected. Indeed, some attempts in the past have been haphazard and based on a lack of understanding of how the coast works naturally. As a result, they have either failed or merely shifted the problem elsewhere.

Managed realignment of the coast at the RSPB Reserve at Nigg Bay on the Cromarty Firth. Two breaches were made in the seawall in 2003 to allow the land behind, formerly used for rough grazing, to be flooded periodically during high tides. This land is now returning to salt marsh and mudflat that not only provide important habitats for plants and birds, but also act as a natural form of sea defence, reducing the energy of the waves.

There is now growing recognition of the need for a more adaptive approach to coastal defence, to work with natural processes and to manage the coast in an integrated way. This is based on understanding the role of landforms and habitats, such as beaches and salt marshes, as a natural form of coast defence, and the need to maintain sediment supply to them. Coast defences frequently reduce, or cut off, sediment supply from eroding stretches of coast or interrupt sediment movements along the coast, so that erosion is transferred elsewhere. One solution is to recharge a beach artificially with sand from offshore. This has proved successful, for example at Portobello near Edinburgh, and is currently being implemented at Aberdeen. But this method is unlikely to prove successful on more exposed, open coasts.

Another approach involves 'managed realignment', which means removing all or part of existing defences, moving them landwards and allowing the coast to rebuild natural beaches and salt marshes. This creates space for natural forms of coast defence in order to protect areas further inland. Such an approach has potential in areas where land has been reclaimed from the sea, but it will require acceptance that some other areas will also have to be sacrificed to provide a sediment supply. Managed realignment has been implemented in parts of England and a trial demonstration is currently underway at the RSPB Reserve at Nigg Bay on the Cromarty Firth. Such an approach will not be applicable where there is high capital investment in property, industry or infrastructure at the present coastline, and high-cost sea defences will still need to be maintained in such areas. However, hard structures will not provide the solution elsewhere. Lower-cost options, such as managed realignment, will inevitably require some difficult choices in order to make space for the coast to readjust, and in many cases society will have to accept that the softer and more vulnerable parts of the coast will move inland, with loss of some properties and parts of links golf courses. In the long run, managing the coast from a dynamic perspective will be essential, whereas trying to play Canute will not work or be economical in any but the highest-value areas.

Future climate surprises: abrupt climate change

Within the context of our relative lack of understanding about the key processes that affect climate change, and their interrelationships, one issue of particular concern is whether the climate will continue to warm steadily until natural checks or reductions in greenhouse gas emissions eventually restore some balance, or whether thresholds will be crossed resulting in abrupt, dramatic and unexpected changes. What sort of surprises might the climate have in store for the future? It is apparent from the geological record that abrupt changes in climate have occurred in the past as a result of natural processes. To what extent are human activities nudging the climate system towards thresholds for similar changes in the future? The answer is that we do not know. However, looking back in the geological record and examining extreme natural changes in the past provides some clues. Two particular examples, both of which have received prominent media attention, are explored: changes in the thermohaline circulation in the North Atlantic and extreme global warming due to massive release of methane from the sea floor.

The thermohaline circulation (see chapter 2) transports a substantial amount of heat northwards in the Atlantic Ocean, which helps to moderate the climate of Scotland by several degrees, particularly in winter. As they move north, the surface waters cool and become increasingly salty as a result of evaporation. The cold, dense, salty surface waters then sink to great depths in the North Atlantic off Greenland and Labrador and flow southwards, maintaining the circulation. This circulation is potentially sensitive to climate change and to inputs of freshwater into the North Atlantic. Freshening of the surface waters would reduce their density and slow their sinking, and hence weaken the circulation. Such weakening could progressively increase to a tipping point, or threshold, where the circulation closed down abruptly. This appears to have happened in the past following massive outburst floods of freshwater from glacial lakes along the southern margin of the last ice sheet in North America. For example, at a time when the climate was warming at the end of the last glaciation around 13,000 years ago, one such event significantly weakened or shut down the circulation, which in turn chilled the North Atlantic region, interrupted the warming and led to the readvance of glaciers in Scotland (Loch Lomond Readvance – chapter 4), where mean annual temperatures fell by over 15°C. Concern has been expressed that this could happen in the future as a result of global warming, particularly if there is extreme greenhouse-gas forcing of the climate or the addition of large amounts of freshwater into the North Atlantic. Some models suggest that warming of 3°C by the end of this century could significantly increase the risk.

If such a dramatic event did occur, it would not involve an instant ice age as portrayed in the disaster movie *The Day After Tomorrow*. However, there would be a significant impact on the climate in Scotland. Projections based on modelling suggest that the cooling would eventually exceed global warming, and the climate of Scotland could rapidly get colder over a matter of a few decades, particularly if emissions of greenhouse gases started to decline and the circulation remained shut down. Conditions would become substantially colder than during the Little Ice Age when Thomas Pennant and other travellers in the seventeenth and eighteenth centuries noted that snow lay in the mountains throughout the summer and the Rev. Robert Walker was able to skate on Duddingston Loch in *c.*1795 (see chapter 4). If such conditions persisted, then small glaciers might start to re-form temporarily in the Highlands until the circulation switched back on again. Probably, too, there would be an increase in extreme events, such as storms and floods.

Currently, however, the probability of the circulation shutting down abruptly later this century is generally considered to be low, but it is not known how low. Although the circulation has shut down in the past, this has been the result of the massive and rapid influx of glacial meltwater into the North Atlantic. This is improbable in the future, unless there is a dramatic collapse of the Greenland Ice Sheet due to some unforeseen factors, and a more likely scenario is a steady freshening of the surface waters of the ocean. This is already underway as a result of increased discharge of rivers into the Arctic Ocean, associated with greater precipitation at higher latitudes as global temperature increases, and the input of freshwater from retreating mountain glaciers. The extent of Arctic sea ice, which is made of mainly freshwater, is also contracting and reducing the production of dense, salty water.

As a result of such changes, the circulation may gradually weaken in the next hundred years or so, rather than shut down. Such weakening could lead to seasonal cooling, especially in winter, in Western Europe, but any overall cooling in Scotland is likely to be offset by continued global warming. At present, there are indications of changes taking place in the Atlantic: the circulation has slowed, the waters at high latitude have freshened and the flow of deepwater southwards from the northern North Atlantic has slowed and freshened. However, it is not clear if these effects are simply part of a natural cycle or an early indication of more significant changes to come.

The second example concerns the release of methane from the sea floor. At the beginning of the Eocene, around 55.8 million years ago, global temperatures rose by 5 to 10°C over a period of 10,000 years and did not return to previous levels until some 100,000 years later. Within a few thousand years, also, sea surface temperatures rose by 5°C in the tropics and as much as 9°C at high latitudes; at the same time, deep ocean temperatures increased by 4 to5°C. This is one of the most striking episodes of global warming in the geological record and was accompanied by great changes in the distributions of plants and animals on land and in the sea. It coincided with substantial emissions of greenhouse gases, and concentrations of carbon dioxide in the atmosphere are thought to have been several times greater than modern pre-industrial levels. Possible sources of these gases may have been the extensive volcanism in the North Atlantic region, including the Hebridean volcanoes (see chapter 3); the generation and release of methane produced by volcanic sills which injected hot magma into organic-rich sediments in the same region; and the release of a huge volume of methane from the ocean floor. Current evidence suggests that the last of these is the strongest candidate. It is thought that gradual warming of the deep ocean caused melting of deposits of gas hydrates (ice-encapsulated methane) buried in marine sediments on the continental slopes of the ocean floor. This released large amounts of methane, a potent greenhouse gas in its own right. The methane combined with dissolved oxygen in the seawater to produce large amounts of carbon dioxide that escaped to the atmosphere along with unreacted methane. Hydrates are only stable in a narrow range of pressures and temperatures. If a small amount of warming of the deep ocean disrupted the balance, it could have started a positive feedback – as more methane was released, more gas escaped to the atmosphere, resulting in more warming.

Could this happen in the future? Could global warming trigger such changes in the deep ocean, with results comparable to those in the Eocene? The geological record suggests that we should at least be aware of the possiblity and its enormous implications. For Scotland, it would mean a return to a subtropical climate and substantially higher sea levels as the Greenland and Antarctic ice sheets melted and the ocean waters expanded in a significantly warmer world. A possible side effect of the release of gas hydrates might also be the destabilisation of marine sediments on continental slopes, leading to large submarine slides and tsunamis. This may have been a contributory factor in triggering the Storegga slides on the Norwegian continental slope around 8,100 years ago which were mentioned in chapter 4. There are large accumulations of sediments on the continental shelf and slope to the north and west of Scotland, but any risk to their stability is unknown.

In terms of greenhouse gas emissions, it is salutary that some estimates of the total amount of carbon released in the Eocene are comparable to those for the amount that would be released, but at a much faster rate, from anthropogenic sources in the next few centuries if present trends in the use of fossil fuels continue unabated. The warning in the geological record about the possible effects is clear. However, one important lesson from the Eocene is that the climate system eventually recovered as carbon dioxide was removed from the atmosphere by rock weathering and absorption in the ocean. The Earth's natural systems, therefore, have corrective mechanisms, but they may take a very long time (at least in human terms) to reverse extreme global warming.

These two examples illustrate the kinds of high-impact, low-probability surprises that could occur as a result of human interference with the climate system. However, we do not know how low the probabilities are, nor the thresholds for such events in terms of higher temperatures and greenhouse gas levels in the atmosphere. This comes back to the question of whether there are attainable 'safe', or low-risk, levels at which greenhouse gases can be stabilised to prevent dangerous climate changes. Policy makers, and society in general, also require a clearer assessment from climate scientists about what changes can be ruled out as improbable, rather than simply ruled in as possible. Such risk-based forecasting is essential for making informed decisions and developing practical responses, for example in terms of reducing emissions of greenhouse gases or planning for sea-level rise. Although catching the public attention, exaggerated headlines about disaster scenarios are unhelpful in encouraging wider debate about action to address the more likely impacts in the shorter term from greater climate variability and more frequent extreme events, such as heat waves, cold winters, floods and storm surges.

CHANGES OVER THOUSANDS OF YEARS – HOT-HOUSE OR ICE-HOUSE?

It is clear that climate change influenced by human activities will persist throughout the present century and beyond. What will happen after that, beyond the twenty-first century and thousands of years into the future after fossil fuels have been exhausted or replaced and the level of greenhouse gases in the atmosphere stabilised? Will the present interglacial be extended or come to an end and the glaciers return? We can look at future scenarios using climate models, and we can also look back into the past to see if there are analogies for the future in the geological record. However, the future will not be exactly like the past and there are different interpretations of how the climate might evolve.

In the absence of human activities and their impact on climate, it is reasonable to expect a continuation of the series of glacial and interglacial cycles that have characterised the Ice Age, driven by the Earth's orbital variations and their effects on insolation (see chapter 2). We are currently in an interglacial that began 11,500 years ago. Looking back at the geological record of the last 500,000 years, such warm intervals are exceptional and have occupied only about 10 per cent of the time. In the 1970s, it was thought that because several previous interglacials had lasted about 10,000 years, the end of the present interglacial was imminent. This seemed to be supported by the cooling which began after the warmest part of the Holocene around 6,000 years ago. However, it is now realised from the climate records in the ocean floor cores and ice sheet cores that not all interglacials have lasted for the same length of time. Looking back into the past at the more detailed records now available, the most comparable time period to the present in terms of orbital variations and insolation patterns is thought to be an interglacial that began about 430,000 years ago. It lasted for an exceptionally long time, about 28,000 years, twice as long as other interglacials; because the orbital forcing was weak due to the low eccentricity of the Earth's orbit, the climate may have 'skipped a beat'. If this period is a good analogue for the current interglacial and the near future, then we could expect another 16,000 years of interglacial conditions irrespective of global warming.

Some climate models then suggest that small glaciers might form again in the Highlands in 25,000 years' time under a cooler climate similar to that of northern Scandinavia today. Semi-natural deciduous woodlands would be replaced by birch or birch–pine woodland and more extensive heathland and grassland cover. Periglacial processes associated with seasonal freezing and thawing of the ground would be more active and extensive in the uplands. Global sea level would drop as glaciers expanded in the Arctic and elsewhere. The climate would then become more continental (colder and drier) like northern Sweden and Finland today. Subsequently, intense glaciation would occur around 60,000 years from now when an ice sheet would expand to cover much of Scotland, as happened during the last glaciation. The climate would then ameliorate in around 70,000 years' time, followed by a return to full glacial conditions in 100,000 years. Conditions comparable to those of today would not return until 120,000 years into the future.

Other climate models, based on the projected changes in insolation and with levels of carbon dioxide in the atmosphere varying naturally as in the past, suggest that we may be in an even longer interval before the onset of the next glaciation in about 55,000-60,000 years' time and the next glacial maximum in 100,000 years' time. This is because the eccentricity of the Earth's orbit is near its minimum and the size of seasonal variations in insolation will be very small over the next 100,000 years.

Hence the natural trend is for the Earth once more to enter a period of glaciation, although the timing is uncertain. How might this be affected by human intervention through greenhouse gas emissions? If the latter remain high, and atmospheric concentrations are eventually stabilised at around twice their present level, then the scenario of a prolonged present interglacial may be reinforced, particularly in the context of small seasonal variations in isolation. Human effects on the climate, therefore, could potentially be very long lasting.

While it may be reassuring that we do not appear to be heading imminently into the next glaciation, future generations will nevertheless need to adapt to some major environmental challenges. For example, if the analogy with the past holds, sea levels could be significantly higher, regardless of human influences on the climate. During the interglacial that began 430,000 years ago, there is evidence that sea level was 13 to 20 metres higher than at present, which would suggest that the Greenland and West Antarctic ice sheets had largely melted. As discussed in the previous section, global warming enhanced by greenhouse gas emissions could hasten such dramatic changes. Eventually, however, in some millions of years of time, plate movements will shift the northern hemisphere landmasses southwards and the Antarctic continent northwards away from high latitudes, melting the polar ice sheets and ending the present Ice Age.

Did early human activities prevent a new glaciation?

A recent and controversial idea proposed by the American climate scientist, W F Ruddiman, is that human impacts on the climate may even have begun much earlier than previously thought. A natural cooling process should have started around 8,000 years ago, leading to large-scale glaciation in North America about 4,000 to 5,000 years ago, but was offset by the effects of greenhouse gas emissions from human activities long before the onset of the industrial era. These activities were the extensive early clearance of forest for agriculture in the Middle East, Europe and southern China. The effects were enhanced by cultivation of rice in waterlogged fields in south-east Asia. This generated methane, which, as a greenhouse gas, is twenty-one times more powerful than carbon dioxide. According to this hypothesis, early human activities delayed the initial stages of the next glaciation and helped to maintain a climate in which civilisation has flourished.

CHANGES OVER MILLIONS OF YEARS – PLATE MOVEMENTS AND SCOTLAND'S CONTINUING JOURNEY

Even further ahead, millions of years into the future, Scotland will continue its journey across the surface of the globe. It is possible to run current plate movements in fast forward and to speculate on what the future geography of the planet might look like. A series of future projections has been developed by Christopher R Scotese at the University of Texas.

In the next 50 million years, the Atlantic Ocean will continue to widen. Africa will collide with Europe and Arabia, producing a mountain range of Himalayan proportions from Spain, across Southern Europe, through the Middle East and into Asia. At the same time, the Mediterranean Sea and the Red Sea will disappear. Australia will move north and collide with South-east Asia, and California will move northwards up the coast to Alaska. Eurasia will rotate clockwise, moving Scotland and western Europe northwards and Siberia southwards.

Beyond that, prediction of plate movements is more difficult. One scenario involves the onset of subduction along the eastern coasts of North America and South America. According to this scenario, in about 100 million years from now, the present Mid-Atlantic Ridge will be subducted and the sea floor beneath the Atlantic Ocean will begin to narrow, drawing North America and Europe closer together again. In 150 million years' time, the Atlantic Ocean will have contracted significantly as a result of subduction beneath the Americas. Similarly, the Indian Ocean will be smaller due to northward subduction of oceanic crust into the Central Indian Trench. Antarctica will have collided with the southern margin of Australia. Scotland will be located on the northern margin of a vast supercontinent comprising the present Europe, Asia and Africa.

Looking even further ahead, 250 million years into the future, the Atlantic Ocean will have closed. North America will have collided with Africa, but in a more southerly position than previously, forming a new supercontinent, like Pangaea 250 million years ago. A new mountain range will have formed along the boundary. South America will be wrapped around the southern tip of Africa, with Patagonia in contact with Indonesia, enclosing a remnant of the Indian Ocean. Antarctica will once be again at the South Pole and the Pacific Ocean will have widened, encircling half the Earth. Scotland will continue to lie on the northern margin of the supercontinent.

Eventually, plate movements will slow and finally cease as the Earth's internal heat engine runs down in half a billion to one billion years' time. Scotland's journey will then be at an end.

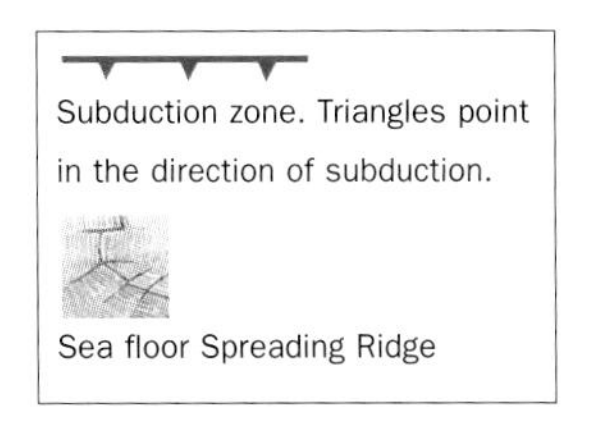

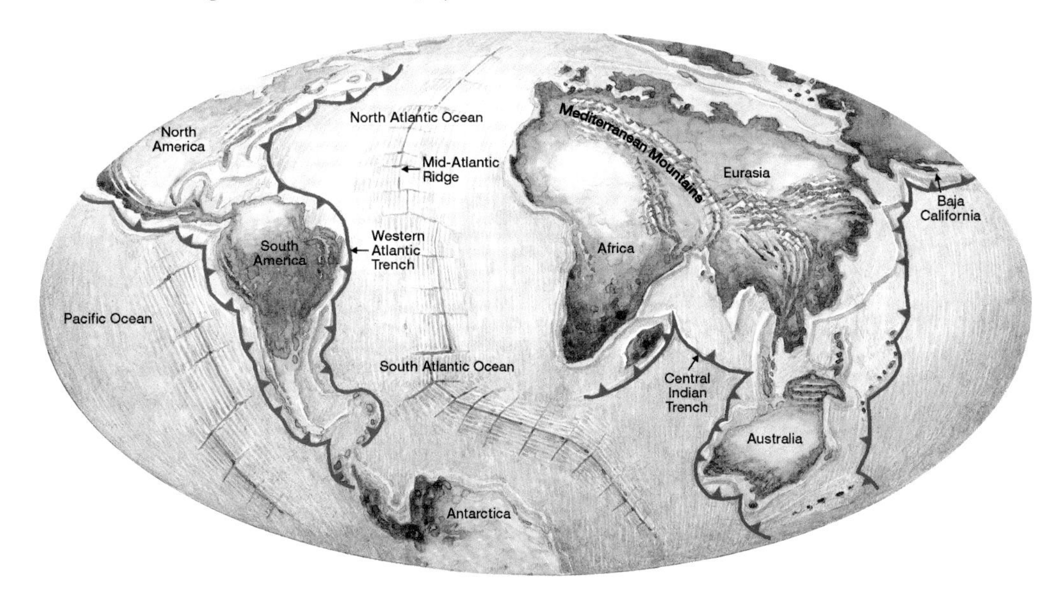

The possible future appearance of the Earth 50 million years into the future, based on projected movements of the Earth's plates derived by Christopher R Scotese. Subduction zones have developed along the eastern coasts of North America and South America, the Atlantic Ocean will soon begin to close and the Mediterranean Sea has disappeared following the collision of Africa and Europe.

CATACLYSMIC EVENTS

Cataclysmic events have affected the Earth throughout its history. Some of the most dramatic effects have resulted from the impacts of extraterrestrial bodies. Today, about 170 extraterrestrial impact craters are known; more are probably buried under younger rocks. One of the most spectacular occurred 65 million years ago, when a large asteroid 10 kilometres across smashed into the Earth. The impact site is the 200-kilometre-diameter Chicxulub crater off the coast of Mexico. The effects of this impact are believed to have finally wiped out the dinosaurs (see chapter 2), although emissions of dust, carbon dioxide and poisonous gases from vast flood basalt eruptions in the Deccan Traps of India may earlier have altered the climate and increased the environmental stress. A similar impact by an asteroid or comet has been suggested for the mass extinction at the end of the Permian 251 million years ago, when over 90 per cent of marine species and 50–70 per cent of land species disappeared off the face of the Earth. However, the evidence for such an impact has been questioned and eruption of massive flood basalts, this time in Siberia, may have created extreme conditions for life on Earth.

Major volcanic eruptions may also have significant global effects. As recently as 1815, the eruption of Tambora in Indonesia, the largest known eruption in the past two millennia, produced a succession of stormier winters and cool, wet summers, including the 'year without a summer' in 1816 when crops failed in Europe and North America. Eruptions of this magnitude are estimated to occur with a frequency of 1 to 2 per 1,000 years.

Eruptions of other 'supervolcanoes' are known in the geological record. These would have had direct impacts on a continental scale through ash fall, and indirect effects globally through 'volcanic winters' as the dust and gas injected into the atmosphere blocked out the sun for years. One of the largest in the world underlies Yellowstone National Park. Here, vast volumes of material have been blasted out three times in the last 2 million years. Such eruptions appear to occur on a 600,000-year cycle, and some geologists predict that the next explosion is 40,000 years overdue. Typically, such super-eruptions occur somewhere in the world every 50,000–100,000 years and involve the ejection of over 500 cubic kilometres of material. The largest in the last 2 million years was at Toba in Indonesia, 75,000 years ago. This blasted out 2,800 cubic kilometres of volcanic ash, and was over 1,000 times greater than the eruption of Mount St Helens in 1980. The Toba eruption may have caused global temperatures to cool by about 3–4°C; a global winter lasting six years; and, possibly, 1,000 years of reduced temperatures. On the basis of genetic evidence, this eruption has also been linked to a decrease in the population of *Homo sapiens* from over 100,000 to less than 2,000.

Another high-magnitude geological threat is from tsunamis caused by earthquakes below the sea floor and/or massive submarine landslides. The Indian Ocean tsunami on Boxing Day 2004 shows the degree of devastation such events can cause. Sonar images of the sea floor off Hawaii reveal the deposits of massive landslides that have slipped off the sides of the volcanic islands. Some geologists predict that a similar hazard is waiting to happen on the volcanic island of La Palma in the Canary Islands. A sudden collapse of this volcano would send a tsunami racing across the Atlantic towards the eastern seaboard of North America and the effects would probably also be felt in Scotland. Closer to home, a tsunami hit the east coast of Scotland 8,100 years ago following submarine slides on the Norwegian continental slope. This indicates that such events are not confined to seismic areas.

These examples give a flavour of the cataclysmic natural disasters which have intermittently impacted the Earth during the course of geological time. Fortunately such high-magnitude events are infrequent. We do not know when similar events will next occur, but from the geological record we do know that they will happen in the future and that Scotland, wherever it has moved on the surface of the planet, will not be exempt from the effects.

In their book *The Life and Death of Planet Earth*, Peter Ward and Donald Brownlee argue that the planet has a finite lifespan of about 12 billion years and that the future will be a re-run of the past, but in reverse. The 3.4 billion years that life has already existed is longer than it will survive in the future as the Earth's life support systems run down. All plant and animal life is likely to become extinct between about 500 and 800 million years from now, replaced by primitive microbes which in turn will face a similar fate. Ultimately in the story of the Earth, the Sun will eventually die as it runs out of fuel. Before doing so, it is expected to expand and become a red giant, exposing the Earth to searing temperatures. It was formerly thought that this expansion would engulf Mercury, Venus and then the Earth in about 7.5 billion years' time, but the Earth could just escape as the Sun will lose mass and the radius of the Earth's orbit will expand. However, the planet would be left as a scorched lump of sterile rock. At present the Earth lies in a narrow inhabitable zone of space. As the Sun expands, this zone will move further away. Will our descendants be able to prolong life by creating colonies in space or even moving the Earth by altering its orbit?

CONCLUSION

In revealing the course of Scotland's journey, we have peered into the "abyss of time", so eloquently expressed in John Playfair's words, and seen a remarkable story in the landscape and its rocks and landforms. The origins of the earliest rocks can be traced back to some 3,000 million years ago. The face of Scotland has been constantly changing over this vast enormity of geological time; it is still changing today and will change in the future. On a grand scale, such change results from the inexorable movements of the Earth's plates, which have driven Scotland across the surface of the globe, opened and closed oceans and joined disparate fragments of crust to form the geological entity that is now Scotland. Continents have collided and split apart, mountain ranges have been raised and worn down to their roots and volcanoes have flourished then expired. Variations in the climate have produced ice ages and the cyclic advance and retreat of glaciers. Yet other changes, at a much smaller scale, have reduced even the hardest rocks to sand and moved the grains along ancient and modern beaches. All of these changes are inscribed today in the rocks and landforms. They have endowed Scotland with a truly astonishing geodiversity and an Earth heritage of exceptional value.

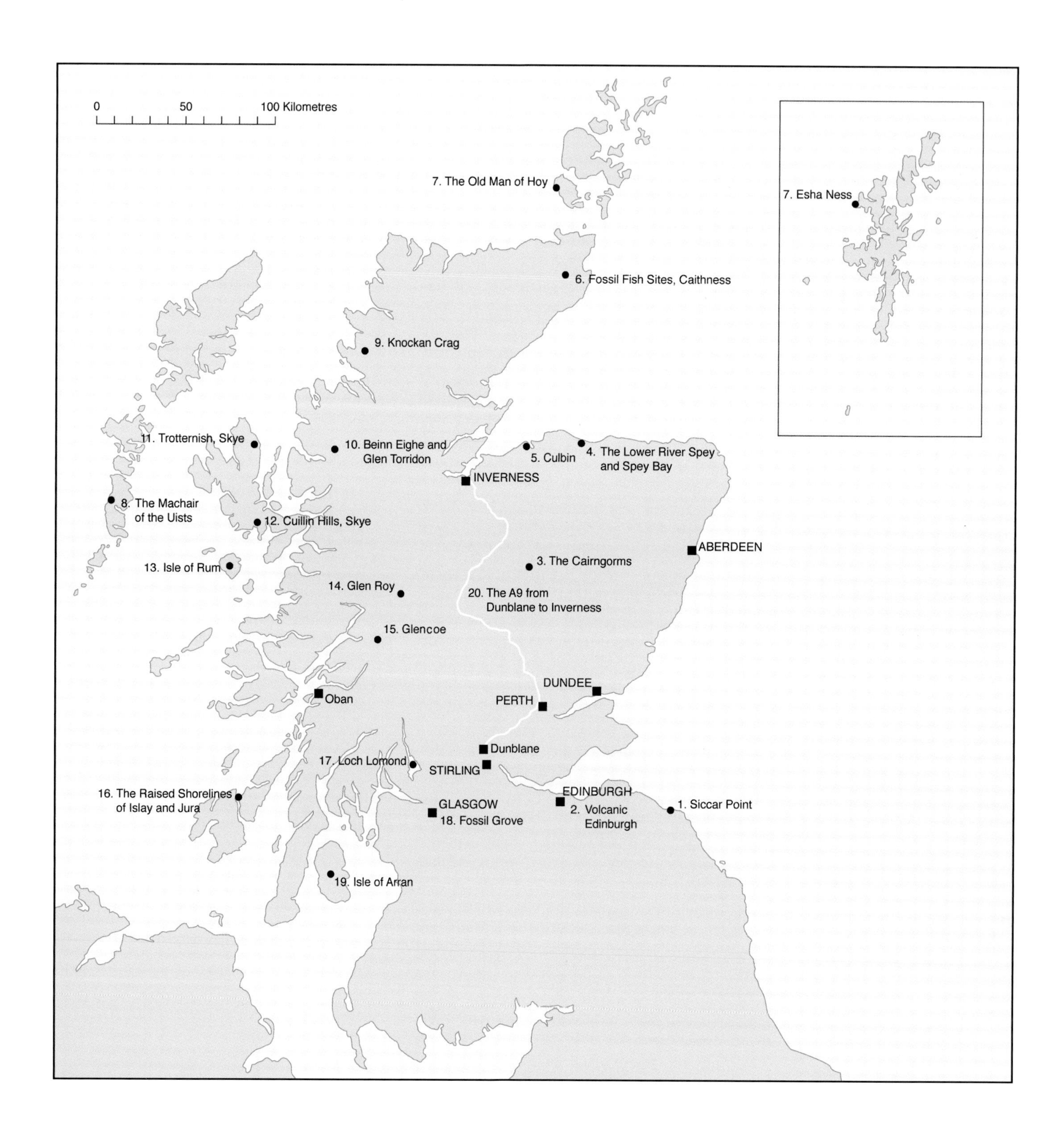

0
50
100 Kilometres
7. The Old Man of Hoy
7. Esha Ness
6. Fossil Fish Sites, Caithness
9. Knockan Crag
11. Trotternish, Skye
10. Beinn Eighe and Glen Torridon
5. Culbin
4. The Lower River Spey and Spey Bay
INVERNESS
8. The Machair of the Uists
12. Cuillin Hills, Skye
ABERDEEN
13. Isle of Rum
3. The Cairngorms
14. Glen Roy
20. The A9 from Dunblane to Inverness
15. Glencoe
DUNDEE
Oban
PERTH
Dunblane
17. Loch Lomond
STIRLING
16. The Raised Shorelines of Islay and Jura
EDINBURGH
GLASGOW
2. Volcanic Edinburgh
1. Siccar Point
18. Fossil Grove
19. Isle of Arran

CHAPTER 6

PLACES TO VISIT

Go to the mountains to read the immeasurable course of time that must have flowed from these amazing operations [of the Earth] which the vulgar do not see and the learned seem to see without wonder.

James Hutton, *Theory of the Earth*, 1788

Geology is essentially a field science. In the words of the distinguished geologist H H Read, "the best geologist, all other things being equal, is the one who has seen the most rocks". So whatever the enthusiast learns about the shaping of Scotland's landscapes from books, the television or other sources, their appreciation will be considerably enhanced by seeing rocks and landforms at first hand. This chapter describes twenty localities from the many thousands that exist throughout the country. This is not to say that these are the only places worth visiting to see Scotland's rich and varied Earth heritage, but they represent a good geographical spread and representative range of interests – from the very oldest rocks of the North-west Highlands to the most recent river and coastal landforms. So we hope that all tastes have been catered for in our selection of field sites. Sources of additional information on the sites can be found in the section on further reading on pp. 305–308.

The sites vary in terms of ease of access – some are immediately accessible from the roadside, whereas others require a well-planned expedition. But our emphasis is largely on those that can be readily accessed and where the interests should be apparent to the non-specialist. Regardless of location, the visitor is asked to follow two codes – the Scottish Outdoor Access Code and the Geological Code. These are described on the following pages.

Opposite. Map of Scotland with the locations of the twenty sites.

1. SICCAR POINT

This section of the Berwickshire coast has been described as, quite simply, the most important geological site in the world. As with many of the key historical sites, it is closely associated with James Hutton, the founder of modern geology. At a time when geologists of the late eighteenth century struggled to understand the world around them, a handful of sites were crucial in unlocking the secrets of the past. This site, more than any other, was critical to Hutton's understanding of the way in which the Earth worked. It is no exaggeration to say that it is a place of pilgrimage for geologists worldwide. Almost every geology student from St Andrews to Sydney will have heard of Siccar Point.

Siccar Point is located south-east of Edinburgh, between Eyemouth and Dunbar. Travelling southwards down the A1, take the A1107 turnoff to Coldingham and in 1 kilometre, a minor road towards the coast. Follow the road towards the vegetable processing factory and park near a gate on the left with a small white sign to Siccar Point. Proceed along the edge of the field, past the ruined St Helen's Church, and then along the clifftop to an interpretation board above Siccar Point. The way down to the coastal exposures is very steep. It is very easy to lose your footing, so care is required. The descent is unsuitable for young children. The exposures, however, can be viewed from the clifftop.

Siccar Point is one of Hutton's unconformities described in chapter 3. These structures arise where rocks from one age sit on top of rocks of an older age, but with a significant time gap between them. In this case, the older Silurian rocks have been upended and are almost vertical. The younger Devonian rocks overlying them, by contrast, dip gently. The plane between the two is known as an 'unconformity'. This undulating contact represents the passage of many tens of

The Scottish Outdoor Access Code: your rights and responsibilities

Know the Code before you go

Everyone has access rights over most land and inland water in Scotland, established by the Land Reform (Scotland) Act 2003 which came into effect in February 2005. You can exercise these rights, provided that you do so responsibly, for recreational and educational purposes, and for crossing land or water. Access rights do not apply to motorised activities, hunting, shooting or fishing, nor if your dog is not under proper control. The Scottish Outdoor Access Code provides guidance on how to exercise access responsibly. It also guides land managers in their responsibilities towards those accessing their land.

The key principles of the Code are:

- **Take responsibility for your own actions**
  If you are exercising access rights, remember that the outdoors cannot be made risk free and act with care at all times for your own safety and that of others.
  If you are a land manager, act with care at all times for people's safety.

- **Respect the interests of other people**
  Acting with courtesy, consideration and awareness is very important. If you are exercising access rights, make sure that you respect the privacy, safety and livelihoods of those living and working in the outdoors, and the needs of other people enjoying the outdoors.
  If you are a land manager, respect people's use of the outdoors and their needs for a safe and enjoyable visit.

- **Care for the environment**
  If you are exercising access rights, look after the places you visit and enjoy, and leave the land as you find it.
  If you are a land manager, help maintain the natural and cultural features that make the outdoors attractive to visit and enjoy.

Specific points to remember when taking access to the outdoors:

- if you have a dog with you, keep it under proper control at all times;
- if you are going to an area where shooting or deer stalking takes place, find out whether your planned route will be affected and take account of advice on alternative routes; avoid crossing land when shooting or stalking is taking place;
- do not intentionally or recklessly disturb or destroy plants, eggs, birds and other animals, or geological features;
- take extra care to prevent damage in more sensitive natural habitats and to avoid disturbing more sensitive birds and animals, particularly during the breeding season;
- follow any voluntary agreements between land managers and recreational bodies, or requests made by local authorities, Scottish Natural Heritage or other public bodies; take your litter away with you.

For more information on access rights and responsibilities visit: www.outdooraccess-scotland.com
For information on access during the stalking season visit: www.hillphones.info

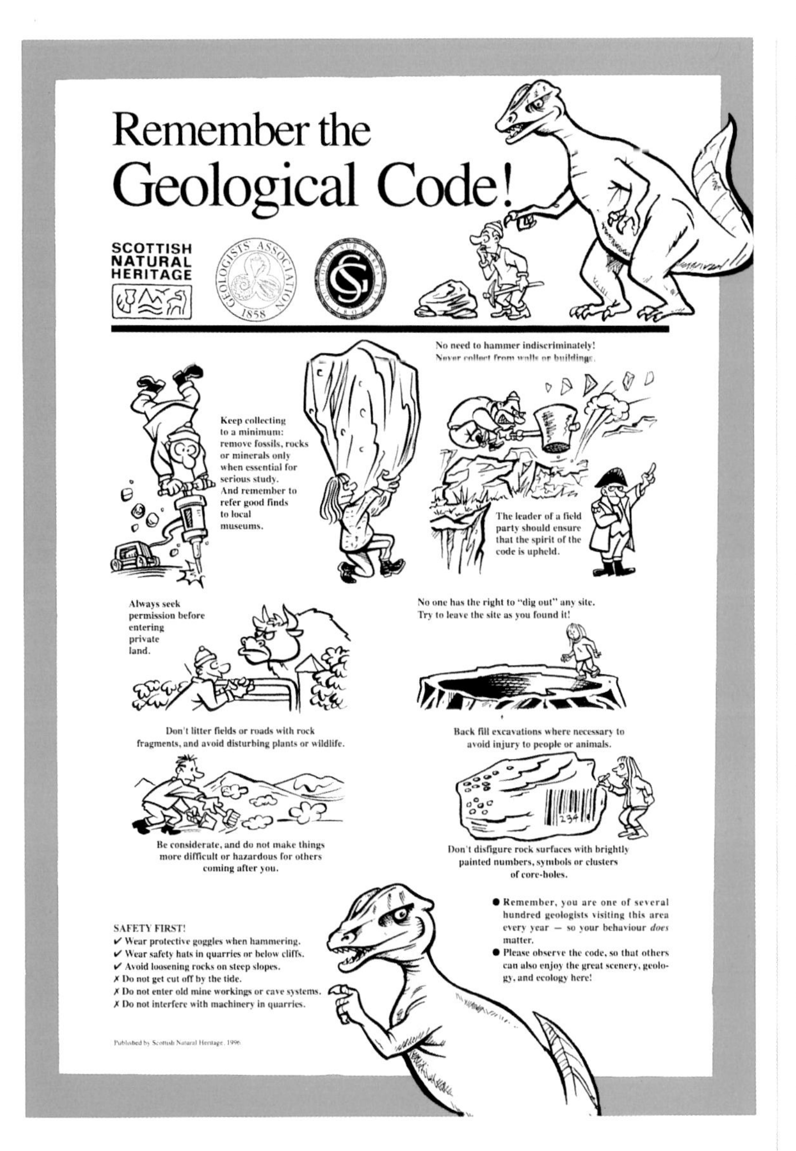

millions of years between the time that the underlying strata were upended and the overlying sandstones were deposited on top.

James Hutton visited Siccar Point in 1788, accompanied by two of his friends, John Playfair and Sir James Hall. He later wrote:

> Having taken the boat at Dunglass Burn, we set out to explore the coast. At Siccar Point, we found a beautiful picture of this junction washed bare by the sea. The sandstone strata are partly washed away, and partly remaining upon the ends of the vertical schistus; in many places, points of the schistus are seen standing up through among the sandstone, the greater part of which is worn away. Behind this we have a natural section of the sandstone strata, containing fragments of the schistus. Most of the fragments of the schistus have their angles sharp; consequently they have not travelled far, or been worn away by attrition.

John Playfair, Hutton's friend, biographer and, in this case, travelling companion, shared the moment of discovery and wrote an account of the trip that has become one of the classics of scientific literature:

> On us who saw these phenomenon for the first time, the impression will not be easily forgotten . . . We felt ourselves necessarily carried back to a time when the schistus on which we stood was yet at the bottom of the sea, and when the sandstone before us was only beginning to be deposited, in the shape of sand or mud, from the waters of a superincumbent ocean . . . The mind seemed to grow giddy by looking so far back into the abyss of time; and while we listened with earnestness and admiration to the philosopher who was now unfolding to us the order and series of these wonderful events, we became sensible how much further reason may sometimes go than imagination may venture to follow.

The unconformity at Siccar Point with the gently dipping Old Red Sandstone sitting on top of the upended Silurian strata.

What prompted these extravagant words was the realisation that the relationship between the underlying upended Silurian strata, or using the contemporary term 'schistus', and the sandstones above, could not have formed in the seven days prescribed in the Bible for the formation of the Earth. Indeed, any timescale measured in terms of human existence would be insufficient to accommodate the chain of events that Hutton and his companions had deduced. They could imagine the inexorably slow rate at which rocks accumulated on the sea bed and, subsequently, the time and cataclysmic force required to upend those strata. Uplift to form dry land followed, with later inundation and deposition of the red sandstones above. The sandstones contain pebbles from the upended strata below, which have been rounded and incorporated within their lower layers. These natural processes would have taken many millions of years to occur. The poetic, and scientifically accurate, reference to the 'abyss of time' captures this 'eureka' moment perfectly.

It was said of Hutton that his writing suffered from "prolixity and obscurity", which prevented many from fully understanding the power of his arguments. But in the field, as he examined rock exposures such as those at Siccar Point, his descriptive powers and ability to engage his audience were not in doubt. According to Playfair, Hutton was a slender figure, with penetrating eyes and a long "aquiline nose", quick of wit and passionate about his subject. He was a popular figure and "a brighter tint of gaiety and cheerfulness spread itself over every countenance when the Doctor entered the room".

A noticeboard has been erected at the top of the cliffs to tell the story of Siccar Point. But that narrative is equally clearly told in the rocks below. The visitor will see the strata that Hutton and his companions examined, minus a few corners knocked off by the sea over the last 200 years. The place reeks of history and discovery and of a free spirit working out his version of the truth. If you go nowhere else to see the geological gems of Scotland, go to Siccar Point. Retrace the steps of a true genius and experience at first hand the place that inspired Hutton to develop one of the most important scientific insights of the Enlightenment. He could see "no vestige of a beginning and no prospect of an end" in terms of the time required to fashion these rocks.

As the revolutionary concept of deep time gained greater acceptance after his death, Hutton's legacy was to free subsequent investigators from a straightjacket and allow the Earth and life sciences to develop and blossom. Charles Lyell enthusiastically embraced Hutton's ideas and they later informed his classic book, *Principles of Geology*. Charles Darwin read Lyell's book as he sailed around South America some forty years after Hutton's death. The idea of an "abyss of time" provided the extended timeframe that Darwin's developing theory of evolution required to operate in a credible fashion. And Siccar Point was right at the heart of it all.

2. VOLCANIC EDINBURGH

Known as the 'Athens of the North' for its architectural splendour, Edinburgh is also of note as a city built on the stumps of ancient volcanoes. Although long extinct, these once fiery volcanic edifices now form many landmarks around the city. Perhaps best known of all is the volcanic plug that supports Edinburgh Castle. Towering above Princes Street Gardens, it is a familiar sight to all visitors to Edinburgh. The black crags of basalt represent the core of the volcano that set hard at the end of its last eruption to form a resistant plug. Many millions of years later, the area was extensively glaciated and this mass of basalt, being harder than the surrounding rocks, was shaped by the ice to form the prominent landmark we see today.

The main volcanic activity took place during the early Carboniferous, when Scotland lay close to the Equator. The environment was that of a steamy tropical lagoon, fringed by trees and teeming with fish and shelled sea creatures known as bivalves and gastropods. The tranquillity was repeatedly shattered as a series of volcanoes erupted in the vicinity. The largest of these was to form the area we now recognise as Arthur's Seat in Holyrood Park. This actually consists of two separate volcanic vents, the larger Lion's Haunches vent and, lying to the west, the Lion's Head vent. The Lion's Haunches vent cuts across rocks associated with the Lion's Head, so it must have remained active after the latter fell silent.

The summit of Arthur's Seat is a magnificent viewpoint for Edinburgh's impressive cityscapes and the landscapes beyond. The Pentland Hills form a very prominent feature to the south-west, and the Castle Rock is also visible. Looking northwards and north-westwards, the volcanic hills of Fife are clearly seen, as are the rocks of the Highlands. To the east are the remnants of further volcanic hills in East Lothian.

Molten rock does not always make it to the Earth's surface and may consolidate below it as a separate sheet of igneous rock. This is the origin of Salisbury Crags. A pulse of magma forced its way between two pre-existing layers of sedimentary rock and cooled just below the surface, later to be excavated by the ice. The Crags now occupy a prominent position and are visible from many parts of the city.

A thick lava pile is associated with the Arthur's Seat volcano. Only half the volcanic cone is preserved, underlying Whinny Hill to the east of the main summit. Alternating hard lava flows and softer ashes give the hill a stepped appearance. The oldest lava in this pile of thirteen separate flows is of particular interest, as it seems likely that it emanated from the Castle Rock vent, some 2 kilometres distant. Pulpit Rock is a very prominent feature on the western slope of Whinny Hill. This lump of rock, surrounded by whin bushes, is a small volcanic side-vent from which one of the older lavas flowed.

At the south end of the park, spectacular lava columns, named Samson's Ribs, are also prom-

Below. Around 350 million years ago, the Arthur's Seat and Castle Rock volcanoes were active. This reconstruction shows what Edinburgh might have looked like at that time, with the contemporary view superimposed.

Opposite. Salisbury Crags is a well-known landmark in the city of Edinburgh. Its origins are associated with the Arthur's Seat volcano.

inent features. The road cuttings created to accommodate the circular road that runs around the park, known as the Queen's Drive, provide excellent exposures of the inner core of the old volcano. Jumbled blocks of lava set in a fine matrix of volcanic ash can be seen in these exposures which mark the position of the main pipe or vent of the ancient volcano.

James Hutton studied Salisbury Crags in some detail to advance his understanding of how the Earth works. At this site and a number of others throughout Scotland, he showed that the rock cooled from a molten state. The prevailing idea of the time was that all rocks were precipitated from a long-disappeared ocean that was supposed to have completely covered the face of the Earth. Hutton's Section is now celebrated by an information board, which explains the significance of the site. Hutton is further commemorated by the naming of 'Hutton's Rock', which is also on Salisbury Crags. This exposure is of basalt, cut by a thin horizon of iron oxide, known as haematite. It is reputed that this interesting exposure was left unquarried at Hutton's request. If we accept this explanation, then this is one of the very first acts of geological conservation to be undertaken anywhere in the world.

At Camstone Quarry, eastwards from the precipitous cliff edge of Salisbury Crags, the sedimentary rocks through which the volcano erupted are well exposed. Sandstones and mudstones are also visible, some showing cracking, which indicates that the sediments periodically dried under a hot sun. The lavas were erupted on top of this foundation of sand and mud that had been laid down in a quiet lagoon on the fringe of a tropical coastal plain.

Many of the hills in and around Edinburgh also have a volcanic history. Calton Hill, at the east end of Princes Street, and Corstorphine Hill,

Our Dynamic Earth with Salisbury Crags in the background.

on the western outskirts, are both similar in age to the Arthur's Seat volcano. Blackford Hill, the Braid Hills and the Pentland Hills, located on the south side of the city, represent volcanic activity from an earlier age – the Devonian. The Pentlands are built from lavas of different mineral compositions that vary in colour from pale orange to dark grey. Thick layers of volcanic ash are also present, indicating that the volcanic activity was fairly explosive when these lavas were being erupted around 400 million years ago.

More recently, the landscape of Edinburgh has been modified by the Ice Age glaciers. The bedrock, which is a mix of hard igneous rock and softer sedimentary rocks, was selectively eroded by the ice. The more resistant basalts stood up well to the power of the ice, whereas the weaker sedimentary rocks were worn down. Locally, the igneous rocks protected the sedimentary rocks on their lee sides, and crag-and-tail landforms were produced, with steep west-facing slopes and more gently sloping gradients to the east. Edinburgh Castle Rock and the Royal Mile is a classic example; others include Blackford Hill, Calton Hill and Craiglockhart Hill. Good examples of ice-smoothed bedrock with grooves aligned in the direction of ice flow can be seen at Corstorphine Hill, while on the south side of Blackford Hill there is the historically famous Agassiz Rock, where Louis Agassiz asserted in 1840 that scratches on the rock surface had been produced by glaciers.

Our Dynamic Earth, housed in a futuristic building constructed at the foot of Salisbury Crags, is also worth a visit. As described in chapter 1, this Millennium project interprets the workings of our dynamic planet for the general public.

3. THE CAIRNGORMS

The Cairngorm Mountains display an exceptional diversity of landforms, revealing the evolution of a granite mountain landscape over the last 400 million years. The area contains the largest expanses of high plateaux above 1,000 metres in Britain, most of the highest summits in Scotland and a wealth of spectacular glens, corries and other features sculpted by the Ice Age glaciers. What is particularly remarkable is the combination of landforms that originated before the last glaciation – tors, weathered bedrock and plateau surfaces – with those shaped by the glaciers. Together, they form one of the world's outstanding examples of a mountain landscape of selective glacial erosion, showing how the erosive power of the glaciers was very effective in particular areas but minimal in others. In addition, the area contains a great range of other

Glen Quoich and the southern Cairngorms, showing the high plateaux with their rounded summits, the glacially steepened flanks of the upper glen and the alluvial floor of the middle glen.

glacial, periglacial, river and slope landforms. In view of the exceptional significance of their Earth history and landforms, the Cairngorm Mountains are included in the UK Tentative List of World Heritage Sites and they became the core of Scotland's second National Park in 2003.

The main elements of the formation of the Cairngorms landscape are described in chapters 3 and 4, including many specific examples of the features present. Granite underlies most of the central mountain massif, apart from an area in the west. The properties of the granite, in particular the jointing and other lines of weakness, have strongly influenced the detailed patterns of weathering, erosion and landscape evolution. The rocks that surround the granite are of Dalradian age, comprising ancient muds, sands and limestones, which were greatly altered when the Caledonian Mountains were formed.

A mass of molten granite rose to within 4 kilometres of the Earth's surface as it then existed towards the end of the Caledonian mountain-building period. Blocks of Dalradian rock were prised away as the granite ascended through the crust. These lumps of foreign rock, or xenoliths, were been affected by heat from the molten magma and are much altered from their original state. Recent work by the British Geological Survey has established that within about 20 million to 30 million years after it cooled around 427 million years ago, erosion 'unroofed' the granite, removing the layers of rock between it and the surface of the Earth at that time. It seems likely that the Cairngorm granite then remained exposed to the elements until the present day, without being buried by younger rocks or inundated by the sea.

The granite of the Cairngorms comprises three principal minerals: clear quartz, pink feldspar and flecks of black mica. In fist-sized specimens, the predominant colouration of the rock is pink. The composition of the granite is singularly uniform across the intrusion, although it is possible to identify different rock textures. Some rocks are uniformly fine grained, whereas others are coarser grained with large feldspar crystals. The granite of Glen Avon is generally accepted to represent a separate pulse of magma, although integral to the intrusion as a whole. The area is well known for semi-precious Cairngorm Stones, a variety of quartz in which the crystals have a smoky appearance. Exposure to natural radiation is thought to be responsible for this alteration. They have been widely collected over the years and are now comparatively rare.

The broad outlines of the present landscape began to appear as the rocks overlying the granite were broken up and removed during Devonian times by processes of weathering and erosion which greatly lowered the high Caledonian Mountains. As the granite was exposed around 400 million years ago, the weathering and erosion selectively exploited zones of weaker rock and, over time, the early rivers excavated the precursors of the present glens along these zones of weakness. By Cretaceous times, the huge mountains had been reduced to a landscape of low relief near sea level. Later, during the Palaeogene and Neogene, the area was uplifted by crustal movements associated with the opening of the North Atlantic Ocean. The surface layers of the granite were also decomposed and eroded under the warm, humid climate conditions. The uplift, weathering and erosion produced a landscape of plateau surfaces and rolling summit hills. Subsequently, during the Ice Age, glaciers selectively eroded the mountains, carving some of the existing valleys deeper and breaching the divides between other valleys. At the same time, ice and frost etched out the corries and helped to

Opposite. The Cairngorms from the north, showing Coire an t-Sneachda (left) and Coire an Lochain (right) cut into the northern edge of the plateau leading south to Ben Macdui. Loch Lomond Readvance boulder moraines (outer limits arrowed) occupy the corrie floors. The adjacent slopes have been extensively modified by solifluction, and garlands of boulder lobes are particularly well displayed in Lurcher's Gully (right). The deep trough of the Lairig Ghru (top right) lies largely hidden, but the truncated spur of the Devil's Point forms a prominent landmark.

expose many of the tors. The melting of the last glaciers left behind abundant deposits in the glens and corries, along with meltwater channels around the fringes of the mountains. Periglacial activity on the higher slopes has slowly moved the soil downhill, forming sheets and lobes of often bouldery debris; rivers have cut into the glacial deposits and the underlying bedrock, producing terraces, fans, and gorges; and slopes have been gullied during heavy storms.

With so much to see over a very large area, the visitor has many options and only a few are highlighted here. There is no 'best place' to see the granite of the Cairngorms. It can be examined wherever solid rock reaches the surface. However, Glen Avon is a good place to see the blocks of 'foreign' Dalradian rock incorporated into the granite. For the 'look and see' visitor, there are good overall viewpoints in the Dee Valley, Strath Don, Glen Feshie and Strath Spey. Starting in the south, the A93 over the Cairnwell north to Braemar offers excellent views of the great granite dome and the summit plateaux, with the tors of Beinn a' Bhuird and Ben Avon on the skyline. Driving west from Braemar, stop at the viewpoint opposite Glen Quoich to get a better impression of the juxtaposition of the glens and the plateaux, as well as the large alluvial fan of the River Quoich. Further west, the Linn of Dee is carved in Dalradian rocks and separates two alluvial reaches of the River Dee.

To the north, the road over the Lecht from Strath Don gives excellent views across the eastern Cairngorms to the plateau and tors on Ben Avon; also, notice how small the Don is in relation to the size of its valley, the result of having its headwaters captured by the Avon. On the west side of the massif, in Strath Spey, the roads up Glen Feshie allow an impression of one of the most dynamic rivers in Britain, with a constantly moving bed of gravel, as well as glimpses of the mountains. The only easy way to experience a flavour of the high mountains is from the ski road from Aviemore to Coire Cas, where the visitor can take the funicular railway to the upper slopes of Cairn Gorm to enjoy a wider vista, although there is no access out on to the mountain. En route, the flash-flood deposits of the Allt Mór and the meltwater channels on the northern flanks of the hills are clearly evident.

For the walker, on the south side of the Cairngorms, the side glens of the Dee, especially the Quoich and Lui, give low-level walks in towards the mountains, and for the more energetic onto the plateaux of Ben Macdui and Beinn a' Bhuird, where a range of glacial and periglacial landforms can be seen. In the lower parts of these glens, there are terraces where the rivers have cut through the glacial deposits, and a variety of river gorges and potholes at the Linn of Lui and Linn of Quoich. An ascent of Morrone, south of Braemar, gives excellent views of the southern part of the massif from Beinn Bhrotain to Ben Avon, as well as upper Deeside.

Similarly, on the north side of the mountains, the paths through the Rothiemurchus Estate into the Lairig Ghru and Gleann Einich and through the Glenmore Forest Park give a better impression of the juxtaposition of glacial erosion and deposition, the glacial and modern river channels, and the old plateau surfaces. Some of these paths lead into remote and challenging terrain, although the lower sections involve only moderate amounts of ascent on reasonable tracks. The paths in Glenmore Forest Park are relatively easy. There is a good view of the River Feshie fan at its junction with the River Spey from the public footpath at Speybank. Three short, self-guided walks with leaflets, *Trails Through Time*, have been produced by SNH to introduce visitors to some of the landforms in Glenmore and the area near Kincraig. There are visitor centres at Inverdruie (Rothiemurchus Estate), Glen More (Forestry Commission) and Coire Cas (Cairn Gorm Mountain). Seasonal ranger services are provided at these centres and at Mar Lodge (National Trust for Scotland).

There are walks from the Coire Cas car park into the Northern Corries of Coire an t-Sneachda, Coire an Lochain and Lurcher's Gully, where corries, moraines and periglacial features are present in the harsh, high-altitude environment. In good weather, a high-level mountain walk can be made by experienced hillwalkers from the Coire Cas car park via Sròn an Aonaich, Cairn Gorm and around the headwalls of Coire an t-Sneachda and Coire an Lochain and then completing a circuit back to the car park at a lower level. From the summit area of Cairn Gorm, there are superb views in clear weather of the plateau surfaces, the tors on Beinn Mheadhoin, the boulder lobes of Lurcher's Gully and the meltwater channels on the lower slopes of the Northern Corries; also just east of the summit there is a good example of a small tor with sheet jointing. From the top of the Fiacaill a' Coire Chais, there are good views of Coire an t-Sneachda, including the boulder moraines on its floor. Continuing westwards, periglacial features are common, notably solifluction sheets and lobes and wind-eroded vegetation. From the path on the west side of Coire an Lochain, boulder moraines can be seen on the corrie floor. For those walking to Ben Macdui, a diversion can be made to see the glacial trough of Glen Avon, a spectacular example of selective glacial erosion.

The Cairngorms are notorious for severe weather at any time of year and all excursions require careful planning, use of a map at a scale of 1:50,000 or greater, competent navigation and appropriate clothing and footwear.

4. THE LOWER RIVER SPEY AND SPEY BAY

The lower River Spey and the coast at Spey Bay provide an exceptional opportunity to see a highly dynamic river and coastal environment. The interaction of such an active river in its lower reaches with a coastal shingle plain that has emerged from the sea is unique in Britain.

The lower River Spey is unusual for a lowland river in that it remains highly dynamic right down to its mouth. This dynamism is apparent in the braided form of the channels, their continuously shifting nature and the rapid rates of change in the position of individual channels, gravel bars and islands. The river has a high energy and is able to alter its gravel bed, particularly during floods. It moves a large amount of sand and gravel down to the coast at Spey Bay, which are then reworked by the sea into a gravel bar and beach. The supply and throughput of this sediment are an essential part of maintaining the dynamic nature of the river channels and the stability of the gravel beach. Beyond the margins of the presently active river floodplain is a wooded area that has been occasionally occupied by the river in the last 200 years, notably during the Muckle Spate of 1829 (see chapter 4).

The coast at Spey Bay is notable for the scale of the active gravel beach, extending from Portgordon to near Lossiemouth, and for the largest area of raised gravel ridges in Scotland. The latter are particularly well developed west of Kingston where a suite of ridges, many unvegetated, extends 800 metres inland to the base of Binn Hill. These ridges owe their existence to the supply of material from the River Spey and formed because relative sea level has fallen during the last 7,500 years. Hence, the dynamic processes of the river and coast are closely linked in this area. This is clearly evident at the mouth of the Spey, which forms a constantly changing environment as gravel from

The River Spey is exceptionally active right down to its mouth at Spey Bay. The river channels and gravel bars frequently change their positions during floods.

the river is carried westwards by longshore drift, forming a spit and diverting the river mouth westwards. Eventually the river breaches the spit and returns to a more central outlet at Tugnet. Further west, the active shingle bar has been extending westwards towards Lossiemouth at a rate of over 30 metres a year over the last hundred years. To the east, however, the coast has been receding landwards at about 1 metre per year.

The presence of the river and dynamic coastline have proved both beneficial and hazardous to human activities in the area. In the eighteenth century, the port of Garmouth was a centre for the processing and export of timber, while Kingston was established as a new settlement at the mouth of the river by a company from Kingston-upon-Hull. Kingston became a prosperous centre for timber export and shipbuilding, including clipper ships for the India tea trade. Timber was floated down the Spey from Rothiemurchus Forest, as many as 20,000 logs at a time, with gangs of men going along the banks to push them off where they became stuck. The economic base of both Garmouth and Kingston declined after 1860 due to increased use of iron and steel in shipbuilding. However, the dangerous nature of the estuary and the shifting course of the river, which made activities increasingly difficult and prevented development of a proper harbour, played an important part. The village also suffered during the Muckle Spate when many houses were destroyed. The westerly growth of the spit has contributed to flooding and erosion in Kingston and in recent years it has been artificially breached to reduce the flood risk. Various bank stabilisation and flood alleviation measures have also been implemented along the river, as well as modifications to improve the salmon fishing. The flat land of the river

The supply of gravel from the River Spey has contributed to the formation of the best dynamic gravel spit and beach in Scotland at Speymouth.

floodplain at Garmouth, and raised beaches at Spey Bay, have provided sites for golf courses, although both are at risk from flooding and erosion by the river and the sea. West of Kingston, the unvegetated shingle ridges are used as a Ministry of Defence rifle range; elsewhere, they are largely concealed by commercial forestry plantations.

The area is also noted for its diverse birdlife and habitats. There is a Scottish Wildlife Trust Reserve at Spey Bay with car parks and interpretation boards at both Kingston and Tugnet. The old salmon-fishing station at Tugnet has been converted into a wildlife centre run by the Whale and Dolphin Conservation Society. There are good views of the river from the footpath which crosses the old railway viaduct near Garmouth and from the Speyside Way which passes along the east side of the river.

5. CULBIN

Culbin is probably best known for the history of human struggle and eventual success against the forces of nature, in this case blowing sand and dunes overwhelming coastal settlements and fields. However, this human story is only one element of a much bigger picture of coastal evolution recorded in the landforms, a picture that is still changing today. Culbin is exceptional at a European level for the scale, diversity and dynamism of its coastal landforms. At one time, Culbin was the largest area of bare coastal sand dunes in Britain.

As with many other beach complexes (see chapter 4), the history of Culbin starts during the latter stages of the last glaciation when large quantities of sand and gravel from the melting glaciers were carried down to the sea by the Rivers Nairn and Findhorn. As sea level rose to a high point, some 7,500 years ago, this material was moved towards the coast and redistributed by the dominant westerly drift of material along the coast. Then, as relative sea level fell again, the waves reworked the gravel into staircases of ridges that later became buried with blown sand and dunes.

The gravel ridges that provide the platform for sand-dune development are locally visible underneath the sand where they form a corrugated surface. They lie seawards of a prominent abandoned cliffline which runs more or less continuously along the coast and which marks the coastline as it was around 7,500 years ago. The Culbin sand dunes extend over a distance of some 14 kilometres between the mouths of the Findhorn and Nairn rivers and include a range of forms unequalled in Britain. The sand dunes moved inland in a south-westerly direction, overwhelming the land in front of them. The massive Lady Culbin dune, east of Buckie Loch, is 30 metres high and is estimated to have moved by up to 100 metres a year at its peak of activity. Over the last few millennia, the Culbin dunes have undergone alternating periods of activity and stability; soils that formed on the sand surface during stable periods were later buried when sand movements were reactivated. Periods of sand movement date back to around 5,000 years ago, but the most dramatic events are those recorded in more recent times over a period of about 400 years from the thirteenth century onwards, culminating in the huge sand blow of 1694.

A combination of strong winds and the removal of marram grass for thatching resulted in a devastating sand blow in October 1694, burying fifteen farms, fields, crops, livestock and houses, as well as Culbin House and its church and gardens. However, there seems to have been no loss of human life. A contemporary account describes the scene: "It came suddenly from the west on an October day that year, a high cloud of sand, two miles in width like a river flowing at great speed. Men reaping barley had to run from the fields and within a few hours the barley was smothered . . . People stayed in their houses overnight only to find that their doors and windows were completely blocked by sand the next morning . . . The following day there was nothing to be seen but sand, not even the tops of trees or the chimneys of the laird's house" (quoted in David Thomson, *Nairn in Darkness and Light*, 1994). In more recent times, there have been occasions when the chimney pots of houses and the tower of the church appeared out of the sand, and local children are reminded not to trip over the church spire as they walk through the dunes.

There have been many attempts to constrain this turbulent sand activity, particularly since the 1840s. The most successful and extensive were by the Forestry Commission between 1922 and 1963. The sand was first covered with a thatch of dead wood, then marram grass was planted, followed by conifers, notably salt-tolerant Corsican pines. The surface of the sand dunes is now so stable that in places it is covered with mosses.

Another remarkable feature at Culbin is the highly dynamic system of bars along the modern coast. At the eastern end, the Buckie Loch sand spit has extended westwards at a rate of 22

The recent development of the Bar at Culbin is shown by the sequence of shingle ridges. Each recurve represents a stage in the extension of the bar westwards as new shingle is carried along the shore and deposited.

metres per year for the last 130 years. To the west, the Bar is a superb example of a gravel spit, comprising a spectacular series of recurved gravel ridges, with tidal lagoons and salt marsh behind. It has been moving westwards at a rate of about 15 metres per year. The gravel has been transported along the shore by the action of waves hitting the coast obliquely from a north-easterly direction. As a result, there are remarkable hooks, or recurves, at the end of each ridge marking the final point of deposition before a new one began to form. All of the gravel ridges have a smooth seaward face fashioned by the waves, and an irregular, crenulated inland face due to the hooked ends and washover of material in storms.

Access to Culbin Sands is easiest from the existing car parks. At the west end, a car park on the Loch Loy road east of Nairn gives access to the RSPB reserve, the salt marsh and a view of the Bar. At the east end, there are two car parks, at Wellhill and Cloddymoss, giving access to the dunes and forest and eventually to the beach. Although the trees obscure the views, there are still plenty of places to walk through the dunes and it is possible to climb to the top of Lady Culbin dune. There is also access along the beach at the Findhorn end to see the scale of the dunes where they are actively being eroded by the sea at an average rate of up to 1 metre per year.

6. FOSSIL FISH SITES, CAITHNESS

The Old Red Sandstones of Caithness have been extensively worked over the years for flagstones. This has left an industrial legacy of many quarries, most now disused. These are also a fossil-hunters' paradise. It was the quarrymen of Hugh Miller's time in the middle of the nineteenth century who were first to notice the prehistoric creatures that stared out from the freshly split slabs of sandstone. Hugh Miller, a stone mason by trade and later a prolific writer on geology, was intrigued by the variety of plants and animals he found entombed in the rock and began his life's work as a palaeontologist shortly after he started work in the sandstone quarries of the Black Isle. The locations described below lie to the north of Miller's beloved Cromarty, but are equally productive for collecting new fossil material.

Please only collect from the loose material and do not hammer the rock faces. This should not detract from your visit to the sites, as scree material is abundant. Please limit what you remove from any of the locations to two specimens at most. There are many other fossil-hunters who will want to visit these sites after you, so please leave the places as you would wish to find them.

This is a composite 'place to visit' made up of two sites. The first is the classic Achanarras Quarry. This was formerly a National Nature Reserve, but retains protected status as a Site of Special Scientific Interest (SSSI). A permit is required to collect from this site and this should be obtained from the SNH office in Golspie. The quarry is located just off the A9, near the Mybster cross-roads, around 20 kilometres north of Latheron. Travel west along the B870 and turn right up a track after about 1 kilometre. Proceed on foot from the locked gate near the disused croft. The disused quarry is flooded and the edges should be approached with care. The main fish-bed horizon is permanently under water, but the

Pterichthyodes milleri are common at Achanarras Quarry. This fossil takes its name from the celebrated palaeontologist, Hugh Miller.

site has a huge amount of spoil left by the quarrymen that can be searched for fossils. Although these spoil tips are not high, they are by their very nature unstable, so great care must be exercised in examining this material. A range of types of fish has been identified from this site, including perhaps the most common, *Dipterus*, the lung fish. Other genera include *Osteolepis*, *Coccosteus* and *Pterichthyodes* – the winged fish of Hugh Miller.

There were periodic mass mortalities of the fish population. This goes some way towards explaining why the remains of these 385-million-year-old fossil fish are largely contained within specific geological horizons that are known as fish beds, rather than having an even distribution throughout the sandstone strata.

The second site lies just to the west of Thurso, near Scrabster. Holburn Head Quarry is located a brisk twenty-minute walk from the end of the metalled road in Scrabster. The quarry is disused, although, as with any site, danger lurks for the unwary. The edge of the quarry abuts a vertical sea cliff, so keep clear of this northern boundary of the site.

The main fish bed actually lies beneath the quarry floor, but it is exposed in a limited area where numerous fish fragments can be collected. Again, there is a great deal of loose material among which to search for fossils. The dominant species to be found here is *Osteolepis panderi*, commonly up to 12 centimetres in length. These fish have fins, prominent scales and skulls that have similarities to later amphibians. They belong to the group that gave rise to land-living tetrapods, or four-legged animals, and eventually to humans.

7. THE OLD MAN OF HOY (ORKNEY) AND ESHA NESS (SHETLAND)

The sea cliffs on the west coast of Hoy in Orkney and at Esha Ness on the north of Mainland Shetland are excellent places to see rock-coast landforms and experience the power of natural forces.

The cliffs from St John's Head southwards to Rora Head and Rackwick on Hoy are the third highest in Scotland, rising to 335 metres. They are formed of Old Red Sandstone. The beds are horizontal and have both harder and softer layers and a number of vertical fractures and faults which the waves have exploited to produce a tremendous variety of landforms: stacks, arches, caves, geos, over-hanging cliffs and shore platforms. These are some of the best examples in Europe of Old Red Sandstone cliffs and other features of coastal erosion, including the renowned Old Man of Hoy, one of the highest and most spectacular sea stacks in Britain. The cliffs appear solid and enduring. However, the Old Man has a finite life, having already lost one leg, and there are cracks in the rock showing that the stack will lose part of its top relatively soon. One day it will be reduced to a heap of rubble at sea level. Meanwhile, other stacks are being formed along the coast at Rora Head as the roofs of caves collapse, geos form and pillars of rock become isolated from the cliffs. These, too, will eventually collapse to form a boulder beach. So, several different stages in the formation of coastal landforms can be seen here.

The rocks also have a story to tell. Near the base of the Old Man, a series of volcanic rocks is among the oldest on the island and includes some thick lava flows. Below the lavas are a series of sandstones rich in ash. These accumulated as ash from nearby volcanic eruptions fell from the sky. Above the lavas, a thick sequence of sandstones was laid down. Some of the sandstone horizons are wind-blown deposits; others were laid down

The Old Man of Hoy is the most spectacular feature of the towering cliffs of the west coast of Hoy. This sea stack will eventually be eroded away, but others are already being formed around Rora Head.

The cliffs at Esha Ness, Shetland, display a spectacular range of rock coast landforms.

by braided streams and rivers. So a picture can be built up of the various environments that existed during Old Red Sandstone times when Scotland was a few degrees south of the Equator (see chapter 3).

While on Hoy, also take the chance to see the glacial and periglacial features at the north end of the island. For the energetic, Ward Hill can be ascended from Sandy Loch to the north-west. Turf-banked terraces formed by solifluction, as well as vegetated stripes and sand deposits formed by the wind, can be easily seen. Also worth a visit are the end-moraine ridges at Dwarfie Hamars, formed by a small glacier that occupied a shallow corrie in the hillside at the time of the last phase of glaciation on Orkney. The Dwarfie Stane is a large glacially transported block of sandstone resting on the moraine. Carved into it is a large chamber, thought to be a Neolithic tomb and apparently constructed for a man of small physical stature who mined the haematite at the north end of the island. Further along the road, near Rackwick, a large glacial outwash terrace fills the floor of the glen. This was formed by the rivers from a melting glacier.

To reach Hoy, take the car ferry from Houton on the Mainland Orkney shore of Scapa Flow across to Lyness. Then take the road north towards Quoy and then across the glen to Rackwick where there is parking and a clearly marked path along the cliffs to the Old Man and other features. The ferry to Orkney from Scrabster, in Caithness, gives superb views of these cliffs from the sea.

On Shetland, the cliffs at Esha Ness are accessed from the main road north from Lerwick and along the B9078 past Hillswick to the car park at the road end by the lighthouse. The cliffs, up to 50 metres high in the south but lower in the north, are cut in resistant andesite and basalt lavas and display spectacular geos, blowholes and stacks. In stormy weather, the power of the Atlantic Ocean is evident; waves crash over the cliff tops, stripping off the soil and vegetation and, at Grind of the Navir, throwing huge boulders and blocks of rock up into storm-beach ridges 50 metres inland. The Cannon, as the name suggests, is a place where water shoots horizontally out of the cliff some seconds after the waves have hit the cliff. To the south, the spectacular arches on Dore Holm and the Skerry of Eshaness can also be seen. To the north at the Villians of Hamnavoe, the cliffs are formed in andesitic tuffs and andesite lavas and display the most impressive effects of wave power in Scotland. Here, storm waves have stripped vegetation and soil off the top of cliffs up to 30 metres high, formed boulder beaches up to 20 metres above sea level and scattered boulders up to 100 metres inland. There are also examples of caves, natural arches and a blowhole.

For those interested in upland processes, it is worth continuing northwards, past the fjord of Ronas Voe, to visit Ronas Hill. This is the highest hill in Shetland at 450 metres and, like Ward Hill on Hoy, is noted for its periglacial landforms and processes, in particular, superb wind-patterned vegetation and solifluction terraces. Such features are normally developed only at much higher altitudes on mainland Scotland and reflect the extreme exposure of the area.

8. THE MACHAIR OF THE UISTS

Enduring memories for the visitor to the west coasts of North Uist, Benbecula and South Uist are the wonderful sandy beaches and the profusion of flowers on the sand plains inland, with cattle grazing, ground-nesting birds fledging their young and crops of oats and rye growing on narrow strips of cultivated land. This is quintessentially the 'mecca of the machair', a focal point in the Western Isles for anyone interested in the interaction between human communities and the natural history of the land. A patchwork of sand, water, rock, and peat forms the very special and unique landscape of the machair.

Machair occurs on other islands in the Western Isles, in the Inner Hebrides – notably on Coll, Tiree, Colonsay and Oronsay – and also on the north-west part of the mainland, but nowhere is it more extensive and more significant than along the west coast of the Western Isles, where it stretches almost continuously from south Harris to the southern extremity of Barra: a distance of some 110 kilometres.

For many generations the sand plain has been cultivated on a rotational basis by the crofters to produce potatoes and local varieties of oats and rye. It is also used for cattle grazing in winter. Seaweed washed onto the beaches during stormy conditions is used to fertilise the soil. The

Rueval, south of Geirnish, gives an excellent view over the machair of South Uist. There is a road up the hill to the viewpoint.

machair demonstrates, perhaps more than anywhere else, the links between physical processes, natural heritage and traditional land management.

The machair has developed over the last few millennia and has always been a dynamic environment, as described in chapter 4. As sea level has risen since the end of the last glaciation, sand has been carried to the coast from the shelf offshore to the west. This sand has been blown inland to form dunes and machair behind the fringing beaches. Reduction in the supply of material from offshore has led to the natural recycling of sand from the dune fronts and its re-deposition inland, which has until now helped to maintain the machair.

The future of the machair is, however, uncertain. The amount of new material brought from offshore is much diminished compared with a few thousand years ago. The coastline is now receding in many places, especially during storms. The combined effects of such storms and the continued rise in sea level means that periodic recession is expected to go on for the foreseeable future. Areas of machair behind the blowouts and other low sections of the coast without the natural protection of a fringing barrier of dunes are likely to bear the brunt of erosion and flooding. Localised coast defences may be necessary to protect buildings and infrastructure in some areas; more extensive coast protection would simply transfer the problem elsewhere, would probably be prohibitively expensive and would reduce the sand supply from erosion of the coastal edge that now maintains the machair. All of these factors suggest that there will be a gradual reduction in the area of machair land as the coastline is forced inland, along with more frequent marine inundations.

There are many places from which to view the machair, including West Geirinish, Stoneybridge, Rubha Ardvule/Bornish, Ashernish, Daliburgh and Garrynamonie on South Uist, and Paible and Balranald on North Uist. However, to see the full range of features, there is no substitute for walking from the beach across the sand plains and blackland to the hills behind. An ascent of one of the hills gives a good impression of the variety of machair forms from the shore inland. Good viewpoints in North Uist are South Clettraval and Uneval, and in South Uist, Rueval, Haarsal near Howmore, and Askervein north of Daliburgh. To gain a different perspective, Baleshare in the south and Vallay in the north of North Uist have huge dune fields, unlike most of the rest of the machair area.

Cultivation of the machair, seen here at Hougharry on North Uist, is a traditional activity in the Western Isles. Seaweed is used as a fertiliser and a rotational cropping pattern ensures that the soil does not become overused.

9. KNOCKAN CRAG

Knockan Crag is one of the best places to see the Moine Thrust in an area renowned for its classic geological features. SNH opened a visitor centre here in 2001 to celebrate the great achievements of Benjamin Peach and John Horne in unravelling the mysteries of the Moine Thrust (see chapter 3). The intricacies of the Knockan story had previously been just for the cognoscenti to appreciate. But the narrative at the Knockan Crag visitor centre has been told in such a way that everyone can understand the nature of the debate that took place between the early pioneering geologists as they wrestled with the competing theories to explain the geology of the North-west Highlands. Knockan Crag is at the heart of some of Scotland's finest scenery and geological landscapes. This was recognised in 2004 by the award of European Geopark status for the area by the European Geoparks Network.

To find this historic site, take the A835 from Ullapool and travel northwards for around 20 kilometres and you will arrive at Knockan Crag Visitor Centre. There is parking for around twenty cars and two touring buses. There are toilets, which, like the interpretive displays, are open all year round, but no catering or other retail facilities. The interpretive displays are centred round a turf-roofed, open-air facility, known as the 'Rock Room'. A geological trail starts at the Rock Room and leads the visitor around a route that takes them past exposures of Cambrian strata, up the cliff towards the Moine

Left. Rock art and sculpture help to enhance the visitor experience at Knockan Crag and to tell the geological story.

Right. The Knockan centre has a number of interactive displays which explain the complexities of the geology to all ages of visitor.

Thrust and back round to the car park. The trail is further embellished with rock sculptures, such as a representation of the globe, and fossils associated with the Cambrian rocks.

In addition, SNH has used innovative, state-of-the-art interpretation, including interactive models to represent Scotland's journey across the globe, a multilingual CD-ROM, an eye-catching rock show, poetry etched in rock, and an accompanying leaflet to tell the Knockan story in a new and engaging way. Visitors perhaps with no previous interest or knowledge of the subject are encouraged to open their minds to new and challenging concepts that explain the geological development of this long-disappeared world of colliding continents and ancient shorelines.

The Rock Route, a series of twelve locations, mainly along the Ullapool to Kinlochbervie road, takes the story beyond the immediate environs of Knockan Crag. These locations include key landscapes and rock sections and add to the visitor's understanding of the area. Sites include the Bone Caves where the remains of bears, wolves and reindeer have been found in the cave-floor sediments; the view across to the north shore of Loch Glencoul where Lewisian rocks have been thrust over the younger Cambrian strata; and the famous road cuttings at Laxford where a slice through the Lewisian gneiss is beautifully exposed. The glacial history of the area is also described at Rhiconich, where there are glacial erratic boulders; near Inchnadamph, where there are moraines; and at Strathcanaird, where the depth of the ice sheet that once almost completely covered Scotland is demonstrated. At each location a weather-proof notice board explains in simple terms the geology and landscapes. All the locations on the Rock Route are lay-bys or other places where cars can be parked easily.

10. BEINN EIGHE AND GLEN TORRIDON

Beinn Eighe and Glen Torridon are outstanding for their geology and glacial landforms. This is a classic place to view spectacular Torridonian sandstone mountains rising abruptly from the underlying Lewisian gneiss basement. The area was heavily glaciated during the Ice Age and has many excellent landforms of glacial erosion, as well as moraines dating from the time of the Loch Lomond Readvance.

Beinn Eighe, or, from its Gaelic roots, 'the File Peak', was declared a National Nature Reserve in 1951, the first such designation in Britain. It lies on the southern shore of Loch Maree on the main tourist route from Wester Ross to the far North-west High-lands. This is another important geological area that has been interpreted for the benefit of the interested visitor. There is parking for around thirty cars at the Reserve interpretive centre, which is located on the A832 just to the north-west of Kinlochewe.

The geology of Beinn Eighe is complex, but the displays at the interpretive centre and the mountain trails that lead from it help the visitor to make sense of over 3,000 million years of geological time. The mountain sits astride the Moine Thrust: that line of disturbance in the Earth's crust formed during the final throes of the Caledonian mountain-building period. This cataclysmic event involved colliding continents and the closure of a great ocean (see chapter 3). It was also responsible for the formation of the bedrock of the Beinn Eighe Reserve. The views from the loch shore to other key geological localities, such as Slioch and Gleann Bianasdail, are also memorable and another good reason to visit.

The Beinn Eighe area is perhaps most remarkable for its 'out of sequence' strata. Geological rules normally dictate that the oldest rocks are to be found at the bottom and the

Quartzites of Cambrian age cap many of the ridges on Beinn Eighe

youngest at the top, and this is the case for Beinn Eighe itself, where Cambrian quartzite overlies Torridonian sandstone. However, the outlying hill of Meall a' Ghiubhais flouts that convention. A full explanation of the processes that caused the rocks that build Meall a' Ghiubhais to be transported westwards over younger rock as continents collided can be found in chapter 3. An improbable concept, perhaps, to those unfamiliar with geological processes operating on a global scale, but how else could we explain the fact that older Torridonian sandstone rocks sit above the younger Cambrian strata? The sandstone comprising the upper part of Meall a' Ghiubhais is separated from the Cambrian below by a particular kind of fault, known as a thrust. Overlying strata can be transported many kilometres on the back of a thrust plane. An isolated knoll of rock, separated by faulting as seen at Meall a' Ghiubhais, is known as a klippe.

These contrasting rock types can be picked out from a distance by the characteristic screes they shed: brick red for the Torridonian and ice white for the Cambrian. On the east and south-eastern flank of Meall a' Ghiubhais, the Cambrian rocks have also been intensely folded and faulted, so the normal geological succession from the oldest – the basal quartzites – to the youngest strata, known as the Salterella grits, has been repeated. These episodes of folding and faulting are also associated with the disturbance caused by earth movements along the thrust plane.

The peaks of Ruadh-stac Beag and Creag Dhubh that sit close to the heart of the Reserve are capped by Cambrian quartzite. These rocks were laid down close to the shoreline of the long-

disappeared Iapetus Ocean. This belt of Cambrian strata runs northwards from here to the coast at Durness and also southwards to Skye.

The views from the mountainside across Loch Maree are spectacular. The mighty Slioch towers above the loch and, as explained in chapter 3, this hill reveals how the surface of the Earth looked around 1,000 million years ago. We do not have to guess – we can see it. The lower slopes are carved in Lewisian gneiss. A wide valley has been cut into the gneiss and was later filled by the Torridonian sandstones that now form the upper part of the hill. That wide valley must have been eroded before the sandstones were deposited, and so represents the contours of the Earth's surface as they existed around 1,000 million years ago. There are few places on Earth where such breathtaking vistas of long-lost worlds are available. Loch Maree was gouged out by ice in more recent times along a much earlier line of weakness in the basement rocks, known as the Loch Maree Fault.

The Torridonian sandstones are also magnificently displayed in Glen Torridon. The adjacent hills form part of a sequence of Torridonian sandstones that smothered the ancient Lewisian crust. Around 7 kilometres of sandstones and related rocks were deposited by a series of rivers that flowed eastwards across the ancient continent of Laurentia. The rock tiers on the hillsides represent the layers of sand that were laid down from 900 million years onwards.

Those who continue their journey westwards towards Upper Loch Torridon should stop at one of the viewpoints that overlook the loch. The ground on the northern side of the loch between Liathach and Inveralligin is built from Torridonian sandstones that overlie the Lewisian basement. When the rocks are examined close up, pebbles of Torridonian sandstone can be seen adhering to the ancient Lewisian land surface, so erosion has planed the cover rocks right back to the landscape as it existed around 1,000 million years ago. This is known as an exhumed landscape and has been revealed by the erosive elements of ice and water.

For those interested in landforms, there can be

Above. View over Upper Loch Torridon showing the relationship between the Torridonian sandstones and Lewisian basement. The ground beyond the loch comprises Lewisian gneiss, which is overlain by Torridonian sandstones, which are visible in the far distance. Sandstone completely covered the basement at one time, but that cover has been eroded and the older Lewisian rocks are once again visible.

Right. Coire Mhic Fhearchair on Beinn Eighe is one of the finest examples of a corrie in Scotland. Its steep headwall rising above the loch comprises three prominent buttresses of Torridonian sandstone overlain by Cambrian quartzite.

few finer views than that from the top of Glen Docherty down the glacial trough which follows the line of the fault to Loch Maree, with the ancient mountains of Torridon and Letterewe on either side. On a sunny day the glint of the white quartzite caps and screes on the Torridonian peaks, especially Beinn Eighe, catches the eye. While travelling down Glen Docherty, notice the active debris flows and fans, particularly along the north side of the glen. These slopes are still active during periods of heavy rainfall.

The car park midway down Glen Torridon is a good starting point for two fine walks to see some superb glacial landforms. From the car park there is a good view across the glen to the remarkable Coire a' Cheud-chnoic ('Valley of the Hundred Hills'), which is one of the best examples of hummocky moraine in Scotland (see chapter 4). A footpath leads past the small loch, Lochan an Iasgair, southwards through the moraine mounds and ridges. For a close-up view of one of Scotland's most spectacular corries, then Coire Mhic Fhearchair is recommended. Start at the car park, where Coire Dubh Mór separates Liathach from Beinn Eighe, and follow the well-made path to the lip of the corrie where there are good examples of smoothed rock surfaces abraded by the ice. The scale of the corrie, with a rock basin and loch on its floor and massive rock buttresses at its head, is most impressive.

In the lower part of Glen Torridon, just inland from the head of Loch Torridon, there is a fine example of the end moraine of a Loch Lomond Readvance glacier, which occupied the glen of the Abhainn Thràil. The glacier formed a tongue of ice extending across the floor of Glen Torridon and its former limit is marked by a large end moraine curving across the glen. An outwash fan extends away from the moraine.

It is also worth visiting Beinn Alligin, not only for the views of Liathach but also to see a massive rock avalanche which happened around 4,000 years ago (see chapter 4). A large part of the rock-face below the ice-steepened summit of Sgùrr Mhór collapsed into the corrie below, leaving a prominent scar that can also be seen from the south shore of Loch Torridon. The avalanche formed a tongue of bouldery debris extending over 1 kilometre down the corrie. At a lower level, there is a relatively easy walk from the car park at the side of the road near Torridon House along Coire Mhic Nòbuil where much of the glen floor is covered with hummocky moraine.

The land in this part of Torridon is almost entirely in conservation ownership, with most of Beinn Eighe owned by SNH and Beinn Alligin, Liathach and Glen Torridon owned by the National Trust for Scotland (NTS). As well as the SNH visitor centre near Kinlochewe, there is an NTS Countryside Centre in Torridon village.

11. TROTTERNISH, SKYE

The Trotternish peninsula on Skye is remarkable for its massive landslides and it is one of the few places in Scotland where Mesozoic sedimentary rocks occur. Combined with a visit to the Cuillin, the visitor has an opportunity to see some astonishing contrasts in rocks and landforms in a relatively small area.

The Trotternish escarpment may seem just like any other cliff or escarpment, but the slopes have collapsed to produce some of the strangest landforms in Britain. The easterly facing escarpment, some 28 kilometres in length, has the largest continuous area of landslides in Britain, along with two of the most spectacular examples at the Storr and Quiraing. Great slabs of rock have moved down the mountainside and turned on their ends. These have weathered to produce upstanding pinnacles of many shapes and sizes. The Old Man of Storr is perhaps the most striking, along with the many pinnacles at Quiraing.

The landslides of the Trotternish escarpment arise from a particular set of geological circumstances. The cliffs are formed in a 300-metre-thick series of basalt lava flows. Since they were erupted some 60 million years ago, the lavas have been tilted gently westwards. These lavas are underlain by sedimentary rocks of Jurassic age and a thick layer of igneous rock – a dolerite sill. The underlying sediments foundered under the great weight of the lavas, initiating the landslides (see chapter 4). The landslides form two zones: an inner zone of spectacular slipped blocks formed during the postglacial and an outer zone of subdued hummocky ground consisting of earlier landslides later moulded by the last ice sheet. The inner zone is a maze of tabular and tilted slipped blocks, pinnacles, buttresses and remarkable rock architecture. The Storr landslide has been dated to around 6,500 years ago.

The scree slopes below the cliffs are also actively eroding, principally during periods of prolonged or intense rainfall, and are extensively scarred by the tracks of debris flows. Buried soils show that there have been periods of stability and erosion in the past. Finer material has been blown from the exposed cliff faces and deposited over the top of the escarpment as a fine sandy soil that carries a lush green grass. This is best seen on the ridge around the Storr.

The Storr is also a good place to see an

Above. Some of the landslipped blocks of lava along the Trotternish escarpment have weathered into spectacular pinnacles, epitomised by the Old Man of Storr.

Opposite. The Table is one of many spectacular landslide blocks at Quiraing on the Trotternish escarpment.

example of a periglacial trimline marking the upper limit of the last ice sheet. The upper slopes of the Storr lay above the surface of the ice sheet and are covered by a mantle of frost-weathered rock, whereas the Bealach Beag to the south was covered by the ice and displays ice-moulded bedrock.

The local names given to the landforms reflect their shapes and allude to their use by mythical figures: the Needle Rock and the Old Man at the Storr, and the Needle, the Prison and the Table at Quiraing. It is perhaps also easy to imagine why the great hollow below the Storr is called Coire Faoin (Corrie of the Fairies), especially if visited on a classic Skye day of swirling mists. Indeed, although the views from the Storr and Quiraing are superb on a clear day, a visit on a day with mist and a little wind, and the piercing calls of ravens, is much the more magical and eerie.

There are two obvious places to see the landslides at close quarters. On the Portree–Staffin road there is a car park and good path to the Old Man of Storr and, for the hillwalker, a steep climb up onto the ridge behind, passing on the way through, the Corrie of the Fairies. Further north, take the narrow twisting road from Staffin to Uig. There is a car park at the top and a footpath right into the heart of Quiraing, past the many pinnacles. It is also worth visiting Castle Ewen in Glen Uig, east of Uig, where there are further landslides.

Underlying the lavas of the Trotternish escarpment is a series of Jurassic sedimentary rocks that have their own special appeal. The remains of dinosaurs have been found here, both individual bones and traces of their presence in the form of footprints in the rock. These rocks are primarily of Middle Jurassic age, consisting of limestones, sandstones and shales. These strata are beautifully exposed in a series of coastal exposures adjacent to the A855 road, which runs from Portree to Duntulm at the northern tip of the island. The rocks are for the most part fossil-bearing and a detailed picture of the environmental conditions under which they were deposited has been established. At the beginning of the Jurassic Period, this area was a broad coastal plain. The sea periodically covered this area, depositing thick layers of sands and muds. The climate was tropical and areas of coastal swamp occasionally developed.

An additional feature of interest is at the Kilt Rock waterfall, just to the north of Valtos. Around the time the Skye volcano was active, great pulses of magma were forced between the layers of Jurassic sedimentary rock, forming a feature known as a sill. These bursts of molten magma, which later cooled to form sheets of dolerite, became interleaved with the sedimentary layers. The upper layers of the cliff are formed from beautifully columnar-jointed dolerite rock, with the lower layers cut in sediments of Jurassic age. An impressive waterfall tumbles over the edge of the cliff at the Kilt Rock, adding to the interest and drama of the spectacle. A car park, viewing point, and site interpretation help the visitor to appreciate this popular tourist attraction.

12. CUILLIN HILLS, SKYE

The Cuillin Hills of Skye are among our most dramatic mountain ranges; a paradise for climbers and walkers, as well as Earth scientists. Here, the visitor will see the roots of an ancient volcano, which has been carved by the Ice Age glaciers into one of Scotland's most distinctive mountain landscapes. The hills rise dramatically out of the sea. The combination of sun, rain and passing cloud on the varying colours in the rock and shapes of the hills, provides an ever-varying scene.

The rugged peaks, often lost in low clouds regardless of the season, are steeped in the history of geological discovery. They were first studied in earnest in the 1870s, but it was Professor Alfred Harker's classic memoir, published in 1904, that more fully elucidated their geological history. This was the first time that an ancient volcano had been studied in great detail.

The high ground that forms the mountainous core of Skye is divided in two: the Red and Black Cuillin. This distinction relates directly to the underlying geology. The Red Cuillin are generally rounder in outline and are underlain by granites and related rocks. The Black Cuillin, by contrast, form more jagged and irregular peaks and ridges, built from more basic rocks known as gabbro. Although these rocks look very different in fist-sized specimens, they have a common heritage. It was Harker's detailed survey that revealed the gabbros were formed in the magma chamber of a long-extinct volcano. We now know that this volcano was active over 60 million years ago when the North Atlantic was widening apace. Harker mapped the internal plumbing of this ancient volcano, which we also now know to be the largest and best developed in the country. It is the variety of rock types that have been described from here that provide part of the fascination of the place.

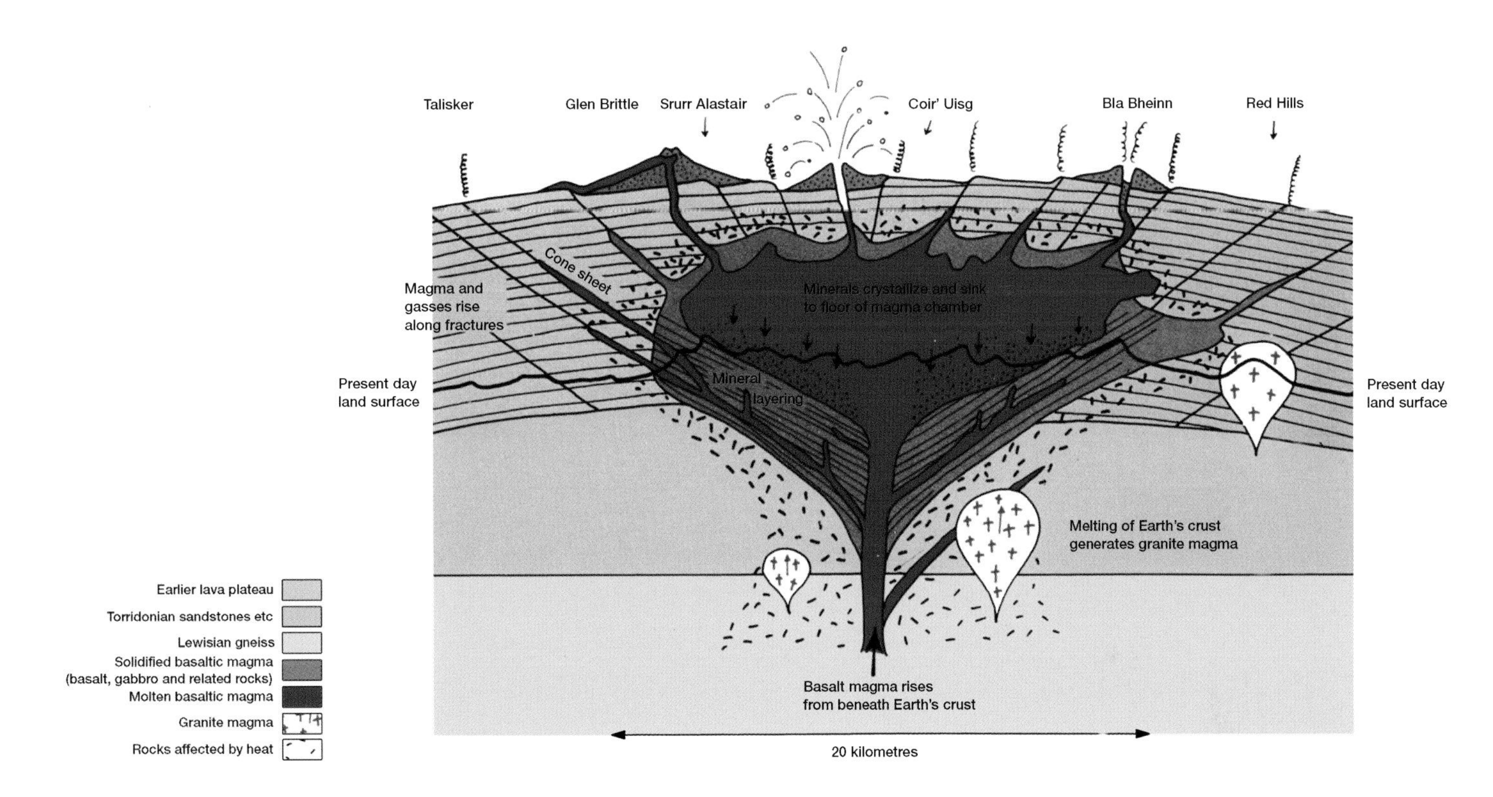

A reconstruction of the Cuillin volcano about 60 million years ago. Since then, the higher parts have been eroded away to reveal details of its 'roots'.

The overall shape of the volcano is that of a funnel. The magma chamber was fed from below as molten rocks were introduced from the lower reaches of the Earth's crust. Many of the gabbros and related rocks are layered, with heavier, black, rock-forming minerals, such as pyroxene and olivine, segregated from the lighter-coloured layers of feldspar. This separation was initiated when the rock was still molten. As the crystals formed in the magma chamber, so they settled to the floor of the chamber like sand grains building up on the sea bed. Over time, great thicknesses of these sedimentary layers built up, and as life was extinguished from the volcano, they solidified to form the 'layered series' that has attracted the attention of geologists over many generations.

The granites of the Red Hills also owe their origin to the same volcano. As the first pulses of basic magma were introduced, so the lower areas of the crust were subjected to greatly increased temperatures; to the point that the surrounding rocks melted. A rock melt of granite composition was created by this process and, being less dense than the surrounding rocks, great balloon-shaped gobbets of this magma rose through the crust. Some reached the surface; others chilled at depth.

The final chapter in the formation of both the Black and Red Cuillin was for the upper 2 or 3 kilometres of rock to be sliced away by the processes of erosion to reveal the lower structures of the volcano. The layered series of gabbros and related rocks, formerly deeply buried in the magma chamber, is now exposed at the surface, as are the granite intrusions.

The magma chamber containing the gabbros and related rocks is demonstrably later than the lavas that build the Trotternish Peninsula at the northern end of Skye. Thick lava flows were erupted over many years to build this part of the island. The molten rock, probably emanating from fissures or cracks in the ground, created an amazing layer cake of individual lava flows that can be followed for many kilometres.

The Black Cuillin are also renowned for their glacial landforms. This is the most spectacular area in Britain for its 'alpine' scenery and superb examples of corries, arêtes and ice-moulded bedrock, as well as the glacially excavated trough of Loch Coruisk. As described in chapter 4, these landforms have been shaped during the course of successive periods of glaciation by ice sheets and mountain glaciers. During the last glacial maximum, the higher peaks rose above the surface of the ice as nunataks. Later, during the Loch Lomond Readvance, an icefield formed in the Cuillin, with smaller glaciers in the corries above Glen Brittle. Some of these later glaciers formed superb lateral and end moraines, notably below Coir' a' Ghrunnda, and a large spread of hummocky moraine at Sligachan.

For the 'look and see' visitor, perhaps the best view of both the Red and Black Cuillin is on the Portree road immediately north of the Sligachan Inn looking southwards. The Red Cuillin, including the imposing peak of Marsco, lie to the east of Glen Sligachan, and the Black Cuillin, dominated by Sgùrr nan Gillean, to the west. The foreground to this panorama is formed by a splendid expanse of hummocky moraine. A path leads southwards from the hotel down the glen

The granites of the Red Hills were formed deep in the bowels of the Skye volcano. Erosion has subsequently exposed them at the surface.

An aerial view of the Black Cuillin from the south, showing the 'alpine' form of the glaciated mountains. These have been carved from the gabbros of the exposed magma chamber of the former Skye volcano. The glacial trough of Loch Coruisk lies at the heart of the range.

towards Loch na Crèitheach. And for the energetic and experienced walker, the ascent of Bruach na Frithe from Sligachan gives one of the more accessible opportunities to reach the Cuillin ridge.

From the west, a closer view of the corries is gained from the road through Glen Brittle, including the moraines in Coire na Creiche. For those able to take a moderately easy walk, there are a number of options: park by the Glen Brittle camp site and walk up to Coire Lagan to see the Loch Lomond Readvance boulder moraines in front of the corrie and the superb ice-moulded bedrock in the corrie itself; to the south a very fine end moraine extends in an arc across the slopes below Coir' a' Ghrunnda. For good viewpoints and a walk from the south, park at the car park just past Kirkibost on the Elgol road and walk over the hill, or park at Elgol and walk around the coast: both paths lead to Camasunary. The serious walker can then continue to Loch Coruisk along the coastal path with the infamous 'Bad Step'. However, an easier way to see the magnificent glacial scenery of Loch Coruisk is to take a boat trip in summer from Elgol.

Walking trips into the Cuillin should not be undertaken lightly. This is a serious mountain area liable to sudden changes in weather.

13. ISLE OF RUM

The wild and rugged landscapes of Rum are carved from some of the most remarkable rocks in Scotland. The history of this island stretches back over 3 billion years and has been punctuated by dramatic events, such as periodic volcanic eruptions and Ice Age glaciations. The Rum volcano was active some 59 million years ago, when a line of active vents stretched down the west coast of Scotland from Skye southwards to Ailsa Craig. Rocks relating to the volcano dominate Rum's landscape, although small patches of rock from earlier episodes of Earth history are also present.

Hallival and Askival are the dominant peaks on the island. They are on the itinerary of all serious geology students who visit Rum. At first sight, these hills look as if they are formed from 'run-of-the-mill' sedimentary rocks, as they comprise regular layers. But on closer examination, these layers turn out to have a more remarkable origin. They are in fact layers deposited on the floor of the magma chamber of the long-extinct Rum volcano. The volcano erupted periodically over the years, blasting ash and water vapour high into the atmosphere. In the quieter, lower reaches of the magma chamber, rock-forming crystals, such as olivine and feldspar, fell to the bottom and built up layer upon layer. Over time, greater thicknesses accumulated and later cooled.

Later, these layers were then uncovered by the forces of erosion, including most recently the Ice Age glaciers, which laid bare the bowels of the volcanic structure. These layered rocks are regarded as being of international importance, as scientific investigators have worked on Rum to develop ideas about the way in which igneous processes operate deep within a volcano. There are more extensive occurrences of such layered igneous complexes elsewhere in the world, such as in the Bushvelt in South Africa and also in Greenland, but none has been more extensively studied.

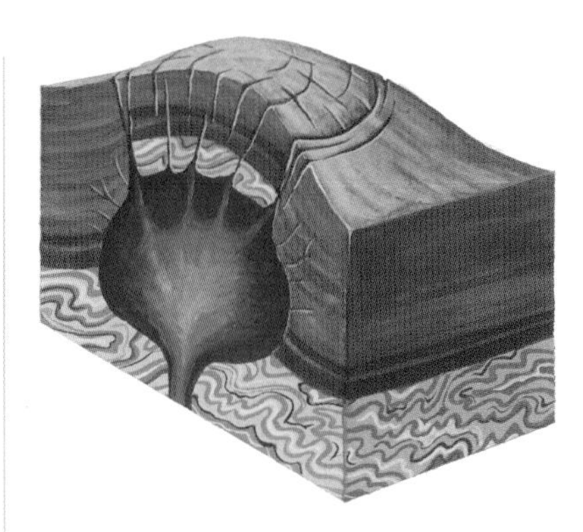

The Rum volcano explodes. The rocks of Rum demonstrate the huge forces that were at work during the Palaeogene. As the volcanic centre developed, so a large dome of rock was created above the upwelling molten rock. Eventually, great cracks developed around the dome, buckling and cracking the strata overlying the magma blister. The 'pressure cooker' eventually exploded. The dome collapsed in on itself into the boiling cauldron of molten rock. Great explosions followed as the pressure that had built up was released. From time to time, the cauldron boiled over, as ash flows and molten material flowed down the slopes of the volcano. The force of these eruptions was such that the shock waves would have reverberated around the world.

The overall structure of the Rum volcano is also remarkable. It began as a blister of molten rock a few kilometres below the surface. As it grew, so the rocks that capped the magma chamber were forced upwards and fractures developed. Eventually, the whole structure failed and the roof collapsed into the magma chamber below. This whole structure is known as a caldera. Erosion has sliced through the upper part of the volcano and only the lower reaches are now left.

It is not clear whether any lavas flowed from the Rum volcano. If they did, they would have been eroded away long since. However, thick lavas do exist on Bloodstone Hill near the western tip of the island, erupted from the nearby volcanic centres on Mull or Skye. The holes created by gas bubbles as the lavas erupted have subsequently been filled by green agates, some with red flecks – hence the name Bloodstone.

Small patches of Lewisian gneisses, similar in age and appearance to those found in the Scourie area of the mainland, are found near Dibidil towards the southern end of the island. These earlier rocks were later overlain by layers of sandstone, collectively many kilometres thick, laid down by rivers that flowed across a barren

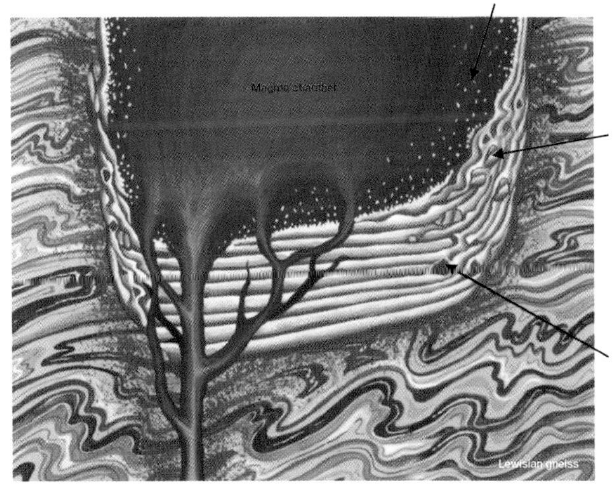

The Rum volcano also had quieter moments. The impressive peaks of Askival and Hallival are carved from a distinctive layered rock called peridotite. The crystals that make up this layered rock accumulated in the magma chamber beneath the Rum volcano; the layers of crystals stacking one on top of another just as beds of sandstone or limestone build up in the seas and oceans.

landscape. These Torridonian sandstones are found predominantly in the north, east and north-west of the island. One small patch of sandstone of Triassic age sits on top of Torridonian strata in Glen Shellesder, just to the south of Kilmory.

The present landscape and landforms of Rum owe much to the impact of repeated glaciation by ice sheets and local mountain glaciers during the Ice Age. These glaciers shaped the detail of the present landscape, carving out the corries on the high peaks and deepening the main glens. They also extensively scoured the bedrock, abrading and quarrying the rock surfaces, forming many fine examples of roches moutonnées and streamlined bedrock, for example around Loch Bealach Mhic Neill and in Glen Harris. The patterns of scratches, or striations, on the bedrock indicate that the ice sheets moved across Rum towards the west and north-west, although the mountains in the southern part of the island deflected the flow of the ice. Erratics from the Moine rocks on the mainland were carried by the ice and deposited on the east coast of the island. The last ice sheet probably covered the whole of Rum, apart from summits over 700 metres, which were nunataks. During the Loch Lomond Readvance, corrie and valley glaciers produced striking moraines in many of the corries, including excellent examples of boulder moraines in the north-west facing corries of the Western Hills. The Rum Cuillin also have many fine examples of glacial landforms, including corries, arêtes and moraines. There are prominent Loch Lomond Readvance moraines on the north side of Ruinsival, in Atlantic Corrie and in Glen Dibidil, but one of the best examples is in Coire nan Grunnd, where the bouldery moraine surface is crossed by the path from Kinloch to Dibidil.

During the later stages of ice sheet glaciation and during the time of the Loch Lomond Readvance, intense frost action on the exposed mountain cliffs and in the soil produced a range of periglacial landforms, including screes, blockfields, solifluction lobes and sorted circles. Some of the small-scale periglacial features remain active today, notably on the Western Hills as described in chapter 4. The upper slopes of Sròn an t-Saighdeir support an extensive relict blockfield and blockslopes (bouldery slopes), and locally the material is sorted into large sorted circles and stripes. On the finer grained soils of Orval and Ard Nev, smaller sorted circles and stripes are currently active. In places, the variable composition of the volcanic rocks has allowed the weaker minerals to decompose and weather out, producing curious rock sculptures, as on Barkeval.

Left. Hallival is a fragment of the magma chamber of the Rum volcano. Rock crystals settled on the floor of the chamber, accumulating as a series of distinct layers.

Right. Glaciers and frost have shaped the hills of Rum. The jagged skyline of the Rum Cuillin forms a backdrop to the more rounded summits and corries of the Western Hills.

During and following glaciation, changes in the relative level of the land and the sea have produced raised shorelines of various ages along the coast of Rum. The south-west coast of the island is famous for a raised shore platform, part of the High Rock Platform of western Scotland (see chapter 4), which is most clearly developed between Harris and A' Bhrideanach. The presence of ice moulding and till on the platform indicates that it pre-dates the last glaciation. At Harris there is an excellent example of a raised gravel beach some 30 metres above sea level. A lower gravel terrace occurs at about 8 metres above sea level. There are also examples of raised beaches at Kilmory and Kinloch.

There is a regular ferry service to Rum throughout the year, departing from Mallaig. During the summer months, there are regular sailings from Arisaig. Boats can also be chartered from Mallaig and Skye. Hostel accommodation on the island is available at Kinloch Castle. There is a campsite at Kinloch and wild camping is permitted throughout the island. Please speak to the local SNH staff for advice. There are also two bothies very close to geological areas of interest mentioned above: at Guirdil and Dibidil. There are no restrictions on access. Rum is a National Nature Reserve and the collection of rocks is not permitted without prior consent from the Reserve office.

14. GLEN ROY

The Parallel Roads of Glen Roy are one of Scotland's natural wonders, a series of lake shorelines traced out along the slopes of the glen and dating from the time of the Ice Age. The origin of these remarkable parallel lines has been a source of great fascination to local people, visitors and scientists alike; they even attracted the interest of Charles Darwin. Today the Parallel Roads reveal a fascinating picture of how glaciers and ice-dammed lakes shaped the landscape of Lochaber.

Glen Roy is one of the most classic landform sites in Britain. It is internationally important for the Parallel Roads and a range of associated landforms in Glen Spean and Glen Gloy. These landforms demonstrate the processes of formation of glacial lakes and their subsequent catastrophic drainage; no other area in Britain provides such a diverse and detailed record. Historically, also, these landforms played a key part in convincing Louis Agassiz of the reality of the former existence of glaciers in Scotland and they provided crucial field evidence that he needed to confirm the Ice Age theory. Following his visit to the area in 1840,

An aerial view of the three shorelines, or Parallel Roads, along the hillside on the west side of Glen Roy. The shorelines are several metres wide and cut into the hard Precambrian bedrock. The hillside is dissected by postglacial gullying and by a prominent debris flow track at the left.

This large, dissected alluvial fan at Brunachan in Glen Roy was formed by deposition of sand and gravel from a tributary river. The river later dissected the surface of the fan, forming the inset terraces. The steep slopes at the front of the fan have been eroded by the River Roy.

he sent his now famous letter from Fort Augustus to Robert Jameson in Edinburgh, announcing the former existence of glaciers in Scotland (see chapter 2).

The story of Glen Roy is encompassed within a much wider legacy of landforms, extending west from Loch Laggan and Loch Treig to near Fort William, and north from Glen Spean to the Great Glen (see chapter 4). As well as lake shorelines in Glen Roy, Glen Gloy and Glen Spean, the landforms include moraines, deltas, alluvial fans, meltwater gorges, lake deposits, river terraces, kettle holes (hollows left by melting blocks of ice) and landslides. These features illustrate geomorphological processes during successive stages of the development of the glacial lakes and their subsequent catastrophic drainage. Several landslides in Glen Roy and Glen Gloy have been tentatively linked to slope instability associated with the sudden draining of the lakes and the rapid unloading of the ground.

Not surprisingly, in view of their striking appearance, the Parallel Roads have long been a feature of interest. According to local legend, they were created for hunting by Fingal, the mythical Celtic hero. Another story is that they were built by the kings of Scotland to assist with hunting when they resided at Inverlochy Castle. By the late eighteenth and early nineteenth centuries, they were already a tourist attraction and were being visited by gentry such as the Grants of Rothiemurchus, as recorded in *Memoirs of a Highland Lady* by Elizabeth Grant. Not surprisingly, they also gained the attention of leading figures in the geological world, generating heated argument as to whether they were old marine shorelines, lake shorelines or had formed by other processes. A marine origin was advocated by Charles Darwin and Charles Lyell. The young Darwin, fresh from his voyage to South America, had been deeply impressed by the uplift of the Chilean coastline by recent earthquakes. However, Louis Agassiz, on his visit to Scotland in 1840, recognised the similarity of the Parallel Roads to the shorelines of glacial lakes in the Alps and thus not only provided a mechanism for their glacial origin, but also used the landforms as evidence in support of the former existence of glaciers in Scotland. Agassiz' ideas were taken forward by Thomas Jamieson twenty years later when he worked out the detailed story of their formation. The basic sequence of events developed by Jamieson has since been elaborated by more modern studies and is described in chapter 4.

The Parallel Roads and associated landforms can be seen at many localities in Glen Roy, Glen Gloy and Glen Spean. The best place to view the shorelines is from the viewpoint on the single-track public road in Glen Roy. On the drive up the glen, lake deposits, or silts, can sometimes be seen in the banks or ditches along the roadside. Further along the road, there are many excellent views of the shorelines, including those in the National Nature Reserve on the west side of the glen, and the massive alluvial fans at Brunachan and Allt na Reinich. There is a small parking area at the end of the road from which the visitor can continue on foot to view the outwash fan at the mouth of Glen Turret, the landslip on the hillside opposite Braeroy and the river terraces between Braeroy and the Burn of Agie. Other areas of interest include the large delta with kettle holes at Fersit, near Loch Treig, and the delta and end and lateral moraines at Roughburn, near the Laggan dam. River terraces formed by drainage of later lakes can be seen on the north side of the road between Roy Bridge and Spean Bridge.

15. GLENCOE

Dark, glowering and even depressing are words sometimes used to describe Glencoe. This place is perhaps best known for the dark deeds perpetrated over 200 years ago on the Clan MacDonald, but its turbulent history reaches back even further than that – to Devonian times, in fact, when Glencoe was a supervolcano. Combine that with the effects of the much later Ice Age glaciations, and this is a landscape truly shaped by fire and ice.

The glen looks quiet now, but around 420 million years ago, this place was a simmering 'pressure cooker' that erupted with catastrophic consequences. Glencoe is famous worldwide as an example of cauldron subsidence – a volcano that has collapsed in on itself. It was one of the first such features to be described, so its historical importance is also of considerable significance. Glencoe has been studied for many years and, as each reassessment is made, the story of the volcano's development and subsequent eruption gets ever more complete. The first investigation was undertaken by C T Clough from the Geological Survey and the results were published in 1909. This work, and a later survey by Sir Edward Bailey, led to the conclusion that Glencoe was a volcano where the upper part of the volcanic superstructure had collapsed into the depths below. Bailey made the analogy with a cork being pushed into a full bottle of wine, forcing liquid out of the vessel. As this block of subsiding rock foundered, so granite magma, which was displaced, rose up along its edge. The perimeter of the foundered block is defined by a fault, which is elliptical in plan. The size of the block that collapsed downwards is immense: around 8 kilometres in length. It is also estimated that it dropped around 1,400 metres. The seismic shock waves generated when these events took place must have reverberated around the world.

This view south-west across Glencoe features Beinn Fhada, Bidean nam Bian and Stob Coire nan Lochan. These hills are built from the rocks that subsided into the Glencoe supervolcano. Later glaciation has shaped the present landscape, forming the glacial trough of Glencoe and the massive truncated spurs of the Three Sisters.

The cauldron collapse was accompanied by a series of major ignimbrite or ash flows seen today in the rocks of Buchaille Etive Mór and Buchaille Etive Beag at the head of Glencoe. The volcanic activity lasted for over 5 million years and at least eight major eruptions took place during that period. The eruption of Mount St Helens in 1980 provided images of devastating white-hot ash flows moving at unbelievable speeds. Although the precise circumstances of this recent eruption and events at Glencoe are different, they conjure up a vivid picture of what the Glencoe eruption might have looked like. The features we see in the landscape today are a direct reflection of the solid geology. The Three Sisters of Glencoe – Aonach Dubh, Gearr Aonach and Beinn Fhada – are cut in the tough, erosion-resistant lavas erupted from the volcano, while the Dalradian rocks into which the volcano was intruded have been eroded into a glacial trough by the Ice Age glaciers.

Glacial erosion has had a profound influence in shaping Glencoe. In the West Highlands, the high snow accumulation and steep pre-glacial glens favoured the development of powerful, fast-

A large postglacial rock avalanche has blocked the entrance to Coire Gabhail, better known as the 'Lost Valley'. The rock mass disintegrated and the debris accumulated as a massive cone of scree with many big boulders. The lighter-coloured scree cone on the right represents a later, smaller slope failure. The river has infilled the floor of the corrie behind the rock debris.

flowing, warm-based glaciers (see chapter 4). During successive cold episodes these glaciers deepened the pre-glacial valleys, carved out corries along the mountain ridges, breached the pre-glacial watersheds and scoured the bedrock over large areas. Rannoch Moor was one of the major centres of ice accumulation and dispersal. Powerful glaciers spilled out from there, excavating the deep trough of Glencoe and forming the massive truncated spurs of the Three Sisters. During the last phase of glaciation, the Loch Lomond Readvance, glaciers extended west from Rannoch Moor to the sea at Loch Linnhe; as they melted back across the moor, they deposited extensive areas of hummocky moraine.

The glaciers left behind many steep mountain slopes in the West Highlands. With the support of the ice removed, areas of weaker rock often deformed or collapsed dramatically as detailed in chapter 4. In a few cases, such as in the Lost Valley, the slope failures have been catastrophic. Here a huge mass of collapsed rock has blocked the entrance to the glen. Modern slope activity is less dramatic but nevertheless impressive as seen in the debris flows and large debris cones along the lower flanks of Glencoe.

There are several places to stop and view the rocks and landforms of Glencoe and the surrounding area. Travelling from the east on the A82, the White Corries Ski Centre car park offers fine views across Rannoch Moor and the expanse of hummocky moraine formed during the Loch Lomond Readvance. Travelling west, the massive truncated spur of Buachaille Etive Mór dominates the landscape. There are active debris flows and fans on its lower slopes, notably below the shallow corrie opposite Altnafeadh. From here, the ascent of Buachaille Etive Mór gives views of the layering in the cauldron subsidence at close quarters, as well as a clear impression of the deep glacial gouge of the glen itself and expansive views across the former ice centre of Rannoch Moor. However, this is a steep and serious hillwalk. Just to the west of Altnafeadh, a short walk behind a car park in a former roadside quarry gives excellent views of the glacially breached watershed of the Lairig Gartain.

Further west, just beyond the narrow upper part of the glen, there are car parks on the south side of the road. Here the full immensity of the landscape is apparent, particularly in the form of the glacial trough and the imposing truncated spurs of the Three Sisters. But there is finer detail, too, in the active alluvial fan at the entrance to Coire nan Lochan and the large debris cones arrayed along the lower slopes of the Aonach Eagach ridge (see chapter 4). From the car parks, it is possible to walk up into the Lost Valley between Beinn Fhada and Gearr Aonach to see the rock avalanche blocking the entrance to this hanging valley. At the bottom of the glen, a car park at the site of the old visitor centre gives views back up the glen to the lavas that make up the massive cliffs of Aonach Dubh.

Beyond Glencoe village is the fjord landscape of Loch Leven, a glacial trough drowned by the sea. This narrows at its western end where there is a large, flat-topped accumulation of sand and gravel. This was formed by a Loch Lomond Readvance glacier, which built an outwash delta into the sea. It now provides a convenient bridging point across the loch.

The land of Glencoe was entrusted to the nation by the notable benefactor Percy Unna in the 1930s. It is owned and managed by The National Trust for Scotland for conservation and informal recreation. There is a visitor centre that has easy to understand models of the geology of the glen and the formation of the volcanic rocks and glacial landforms.

16. THE RAISED SHORELINES OF ISLAY AND JURA

Islay and Jura are world famous for their distinctive brands of malt whisky. However, they have another claim to fame that is rather less well known. The north-west coast of Islay and the west coast of Jura contain some of the finest examples of raised shorelines in western Europe. The range of features present, including raised shore platforms, cliffs, and staircases of unvegetated, raised gravel ridges, and their extent and degree of development are quite exceptional. There is no better place in Scotland to observe such features, and, most unusually, they are even noted on Ordnance Survey maps. However, these are not drive-in sites and the visitor will need to walk some distance or take a boat to see the very best examples.

The area is particularly notable for the classic development of the three shore platforms that occur widely in the Inner Hebrides: the High, Main and Low Rock Platforms which are described chapter 4. In addition, it displays quite

Opposite. Raised shorelines on the west coast of Jura, north of Loch Tarbert. The High Rock Platform and its low backing cliff are clearly developed in the upper part of the photograph, running from top left to middle right. A staircase of vegetated and unvegetated gravel beach ridges covers the surface of this platform to the right of the big gully. These formed at a time of high relative sea level as the last ice sheet was melting. At its seaward margin, the High Rock Platform is truncated by the prominent backing cliff of the Main Rock Platform. The surface of this platform is covered by postglacial gravel beach ridges.

Below. The coast of Islay south of Rubh' a'Mhàil, showing the High Rock Platform and its backing cliff. Along its seaward edge, the platform is cut by the backing cliff of the Main Rock Platform. The surface of the Main Rock Platform to the left of the figure is covered with vegetated postglacial gravel beach ridges.

superb suites of raised gravel beach ridges. Although both sets of landforms are present on each island, the platforms are best seen on Islay and the gravel beaches on Jura.

On Islay, the principal landforms occur along the north-west coast between the headlands of Rubh' a' Mhàil and Rubha Bholsa. Here, the High Rock Platform occurs as a prominent bench at about 35 metres above present sea level. It reaches widths of up to 650 metres and is backed by cliffs up to 70 metres high. The platform surface and cliff are obscured by thick accumulations of till. An impressive end moraine, known as the Coir' Odhar moraine, was deposited on the platform towards the end of the last ice sheet glaciation. At several localities along the seaward edge of the platform, beach gravel overlies the till, indicating a period of high relative sea level as the glaciers were melting. The Main Rock Platform occurs at a lower level, 3–5 metres above present sea level. Its surface is covered by thick accumulations of largely vegetated raised gravel deposited later during the postglacial. The glaciated Low Rock Platform is locally exposed in the intertidal zone.

The same three platforms occur along the west coast of Jura, but this area is particularly noted for extensive staircases of unvegetated beach ridges of quartzite gravel. These are superbly developed, particularly north of Loch Tarbert. The higher beaches rest on the surface of the High Rock Platform, which reaches an altitude of 32–34 metres above present sea level. As many as fifty-five separate ridges have been identified. They were formed at a time of high relative sea level as the last ice sheet was melting, when the land was still depressed from the weight of the ice and sea level was relatively high. As the land rebounded, the rate of uplift outpaced the rate of sea-level rise, so that the relative level of the sea against the land fell, leaving the staircases of raised shorelines as a record of these changes. At a lower level, suites of postglacial gravel ridges cover the surface of the Main Rock Platform. These have formed in the last 7,000 years as relative sea level dropped following the postglacial rise in sea level. As many as thirty-one separate ridges occur in an area to the north of Inver. Local areas of beach and machair, such as those at Corpach Bay, add further diversity to this remarkable coast. An additional feature of interest in western Jura is the presence of a very fine medial moraine. It extends westwards as a belt of boulders from the base of Beinn an Oir towards the coast.

Both areas of coast are very remote and difficult of access. The features on the north-west coast of Islay can be reached on foot from Bunnahabhainn over heather moorland without footpaths. Take the road from just west of Port Askaig to Bunnahabhainn and park there. It is a walk of 5–6 kilometres along the Sound of Islay coast to the headland at Rhuvaal Lighthouse and then another 2–3 kilometres westwards to see the best features. Alternatively, the journey is easier for those with access to a boat.

Similarly, by far the easiest way to get to the west coast of Jura is by private boat charter, probably from Port Askaig on Islay. Otherwise it is a very rough walk over heather moorland across the island, again without footpaths. To access the raised shorelines and medial moraine on the south side of Loch Tarbert, park around Lagg on the main east coast road of Jura and strike westwards over the moorland for about 6 kilometres. To access the platforms and raised beaches on the north side of Loch Tarbert, leave the main road a couple of miles north of Tarbert and strike south-westwards over the moorland for about 8 kilometres.

17. LOCH LOMOND

The Loch Lomond area is noted for the Highland Boundary Fault and a diverse range of glacial landforms. It forms part of Scotland's first National Park, Loch Lomond and the Trossachs.

The line of the Highland Boundary Fault, which runs from Arran to Stonehaven, is clearly seen in the remarkable contrast in topography between the Highlands to the north and the Central Lowlands to the south. The latter have been downfaulted and covered by later sedimentary rocks that have eroded more easily than the harder crystalline rock of the Highlands, leaving an abrupt slope between the two areas. The Highland Boundary fault can be seen very clearly in the Loch Lomond area, running through the islands in the loch and across Conic Hill at Balmaha. The latter is a good vantage point to see the general landscape contrast and the effect of the harder and more resistant rocks to the north and the softer less resistant ones to the south. Here the visitor can stand literally on the Highland Boundary Fault and see it stretch through the islands in the loch.

There are plenty of glacial features to see around Loch Lomond. One of the most striking is the contrast between the 'Highland' form of the valley to the north of Rowardennan and the 'Lowland' form to the south. The former is a classic glacial trough eroded to a depth of 600 metres in the hard Dalradian rocks by the glaciers moving southwards and punching breaches through the pre-glacial watersheds (see chapter 4). To the south, the rocks are less resistant sandstones and conglomerates and the glaciers would have been able to spread out more across a broader pre-glacial valley. Their power was less focused, so that the landscape is much gentler with many low-lying islands in the loch. This contrast can be seen quite easily from many good

Opposite. Aerial view of Loch Lomond from the south, showing the contrast between the northern and southern sections. The Highland Boundary Fault runs through Inchcailloch (centre) and Conic Hill (right) and separates the Devonian sedimentary rocks of the Central Lowlands (foreground) from the more resistant Dalradian rocks of the Highlands to the north.

vantage points on the west or east shores of the loch. Hanging valleys, ice-smoothed bedrock and steep valley sides are all evident in the Highland section, especially at Inveruglas. There are good examples of much smaller-scale landforms of glacial erosion beside the car park at Rowardennan. Here the bedrock by the loch shore has been smoothed and abraded by the glaciers into whaleback forms with striations on their surfaces.

Following the melting of the last ice sheet around 15,000 years ago, the sea was relatively higher for a time and flooded into Loch Lomond from the Clyde estuary and through the Vale of Leven, forming what must have been a spectacular fjord. Marine silts and clays were deposited containing the shells of species now confined to more northerly latitudes. These deposits can sometimes be seen in exposures in the banks along the southern shores of the loch north of Gartocharn. Also in this area, a well-developed shore platform and backing cliff were formed contemporaneously with the Main Rock Platform by similar cold-climate processes (see chapter 4). The platform forms a shallow rock shelf extending for up to 500 metres offshore. Later, during the postglacial rise in sea level, the sea again flooded into Loch Lomond, leaving a raised shoreline around the southern end of the loch. Several of these features can be seen in or near the Loch Lomond National Nature Reserve which can be accessed by a footpath from the car park at the Millennium Hall in Gartocharn.

The southern end of the loch is also where the last glacial episode in Scotland, the Loch Lomond Readvance, was first identified and named (see chapter 4). The moraine marking the limit of the Readvance glacier can be traced to the south of the loch as a ridge running from Shantron Muir and Glen Fruin in the west round to Alexandria, east towards Killearn and then north of Drymen. The best example is south from Croftamie towards Cameron Muir. Drumlin swarms can also be seen south of the River Endrick near Drymen and north of Gartness. Other landforms include the crag-and-tail of Duncryne Hill at Gartocharn, which offers a short walk with excellent views of Loch Lomond.

There is information on the geology and landforms of the area at the National Park Visitor Centre at Balmaha.

18. FOSSIL GROVE

Fossil Grove is one of Glasgow's most ancient attractions, situated amid the tranquillity of Victoria Park in the west end of the city. Although Fossil Grove has much to offer the casual visitor, it is also of considerable scientific interest as the best place to see in situ fossil tree stumps. These provide a unique insight into the ecology of the equatorial rainforests that were beginning to dominate the world over 310 million years ago. At this time, the land that was to become Scotland sat astride the Equator.

There are eleven stumps in all, the fossilised remains of trees known as club-mosses, or *Lepidodendron.* They have a distinctive, patterned bark. What is particularly interesting about this site is that in most occurrences in the fossil record all that remains of these steaming tropical forests that once teemed with a huge variety of plant and animal life, are a few fragments of stem or individual leaves. At Fossil Grove, a patch of forest is preserved as it was in life during Carboniferous time. Scientists have been able to calculate the density of trees that would have grown in the ancient forest and to reconstruct what the area would have looked like during the heyday of the great forest. Some estimates suggest that the density could have been as high as 4,500 trees per square kilometre. These trees grew to impressive heights, sometimes in excess of 30 metres. They had long, narrow leaves and bore cones that allowed individuals to seed the next generation. Their trunks were important in the formation of the coal deposits that are now extensive in the Central Lowlands of Scotland.

The site was first discovered in 1887 during the excavation of a quarry in what was to become Victoria Park. If we accept the preservation of Hutton's Rock on Salisbury Crags as the first act of geological conservation, then the work undertaken to safeguard these exposures must be regarded as a close and laudatory second. A pavilion was erected to protect the fossil remains from the elements and this has recently been refurbished and modernised. The interpretive facilities have also been upgraded, so the story of Fossil Grove is told in a way that everyone can appreciate and understand.

There is on-street parking near the western entrance to the park and admission is free.

The tree stumps at Fossil Grove are positioned as in life, so a clear impression is gained of what this fossil forest would have looked like.

19. ISLE OF ARRAN

The Isle of Arran is often described as "Scotland in miniature"; so called as it includes all of the key elements – spectacular upland scenery, enclosed agricultural lowlands, and engaging seascapes – that make Scotland on the grand scale so special. It is, of course, the geology and landforms that have shaped the landscape. The rocks and landforms of Arran for the most part mimic those of the rest of Scotland. Ancient metamorphic rocks and associated intrusive granites build the higher ground of the island's north end and younger sedimentary strata underlie the softer undulations of the agricultural land. The Ice Age glaciers have formed a classic landscape of mountain glaciation in the northern hills, and associated changes in land and sea levels have left raised shorelines around the coast. There are also sites with a historical resonance: James Hutton made it to these shores, and the north end of the island has its own Hutton's Unconformity.

The northern part of the island is mountainous: an extension of the Highlands. Rocks of the Dalradian – ancient metamorphic rocks –

The Northern Mountains of Arran bear a strong imprint of glaciation. The Ice Age glaciers carved corries and knife-sharp arêtes from the granite bedrock.

surround a central granite core. These Dalradian rocks started out as sandstones, limestones, muds and even lavas on the floor of the ancient Iapetus Ocean. These sediments were later cooked and squashed as the ocean closed and now form banded metamorphic rocks. Imacher Point is one of the best places to observe these rocks.

After a gap of many hundreds of millions of years, the granite of the Northern Mountains forced its way into position from deep within the Earth's crust. Around 60 million years ago, active volcanoes emerged in what was to become northern Arran, although the granites we now see exposed at the surface would have been located deep within the bowels of the volcanic structure. Erosion has cut deep to expose them to the elements. The impressive peaks of Goat Fell, Beinn Nuis, Caisteal Abhail and Beinn Bharrain are all carved in this granite. There are in fact two separate granites that build the Northern Mountains; an outer ring of coarse-grained material and a later inner core comprised of finer-grained granite. Granite bodies are to be found on Ard Bheinn and also in Glen Scorrodale, but these have given rise to less prominent landscape features. Ard Bheinn is also of interest, as it contains a block of chalk rock, similar in age and composition to that found along the Antrim coast in the north of Ireland and further afield in the south of England. The possible implications of this limited exposure of rock are profound: was this part of Scotland at one time covered by a thick layer of chalk that has subsequently been entirely removed by erosion, save for this one small exposure? We can only speculate.

Sedimentary rocks from the Old Red Sandstone, Carboniferous and Permian periods, in conjunction with a variety of igneous rocks related to the Arran volcano, comprise the rest of the island. As a rule of thumb, as the visitor tours the island, the higher ground is underlain by igneous rocks, which are generally harder and more resistant to erosion, whilst the areas of more subdued topography are formed from sediments. The Old Red Sandstone and Permian rocks, also known as the New Red Sandstone, form a kind of sedimentary sandwich with the Carboniferous strata as the meat in the middle.

Near Cock of Arran at the extreme northern end of the island, a small patch of Old Red Sandstone rocks overlies the Dalradian rocks. This is one of the first of many sites that James Hutton used to develop his *Theory of the Earth*. Hutton recognised the juxtaposition of the Old Red and the underlying schists and surmised that this relationship represented a period of uplift and erosion before the deposition of the sandstones above.

Rocks of Carboniferous age are limited in distribution. Some coal seams of economic worth have been worked near the north end of the island and limestones, rich in shelled marine creatures known as brachiopods, have been quarried at Corrie.

The upper layer of the sandwich comprises the desert sands of Permian age. At this time, the land that was to become Scotland was located just to the north of the Equator and we had a climate to match. Great sand dunes were whipped up by the winds that blew across the Permian deserts, creating structures that are characteristic of this type of environment – known as 'cross-beds'. These structures are beautifully displayed on the wave-washed exposures near Corrie on the east coast. These sandstones also retain other indicators of the prevailing environmental conditions. Circular pitted marks on some rock surfaces have been interpreted as powerful lightning strikes that occurred during Permian times. Geologists call these structures 'fulgurites'.

Cross-bedded sandstones near Corrie are of Permian age and were deposited by winds blowing across a Sahara-like desert.

As the Arran volcano erupted and the substantial granites were emplaced across the northern part of the island, other related igneous intrusions were also in the process of forming. Many prominent features, such as the Doon at Drumadoon Point and at Dippin Head, were formed at this time. Great bursts of magma were injected into the Permian sandstones and this molten rock cooled below the surface, only to be exposed later as the upper layers were removed by erosion.

The remaining geological delight of the island is to be found along the south coast near Kildonan. A criss-cross pattern of dykes, which were formed by bursts of magma that shot out in a linear format from the Arran volcano, is a prominent feature of the intertidal foreshore. They intruded into the softer Permian sandstones

Raised shorelines at Dougarie on the west coast of Arran. A raised shore platform and cliffline are cut in bedrock. The platform, part of the Main Rock Platform of western Scotland, is overlain by postglacial raised beach deposits.

and, as erosion has gnawed away at the rock platform, so the harder igneous material has been left in the form of upstanding linear features. It is only possible to view these fascinating features at low tide, as at other times they are obscured by the sea.

Arran also has a spectacular range of landforms. The granite mountains in the north of the island are a textbook example of glaciated mountains, with classic examples of corries and arêtes, especially on Goat Fell, A' Chir and Cir Mhòr, and the glaciated valleys of Glen Rosa and Glen Sannox. The mountains were high enough to act as an ice accumulation centre and at times, the higher summits may have been nunataks. There are several points of access. For Glen Rosa, start at Glenrosa Farm, west of Brodick, where there is parking. For Glen Sannox take the path on the south side of the Sannox Burn. Both paths meet at the Saddle, which is a col lowered by glacial erosion. There are fine views from here of the landforms of glacial erosion. Glacial deposits formed by Loch Lomond Readvance glaciers are present in several of the glens and corries; good examples occur in Upper Glen Rosa, Coire a' Bhradain and Glen Sannox and in Garbh Coire and Coire nan Ceum on Caisteal Abhail. Till from the last ice sheet glaciation covers much of the lower ground on the island. There are also some particularly fine examples of granite erratic boulders transported by the ice. Along the rock platform on the east coast, at Corrie for example, these erratics have been washed out of the till by the sea and now rest directly on the sedimentary rocks.

Raised shorelines surround much of island. They are associated with old cliffs and sea stacks and abandoned caves way above sea level, reminding the visitor of the uplift of the island after the Ice Age. The best places to see them are along the south coast west from Dippin Head, at Blackwaterfoot, and along the west coast between Machrie and Lochranza. The most prominent and extensive feature is a raised shore platform and backing cliff with stacks and caves, part of the Main Rock Platform (see chapter 4), which is covered with postglacial raised beach deposits. In many places the coastal road follows this raised platform, for example between Brodick and Sannox, and Machrie and Lochranza. Along the south coast it is also easy to see the dykes intruded during the volcanic activity of the Palaeogene jutting out across the shore platforms. The raised sea caves at King's Cave, north of Blackwaterfoot, can be accessed from a Forestry Commission car park and footpath beside the main road. This is reputed to have been the hiding place of Robert the Bruce when he had his famous encounter with a spider.

Arran is circumnavigated by a coast road, so there are few sites of interest that are far from the beaten track. The String and Ross roads cross the island, enhancing access to the central parts of Arran. The island is served by a regular ferry from Ardrossan on the mainland, which sails to Brodick on the eastern side of the island.

20. THE A9 FROM DUNBLANE TO INVERNESS

A journey along the A9 from Dunblane to Inverness provides a slice through Scotland's geology from south to north. Older rocks, particularly the Dalradian, are well displayed, as are glacial landforms of much more recent age. This 'place to visit' is strung out over 200 kilometres and, although there cannot be a detailed commentary on all of the sights that can be seen from the carriageway, the main highlights are described. The driver is advised to concentrate entirely on the road ahead and this description is largely for the benefit of any passengers who may be sharing the journey. However, a number of safe lay-by stops are also included, so that everyone is able to appreciate the grandeur of the scenery and the rocks and landforms that build it.

Dunblane to Perth

The journey starts in the Central Lowlands amid rural surroundings. Behind us are the spectacular vistas of Stirling Castle and the Wallace Monument, both perched high on ancient volcanic piles. The wide valley of the Forth, once a broad tidal estuary beyond this point, is now occupied by rich farmland. The A9 begins at the end of the M9 motorway, just to the south of Dunblane, with an upward climb. Road cuttings on the passenger side tell us that we are still in Old Red Sandstone country. Past Greenloaning, the Ochil Hills come into view. These lavas were erupted during Devonian times and shaped into their characteristic undulations by recent glaciations. In particular, the glaciers carved out Gleneagles, a spectacular glacial breach through the hills. On the lower ground, the glaciers left behind extensive deposits of sand and gravel in the form of kame and kettle topography and eskers. Past Auchterarder, the route crosses the flat carselands of Strathearn, which were formed when the sea level was higher 7,000 years ago

The carselands of Strathearn, viewed from the Cairney Brae, were inundated during the postglacial rise in sea level when the Tay estuary extended further west. The Ochil Hills to the south are formed from Devonian lavas.

and the estuary of the Tay extended westwards along Strathearn. It then climbs the Cairney Brae past the exposures of cross-bedded sandstones described in chapter 3. The structures in the sandstone indicate that these rocks were laid down by a mighty river, akin to the Mississippi in scale. There is another cutting at the top of the hill, this time in igneous rock. A linear intrusion of dolerite cuts through red sandstone. We see a section through the dyke and its contact with the sandstones. Travelling towards Perth, the Highlands are clearly visible to the north. Often dusted with snow, whatever the time of year, these are the mountains through which most of our journey will be made. Notice the broadly similar level of the summits. These are the roots of ancient mountains that have been planed down by the forces of erosion, then later uplifted and glaciated.

Perth to Pitlochry

Leaving Perth behind and travelling north, we remain on the Old Red Sandstone. The productive soils, developed on hummocky sands and gravels deposited by the last ice sheet, support arable farms and small settlements. Approaching Birnam and Dunkeld, the scene changes dramatically. The Highland Line is crossed and the traveller has journeyed to another continent. Scotland is fashioned from a number of smaller continental fragments and the Highland Boundary Fault is the junction between two such giant

Hummocky moraine at the Pass of Drumochter.

jigsaw pieces. The hard Dalradian rocks that build most of the Highlands were more resistant to erosion than the rocks further south. So from this point onwards, the landscapes are consistently grander and more rugged. The Hermitage Visitor Centre, managed by the National Trust for Scotland, is a good place to stop for a picnic and stretch your legs. A short walk from the car park will take you to a spectacular waterfall on the River Braan. The glaciers gouged a narrow pathway at the Pass of Birnam at Dunkeld; it is through this stretch of countryside that the 'Silvery Tay' now flows, and later road and rail builders have exploited it to the full. The road runs along the floodplain of the River Tay to Ballinluig. This point marks the confluence of the Tay and Tummel. Between Dunkeld and Pitlochry, the river floodplain is flanked by glacial outwash and river terraces. Towards Pitlochry, the long straight of the dual carriageway swings westwards over the broad expanse of the River Tummel and climbs steadily towards Glen Garry.

Pitlochry to Kingussie

Killiecrankie has strong historical associations, with the Jacobite army defeating the Redcoats over the deeply incised ravine of the River Garry. The road cuttings through the foliated Dalradian rocks are an engineering feat worthy of note. The carriageway is supported on stilts at this point and clings precariously to the side of the mountain slopes. A novel engineering technique, known as smooth blasting, was used to create the uniform rock faces that run the length of the rock cutting. Over the River Garry, and travelling towards Blair Atholl, we are never far from Hutton country – Glen Tilt is another place that has strong associations with the founder of modern geology. A quarry that exploits thick limestones within the Dalradian lies to the south of the carriageway near Blair Atholl.

Look out too for the ice-white Blair Castle, which can be seen through the trees to the north of the road. The landscape here bears the marks of glaciation. Outwash and river terraces fill the floor of the strath. As the road rises towards the Pass of Drumochter, the floor of Glen Garry suddenly becomes filled with hummocky moraines. These were formed by a tongue of ice extending south from the Loch Lomond Re-advance icefield in the West Drumochter Hills. This hummocky moraine continues all the way over the pass and some way down the glen on the north side. In low light, or with a dusting of snow, the moraines are particularly impressive. The intermittent roadside exposures are cut in rather unremarkable schists that build much of the central core of Scotland. Before the summit, there is a fine view of Loch Garry to the south-west. High above its eastern shore, the hillside is crossed by a number of diagonal lines delineating a rock slope failure that has moved only a little way downslope. On the hillslopes on both sides of the Pass of Drumochter, there are excellent examples of active debris flows. The Allt Dubhaig which flows south through the pass is one of the most intensively studied rivers in Scotland, particularly with reference to how rivers transport sediment. Northwards, beyond the hummocky moraines and the limit of Loch Lomond Readvance ice, outwash terraces extend towards Dalwhinnie. At Dalwhinnie, the glacial trough occupied by Loch Ericht follows the line of a fault to the south-west. North of Dalwhinnie, the floor of Strath Spey is filled with mounds and terraces of sand and gravel deposited by meltwaters from the last ice sheet. Ruthven Barracks, near Kingussie, was built by General Wade after the 1715 Jacobite uprising, on a mound above the modern floodplain which often floods in winter.

Kingussie to Inverness

The River Spey, the A9 and the Perth–Inverness railway line all run north-east between Newtonmore and Aviemore in a glaciated strath. This trend follows the line of a basement fault, which was later exploited by the ice to create this convenient corridor through the mountains. The floor of the strath here is filled with sand and gravel that form kame and kettle topography. Loch Insh and the important wetland area, the Insh Marshes, lie to the south of Kincraig. To the east, Ord Ban and neighbouring hills are good examples of large roches moutonnées, with ice-smoothed slopes on their up-ice (southern) flanks and steep, ice-plucked slopes on their down-ice (northern) sides. Rising above the strath to the east, the towering ramparts of the Cairngorms dominate the landscape as Aviemore is approached. The granite body of the Cairngorms cooled in the upper reaches of the Earth's crust about 427 million years ago and since shortly after that time has been a prominent feature of the landscape through the geological ages. In clear light looking back south from north of Aviemore, many of the main features on the northern flanks of the Cairngorms are visible: the line of corries from Cairn Gorm to Braeriach, the great gash of the Lairig Ghru and the rounded upper slopes of the mountains. Looking south-westwards, the Monadhliath Mountains are rarely out of the mist. The rocks that underline the high plateaux and adjacent ground between here and Loch Ness to the west are built from sedimentary layers that were cooked and squashed as continents collided. The resulting schists that now form the core of the Highlands have been sliced through by road engineers and are revealed in impressive road cuttings near Aviemore. Thick, lightly coloured granite veins cut through the schists in a manner that is repeated at many locations between Dunkeld and Inverness. North of Aviemore, there are further glacial landforms of note. Kame and kettle topography continues along Strath Spey and there are fine examples of river and outwash terraces in the Findhorn valley at Tomatin. Further north, the A9 crosses Strathnairn, with views to the south-west of thick deposits of sand and gravel linked to the Littlemill eskers (see chapter 4). Descending from the summit of Drumossie Muir towards Inverness, there are fine views northwards across the Moray Firth, with the massive form of Ben Wyvis on the skyline. The remarkably straight coastline of the Black Isle and northwards to Tarbat Ness follows the continuation of the Great Glen Fault.

FURTHER READING

Much of the literature on the geology and landforms of Scotland is written primarily for the advanced reader or specialist and is published in academic journals. The literature for this audience is voluminous and is not cited here, with the exception of a few key references. These provide a gateway into the more scientific literature, as well as lists of references to more specialist sources for those interested in exploring the subject in more depth. For the general interest reader, the range of publications available on geology and landforms is more limited and what follows is a sample of the books that may be of interest.

CHAPTER 1 – GEOLOGY ENRICHES ALL OF OUR LIVES

Building Stones of Edinburgh. A.A. McMillan, R.J. Gillanders & J.A. Fairhurst. 1999. Edinburgh Geological Society, Edinburgh. (An account of the building stones of Edinburgh that gives the geological provenance of most of the city's significant buildings).

Earth Science and the Natural Heritage: Interactions and Integrated Management. J.E. Gordon & K.F. Leys (eds). 2001. The Stationery Office, Edinburgh. (Includes articles on Scotland's Earth heritage, links with biodiversity, and coastal and river management).

Geodiversity. Valuing and Conserving Abiotic Nature. J.M. Gray. 2003. Wiley, Chichester. (A comprehensive overview of geodiversity and why it is important).

Geology and the Scenery of Scotland. J.B. Whittlow. 1977. Penguin Books, Harmondsworth (Regional accounts of the relationship between the underlying geology and landscape).

Hostile Habitats. Scotland's Mountain Environment. N. Kempe and M. Wrightham (eds). 2006. Scottish Mountaineering Trust, Nairn. (Includes chapters on the geology and landforms of Scotland's mountains).

An Introduction to the Geological Conservation Review. N. Ellis (ed). 1996. Joint Nature Conservation Committee, Peterborough. (Introduction to the selection and conservation of geological and landform Sites of Special Scientific Interest throughout Britain).

A Picture of Britain. D. Dimbleby. 2005. Tate Publishing, London. (A contemporary view of landscapes in Britain).

Scotland after the Ice Age: Environment, Archaeology and History, 8000 BC–AD 1000. K.J. Edwards & I.B.M. Ralston (eds). 2003. Edinburgh University Press, Edinburgh. (A comprehensive overview of Scotland's natural environment and archaeology since the end of the last glaciation).

The Scottish Colourists 1900–1930. P. Long with E. Cumming. 2001. National Galleries of Scotland, Edinburgh. (An account of artists working at the beginning of the last century who were known for their love of the Scottish countryside, amongst other subjects).

Whisky on the Rocks. S. Cribb & J. Cribb. 1998. British Geological Survey, Keyworth. (Describes how the bedrock influences the whiskies of different parts of Scotland).

CHAPTER 2 – HOW THE EARTH WORKS

Cassell's Atlas of Evolution – the Earth, its Landscapes and Life Forms. M. Benton (ed). 2001. Andromeda Oxford Ltd. (A review of the evolution of life throughout geological time and related shifts in the continents).

Earth. Susanna van Rose. 1994. Dorling Kindersley, London. (An introduction to the way the Earth works and the scientists who made some of the key discoveries).

The Earth – an Intimate History. R. Fortey, 2005. HarperPerennial, London. (An ambitious attempt to tell the geological story of the Earth for the general reader).

Earth's Restless Surface. D. Janson-Smith. 1996. Natural History Museum, London. (A description of some of the key processes that have helped to shape the surface of the Earth).

Encyclopaedia of Dinosaurs and Prehistoric Life. D. Lambert, D. Naish & E. Wyse. 2001. Dorling Kindersley, London. (An authoritative account of dinosaurs and all other forms of life that have inhabited the Earth since the beginning of geological time).

Fossils – the Story of Life. S. Rigby. 1997. British Geological Survey, Edinburgh. (A very readable introduction to the study of fossils).

From the Beginning. K. Edwards & B. Rosen. 2000. The Natural History Museum, London. (A very readable account of the way in which the Earth functions and the life that has inhabited the planet).

Frozen Earth. The Once and Future Story of Ice Ages. D.

Macdougall. 2004. University of California Press, Berkeley. (A popular account of the causes and effects of ice ages and some of the scientists involved in the major discoveries).

Glaciers. J.E. Gordon. 2001. Colin Baxter, Grantown-on-Spey. (An introduction to glaciers, explaining their formation and how they shape the landscape).

Glaciers. 2nd edition. M. Hambrey & J. Alean. 2004. Cambridge University Press, Cambridge. (A very readable and well-illustrated account of glaciers).

The Great Ice Age. R.C.L. Wilson, S.A. Drury & J.L. Chapman. 2000. Routledge, London. (A textbook on the Ice Age, climate change and human evolution).

Ice Ages. Solving the Mystery. J. Imbrie & K.P. Imbrie. 1979. Macmillan, London. (A historical account of how our understanding of the Ice Age has developed over time from early beginnings in the 19th century to the remarkable discoveries of the ocean-floor records).

The Ice Age World. B.G. Andersen & H.W. Borns Jr (1997). Scandinavian University Press, Oslo. (A well-illustrated account of the Ice Age and glacial landforms).

James Hutton – the Founder of Modern Geology. D.B. McIntyre & A. McKirdy. 2001. National Museums of Scotland, Edinburgh. (A short biography of one of the influential figures from the Scottish Enlightenment).

Origins – the Evolution of Continents, Oceans and Life. R. Redfern. 2000. Cassell and Co., London. (An authoritative and well-illustrated account of the formation and evolution of the Earth through geological time).

Rock and Mineral. R. Symes. 1988. Eyewitness Guide, Dorling Kindersley, London. (A well-illustrated description of the rocks and minerals found in Britain and how this mineral wealth has been used to enhance our lives).

Rocks, Minerals and Fossils. C. Pellant. 1990. Pan Books, London. (An introduction to the classification of rocks, minerals and fossils – a good starting point for the beginner).

Rock Solid. A. Grayson. 1992. Natural History Museum, London. (An account of the rocks, fossils and minerals of Britain).

Savage Earth. A. Scarth. 1997. Harper Collins, London. (Focuses largely on dramatic events such as volcanic eruptions, earthquakes and other natural disasters).

A Short History of Nearly Everything – Illustrated. B. Bryson. 2005. Transworld Publishers, London. (A very readable account of the development of scientific ideas through the ages).

The Story of the Earth. F. Dunning. 1981. HMSO, London. (One of the first popular accounts of the way in which the Earth works).

The Two-Mile Time Machine. R.B. Alley. 2000. Princeton University Press, Princeton and Oxford. (A popular account of the remarkable records of Ice Age climate change revealed in ice cores from the Greenland Ice Sheet).

The Usborne Book of the Earth. F. Watt. 1992. Usborne Publishing, London. (A book for children about the Earth and the way we use its resources).

Volcano. S. van Rose. 1992. Eyewitness Guides, Dorling Kindersley, London. (A beautifully illustrated book that explores the power of volcanoes and associated earthquakes).

Volcanoes – Power and Magic. D. Obert (ed). 1996. Könemann. (Volcanoes explained in the context of plate tectonics and a world view of the planet).

CHAPTER 3 – SCOTLAND'S JOURNEY ACROSS THE GLOBE

Beinn Eighe – The Mountain above the Wood. J.L. Johnston & D. Balharry. 2001. Birlinn, Edinburgh. (A description of the establishment and management of Britain's first National Nature Reserve).

Geology and Landscapes of Scotland. C. Gillen. 2003. Terra Publishing, Harpenden. (A technical account of the geology of Scotland for the more knowledgeable reader).

The Geology of Britain – an Introduction. P. Toghill. 2000. Airlife Publishing, Ramsbury, Marlborough. (An introduction to the geology of Britain for more knowledgeable readers).

The Geology of Scotland. (4th edition). N. Trewin (ed). 2002. The Geological Society, London. (The most authoritative and comprehensive academic text on the geology of Scotland and a gateway to the key scientific publications of the last hundred years).

The Hidden Landscape. R. Forley. 1993. Pimlico, London. (A lucid description of the geological history of Britain).

Mountain Building in Scotland. S. Blake (ed). 2003. The Open University, Milton Keynes. (A technical account of how

Scotland's mountains were formed).

The Nature of Scotland. M. Magnusson & G. White (eds). 1991. Canongate, Edinburgh. (Describes the nature of Scotland – its geological foundations, flora and fauna and rural development).

The Scenery of Scotland – the Structure Beneath. W.J. Baird. 1990. National Museums of Scotland, Edinburgh. (The first published layman's guide to the geology of Scotland, it remains a valuable introduction to the subject).

Scottish Landscapes. A. McKirdy & M. McKirdy. 1996. National Museums of Scotland, Edinburgh. (An activity book for children that provides an introduction to Scottish landscapes and associated topics).

Scottish Rocks and Fossils. A. McKirdy & M. McKirdy. 1995. National Museums of Scotland, Edinburgh. (An activity book for children that provides an introduction to the rocks and fossils of Scotland).

Volcanoes and the Making of Scotland. B. Upton. 2004. Dunedin Academic Press, Edinburgh. (An explanation of the effect that ancient volcanoes have had on the bedrock of Scotland).

CHAPTER 4 – SHAPING THE LANDSCAPE

Coastal Geomorphology of Great Britain. V.J. May & J.D. Hansom. 2003. Geological Conservation Review Series No. 28. Joint Nature Conservation Committee, Peterborough. (Includes descriptions of Scotland's most important sites for coastal landforms).

The Coastline of Scotland. J.A. Steers. 1973. Cambridge University Press, Cambridge. (A systematic account of the evolution and development of the coastline).

The Evolution of Scotland's Scenery. J.B. Sissons. 1967. Oliver & Boyd. Edinburgh. (An overview of landscape evolution focusing especially on the glacial and postglacial periods).

Fluvial Geomorphology of Scotland. A. Werritty & L.J. McEwen. 1997. In: K.J. Gregory (ed.), *Fluvial Geomorphology of Great Britain*. Geological Conservation Review Series No. 13. Joint Nature Conservation Committee, Peterborough, 19–114. (Describes Scotland's most important sites for river landforms).

Glacial Geology. Ice Sheets and landforms. M.R. Bennett & N.F. Glasser. 1996. Wiley, Chichester. (An introductory textbook on glaciers and glacial landforms).

Glaciers and Glaciation. D.I. Benn & D.J.A. Evans. 1998. Arnold, London. (A detailed and comprehensive textbook on glaciers and how they shape the landscape).

Holmes' Principles of Physical Geology. P.M.D. Duff. 1994. Chapman & Hall, London. (A standard textbook explaining the formation of landforms).

Mass Movements in Britain. R. Cooper. 2007. Geological Conservation Review Series No. 33. Joint Nature Conservation Committee, Peterborough. (Includes descriptions of Scotland's most important sites for landslides).

The Periglaciation of Great Britain. C.K. Ballantyne & C. Harris. 1994. Cambridge University Press, Cambridge. (A comprehensive textbook on periglacial landforms, including detailed descriptions and scientific explanations of many of the landforms on Scottish mountains).

Reconstructing Quaternary Environments. (2nd edition). J.J. Lowe & M.J.C. Walker. 1997. Addison Wesley Longman, Harlow. (A comprehensive textbook that examines the evidence and techniques used to reconstruct environmental change during the Ice Age).

Reflections on the Ice Age in Scotland. An Update on Quaternary Studies. J.E. Gordon (ed). 1997. Scottish Association of Geography Teachers and Scottish Natural Heritage, Glasgow. (A series of review articles describing the shaping of Scotland's landscape during and following the Ice Age).

CHAPTER 5 – FUTURE LANDSCAPES: SCOTLAND IN A GLOBAL CONTEXT

Avoiding Dangerous Climate Change. 2006. H.J. Schellnhuber and others (eds). Cambridge University Press, Cambridge. (Presents the findings from a conference of international scientists).

Climate Change 2007. Intergovernmental Panel on Climate Change. 2007. (http://www.ipcc.ch/). (A comprehensive analysis and overview by international experts of the current state of knowledge on climate change).

Climate Change Scenarios for the United Kingdom. The UKCIP02 Scientific Report. M. Hulme and others. 2002. Tyndall Centre for Climate Change Research, School of Environmental Sciences, University of East Anglia. (Presents scenarios of climate change for the UK).

The Economics of Climate Change. N. Stern. 2006. Cambridge University Press, Cambridge. (A detailed review of the economics of climate change and the policy responses for mitigation and adaptation).

Environmental Change. Key Issues and Alternative Approaches. F. Oldfield. 2005. Cambridge University Press, Cambridge. (A textbook that examines past, present and future global environmental change).

Future Flooding: Scotland. A. Werritty and J.Chatterton. 2004. Foresight Programme, Office of Science and Technology, London. (Examines the extent and nature of river and coastal flood risk in Scotland).

Global Warming. A Very Short Introduction. M. Maslin. 2004. Oxford University Press, Oxford. (Discusses the predicted impacts of global warming and the surprises that could be in store in the near future).

The Life and Death of Planet Earth. P.D. Ward & D. Brownlee. 2004. Henry Holt & Co, New York. (Traces the history of the Earth, its life-support systems and the ultimate fate of the planet).

Plows, Plagues, and Petroleum. How Humans Took Control of Climate. W.F. Ruddiman. 2005. Princeton University Press, Princeton and Oxford. (An account of climate change and human involvement, including the controversial hypothesis that early human activity averted a major glaciation).

The Science and Politics of Global Climate Change. A Guide to the Debate. A.E. Dessler & E.A. Parson. 2005. Cambridge University Press, Cambridge. (An accessible and informative account of the science and the issues).

CHAPTER 6 – PLACES TO VISIT

Since 1991, the *Scottish Geographical Journal*, published by the Royal Scottish Geographical Society (http://www.geo.ed.ac.uk/~rsgs/menu.html), has included a series of articles on 'Scottish Landform Examples'. These include the Cairngorms, Glen Roy, Sands of Forvie, Trotternish and the Carse of Stirling, as well as many others.

The Geographical Association (http://www.geography.org.uk/) has published three Scottish titles in its *Classic Landforms* series, which cover some of the places to visit:

Classic Landforms of Skye. D.I. Benn & C.K. Ballantyne. 2000. Geographical Association, Sheffield.

Classic Landforms of the Loch Lomond Area. D.J.A. Evans & J. Rose. 2003. Geographical Association, Sheffield.

Classic Landforms of Assynt and Coigach Area. T.J. Lawson. 2003. Geographical Association, Sheffield.

Landscape Fashioned by Geology. Titles on *Scotland, Edinburgh and West Lothian, Skye, Cairngorms, Loch Lomond to Stirling, Orkney and Shetland, Arran and the Clyde Islands, East Lothian and the Borders, Fife and Tayside, Northwest Highlands, Rum and the Small Isles, Parallel Roads of Glen Roy, Mull and Iona, Glasgow to Ayrshire, Ben Nevis and Glencoe,* and *Western Isles*. Scottish Natural Heritage. 1992–2006. (Series Editor, A. McKirdy). Scottish Natural Heritage, Battleby. (A series explaining the geology and landscapes of different parts of Scotland for general readers).

The Quaternary Research Association (http://qra.org.uk/) has published technical Field Guides, which describe the glacial landforms and deposits in many parts of Scotland.

Websites developed by Adrian Hall explain the geology and geomorphology of the Cairngorms (http://www.fettes.com/Cairngorms/), Shetland (http://www.fettes.com/shetland/)
Orkney (http://www.fettes.com/Orkney/)
and Caithness (http://www.fettes.com/caithness/), all with excellent photographs.

Two further websites provide helpful sources of information for the non-specialist, as well as information on places to visit and links to other sites:
http://www.scottishgeology.com
http://www.knockan-crag.co.uk/

INDEX

Note: Page numbers in bold indicate a plate; after the page number '*fig.*' indicates a drawing or chart, '*pnl.*' indicates matter in the yellow panels. 'Ma' stands for millions of years.

ACKNOWLEDGEMENTS

PHOTOGRAPH SOURCES AND COPYRIGHT

British Geological Survey, 121, 122, 160 right. Reproduced by permission of the British Geological Survey. © NERC. All rights reserved. (IPR/67-37C and IPR/69-27C); Laurie Campbell/SNH, ii, 117; Neil Clark, 143, 147 left; Cox, R. & Nicol, J. (1869). *Select Writings, Political, Scientific, Topographical, and Miscellaneous, of the Late Charles Maclaren, FRSE.* Edmonston & Douglas, Edinburgh, 87 bottom; Roger Crofts, 183 bottom, 192 top left, 193, 219, 271; Alastair Dawson, 213; Edinburgh University Library Special Collections, reproduced by permission, 86 bottom; Geikie, A. (1895). *Memoir of Sir Andrew Crombie Ramsay.* Macmillan, London, 87 top; Geological Society of London, reproduced by permission, 86 top; Lorne Gill/SNH,vi, viii, 8, 9, 10, 12, 13, 15 both, 16, 18, 19, 21 left, 22, 26 all, 27 both, 28, 33, 35, 36 centre and bottom, 37 both, 39, 46, 48, 51, 53 top, 59 top, middle, bottom right, 89, 97, 98, 100, 103, 104, 107, 108, 109, 110 both, 112, 113, 120, 125 right, 127, 130, 140, 148, 150, 153, 155 left, 196, 202, 224, 254, 257, 259, 270, 273 both, 275, 276, 278, 279, 286 left, 288, 290, 296, 299, 302; John Gordon, 2, 3, 4, 6, 43, 82 right, 85 bottom, 157, 158, 159, 160 left, 161, 163, 164, 165, 166, 174 left, 174 right, 177, 178, 180, 181 bottom, 182, 183 top, 185, 188, left, 188 right, 189, 191, 192, top right, 192 bottom right, 194, 198, 201 left, 201 right, 212, 214, 228 left, 228 right, 234, 237, 240, 269, 277, 291, 292, 303; Irons, J.C. (1896). *Autobiographical Sketch of James Croll, with Memoir of his Life and Work.* Edward Stanford, London, 77; David Jarman, 200; George Lees, 239; Marine Nature Conservation Review/SNH, 14; Patricia & Angus Macdonald/Aerographica, 5, 204, 222, 283, 293; Patricia & Angus Macdonald/SNH, front cover, 20, 34, 36 top, 38, 93, 96, 99, 102, 116, 138, 139, 151, 170, 171, 173, 181 top, 184, 192 bottom left, 203, 206, 210, 211, 223, 229, 261, 263, 264, 266, 272, 283, 286 right, 287, 293, 295, 297, 300; Colin MacFadyen, 119, 133 bottom, 143, 156; John MacPherson/SNH, 29, 235; Fiona Mactaggart/SNH, 187 bottom; Donald McIntyre, 125 left and centre; Alan McKirdy, 94; Moira McKirdy, 21 right; William McKirdy, 24; Steve Moore/SNH, 227; D. Munro for RSPB, 241; National Galleries of Scotland, 61 bottom, 197; National Museums of Scotland, 25, 70 all, 133 top, 134, 155 right, 268; Natural History Museum, London, 256; NASA, 230; National Oceanic & Atmospheric Administration, 85; Our Dynamic Earth, 45, 47, 258; Prestwich, J. (1886). *Geology: Chemical, Physical & Stratigraphical: Volume I Chemical and Physical.* Reproduced by permission of Oxford University Press, 84; Science Photo Library, xii; Scottish Environment Protection Agency, 233; Shairp, J.C. (1873). *Life and Letters of James Forbes.* Macmillan, London, 86 bottom; Nigel Trewin, 129; David Vaughan/British Antarctic Survey, 82 left; George Washington Wilson Collection, reproduced with permission of Aberdeen University Library, 88 top.

OTHER ARTWORK

Benton, M.J. & Spencer, P.S. (1995). *Fossil Reptiles of Great Britain.* Geological Conservation Review, JNCC, Peterborough, 142. Richard Bonson/SNH, 90 right; 91 all, 103. British Geological Survey, 220, compiled by the British Geological Survey and based on digital elevation data supplied by Scottish Natural Heritage. Reproduced by permission of the British Geological Survey. © NERC. All rights reserved. (IPR/69–27) C. M. Coates, Hunterian Museum and Art Gallery, University of Glasgow, 135 top. Aongus Collins/SNH, 253. Craig Ellery/SNH, 41; 92; 126 right; 128; 130, after Trewin N.H. (1986). Palaeoecology and sedimentology of the Achanarras fish bed of the Middle Old Red Sandstone, Scotland. *Transactions of the Royal Society of Edinburgh*, 77, 21–46; 136; 147; 152; 190; 217; 281. Claire Hewitt/SNH, 135 bottom; 154 after E.R. Phillips, 175. Estate of John Clark of Eldin, reproduced with permission, 126, left. Lauder, T.D. (1830). *An Account of the Great Floods of August, 1829 in the Province of Moray and Adjoining District.* Adam Black, Edinburgh, 205. Jim Lewis/SNH, 53; 54; 56 all, after Toghill P. (2000). *The Geology of Britain.* Airlife Publishing; 57; 58; 73 left, after Doyle et al. (1994). *The Key to Earth History.* John Wiley, Chichester; 73 right, after Zachos, J. et al. (2001). Trends, rhythms, and aberrations in global climate 55 Ma to present. *Science*, 292, 685–693; 75, after Raymo, M.E. (1994). The initiation of Northern Hemisphere glaciation. *Annual Reviews of Earth and Planetary Science*, 22, 353–383; 76, after EPICA community members (2004). Eight glacial cycles from an Antarctic ice core. *Nature*, 429, 623–628, Siegenthaler, U. et al. (2005). Stable carbon cycle – climate relationship during the late Pleistocene. *Science*, 310, 1313–1317; 78, after Lowe, J.J. & Walker, M.J.C. (1997). *Reconstructing Quaternary Environments.* Addison, Wesley, Longman, Harlow; 80 left, after Broecker, W.S. & Denton, G.H. (1990). The role of ocean-atmosphere reorganizations in glacial cycles. *Quaternary Science Reviews*, 9, 305–341, Skinner, B.J. & Porter, S.C. (1995). *The Dynamic Earth.* Wiley, New York; 80 right, after Ruddiman, W.F. & McIntyre, A. (1981). The North Atlantic during the last deglaciation. *Palaeogeography, Palaeoclimatology, Palaeoecology*, 35, 145–214, Dyke, A.S. et al. (2002). The Laurentide and Innuitian ice sheets during the Last Glacial Maximum. *Quaternary Science Reviews*, 21, 9–31, Svendsen, J.I. et al. (2004). Late Quaternary ice sheet history of northern Eurasia. *Quaternary Science Reviews*, 23, 1229–1271; 81; 100, after Strachan, R.A. et al. (2002). The Northern Highland and Grampian terranes. In Trewin N. (ed), *The Geology of*

Scotland. The Geological Society, London; 106, after Park, R.G. et al. (2002). The Hebridean terrane. In Trewin N. (ed), *The Geology of Scotland*. The Geological Society, London; 114; 117, after Blake, S. et al. (2003). *Mountain Building in Scotland*, The Open University, Milton Keynes; 118; 120; 121, after Peach, B. and Horne, J. (1907). *The Geological Structure of the Northwest Highlands of Scotland*. Geological Survey, Edinburgh; 123 all, after Stephenson, D. et al. (1999). *Caledonian Igneous Rocks of Great Britain*. Geological Conservation Review, JNCC, Peterborough; 134, after Read W.A. (2002). Carboniferous. In Trewin N. (ed), *The Geology of Scotland*. The Geological Society, London; 141; 145, after Glennie, K.W. (2002). Permian and Triassic. In Trewin, N. (ed), *The Geology of Scotland*. The Geological Society, London; 146, after Hudson, J.D. & Trewin, N.H. (2002). Jurassic. In Trewin, N. (ed), *The Geology of Scotland*. The Geological Society, London; 162, after Sissons, J.B. (1965). *The Evolution of Scotland's Scenery*. Oliver & Boyd, Edinburgh; 167, after Linton, D.L. (1954). Some Scottish river captures re-examined. III. The beheading of the Don. *Scottish Geographical Magazine*, 70, 64–78; 168, after Linton, D.L. & Moisley, H.A. (1960). The origin of Loch Lomond. *Scottish Geographical Magazine*, 76, 26–37; 169, after Clayton, K.M. (1974). Zones of glacial erosion. *Institute of British Geographers Special Publication*, 7, 163–176; 176 left, after North Greenland Ice Core Project members (2004). High-resolution record of Northern Hemisphere climate extending into the last interglacial period. *Nature*, 431, 147–151; 179 left, after Hall, A.M. (1997). Quaternary stratigraphy: the terrestrial record. In: Gordon, J.E. (ed.), *Reflections on the Ice Age. An Update on Quaternary Studies*. Scottish Association of Geography Teachers and Scottish Natural Heritage, Glasgow, 59–71; Sejrup, H.P. et al. (2000). Quaternary glaciations in southern Fennoscandia: evidence from southwestern Norway and the northern North Sea region. *Quaternary Science Reviews*, 19, 667–685; Sejrup, H.P. et al. (2005). Pleistocene glacial history of the NW European continental margin. *Marine and Petroleum Geology*, 22, 1111–1129; 179 right, after Gordon, J.E. & Sutherland, D.G. (1993). *Quaternary of Scotland*. Chapman & Hall, London; 186, after Birks, H.H. & Matthews, R.W. (1978). Studies in the vegetational history of Scotland. V. Late Devensian and early Flandrian pollen and macrofossil stratigraphy at Abernethy Forest, Inverness-shire. *New Phytologist*, 80, 455–484; 187 top, after Gordon, J.E. & Sutherland, D.G. (1993). *Quaternary of Scotland*. Chapman & Hall, London, Thorp, P.W. (1991). Surface profiles and basal shear stresses of outlet glaciers from a lateglacial mountain icefield in western Scotland. *Journal of Glaciology*, 37, 77–89; 207, after Hansom, J.D. & McGlashan, D.J. (2004). Scotland's coast: understanding past and present processes for sustainable management. *Scottish Geographical Journal*, 120, 99–116; 208, after Shennan, I. & Horton, B. (2002). Holocene land- and sea-level changes in Great Britain. *Journal of Quaternary Science*, 17, 511–526; 209; 218; 221; 238, after Dawson, A.G. et al. (2001). Potential impacts of climate change on sea levels around Scotland. *Scottish Natural Heritage Research, Survey and Monitoring Report*, No. 178. Robert Nelmes/SNH, 49 all; 50; 52; 55; 60; 64; 65; 71 both; 90 left; 111 top; 111 bottom, after Stephenson, D. & Gould, D. (1995). *The Grampian Highlands*. British Geological Survey, HMSO, London; 124; 129 top; 199, after Ballantyne, C.K. (1991). The landslides of Trotternish, Isle of Skye. *Scottish Geographical Magazine*, 107, 130–135. Courtesy of Perth Museum and Art Gallery, Perth and Kinross Council, 31. Elizabeth Pickett/SNH, 284 both; 285. Christopher R. Scotese, (http://www.scotese.com/), 247. Richard Tipping, reproduced by permission, 195. *Transactions of the Royal Society of Edinburgh* 12, Plate 6, 22 above. R.K. Greville (1831) *Transactions of the Royal Society of Edinburgh* 12, p.6

SOURCES OF QUOTATIONS NOT IDENTIFIED IN THE TEXT

16: Tabraham C. (2003). *The Illustrated History of Scotland*. Lomond, Edinburgh; 30 and 32: Burns R. (1986) *The Complete Works of Robert Burns*. Alloway Publishing, Ayrshire; 60, 178. Hutton, J. (1795). *Theory of the Earth, with Proofs and Illustrations* Edinburgh; 61: Smellie William (1800). *Literary and Characteristic Lives of Gregory, Kames, Hume and Smith*. Edinburgh; 61 and 83: Playfair, J. (1802). *Illustrations of the Huttonian Theory of the Earth*. William Creech, Edinburgh; 84: Agassiz, L. (1842). The glacial theory and its recent progress. *Edinburgh New Philosophical Journal*, 33, 217–283; 190: Darwin, F. (1887).*The Life and Letters of Charles Darwin Including an Autobiographical Chapter*. Vol. 1. John Murray, London; 157: Alexander, H. (1928). *The Cairngorms*. Scottish Mountaineering Club; 162. Forbes, J. 1846. Notes on the topography and geology of Cuchullin Hills in Skye, and on traces of ancient glaciers which they present. *Edinburgh New Philosophical Journal*, 40, 76–99.